Sources in the History of Mathematics and Physical Sciences

16

Springer
New York
Berlin
Heidelberg
Barcelona
Hong Kong
London
Milan
Paris
Singapore
Tokyo

Sources in the History of Mathematics and Physical Sciences

Vol. 1: G.J. Toomer (Ed.)
Diocles on Burning Mirrors: The Arabic Translation of the Lost Greek Original, Edited, with English Translation and Commentary by G.J. Toomer

Vol. 2: A. Hermann, K. von Meyenn, V.F. Weisskopf (Eds.)
Wolfgang Pauli: Scientific Correspondence I: 1919-1929

Vol. 3: J. Sesiano
Books IV to VII of Diophantus' *Arithmetica*: In the Arabic Translation Attributed to Qustā ibn Lūqā

Vol. 4: P.J. Federico
Descartes on Polyhedra: A Study of the *De Solidorum Elementis*

Vol. 5: O. Neugebauer
Astronomical Cuneiform Texts

Vol. 6: K. von Meyenn, A. Hermann, V.F. Weisskopf (Eds.)
Wolfgang Pauli: Scientific Correspondence II: 1930-1939

Vol. 7: J.P. Hogendijk
Ibn Al-Haytham's *Completion of the Conics*

Vol. 8: A. Jones (Ed.)
Pappus of Alexandria, Book 7 of the *Collection*

Vol. 9: G.J. Toomer (Ed.)
Appolonius *Conics* Books V to VII: The Arabic Translation of the Lost Greek Original in the Version of the Banū Mūsā, Edited, with English Translation and Commentary by G.J. Toomer

Vol. 10: K. Andersen
Brook Taylor's Work on Linear Perspective

Vol. 11: K. von Meyenn (Ed.)
Wolfgang Pauli: Scientific Correspondence III: 1940-1949

Vol. 12: F.J. Ragep
Naṣīr al-Dīn al-Ṭūsī's *Memoir on Astronomy*
(al-Tadhkira fī ᶜilm al-hay'a)

Continued after Index

Elaheh Kheirandish

The Arabic Version of Euclid's *Optics*

(*Kitāb Uqlīdis fī Ikhtilāf al-manāẓir*)

Edited and Translated with Historical Introduction and Commentary

Volume I

With 73 Illustrations

Springer

Elaheh Kheirandish
Harvard University
Cambridge, MA 02138
USA
kheirand@fas.harvard.edu

Sources Editor:
Gerald J. Toomer
2800 South Ocean Boulevard, 21F
Boca Raton, FL 33432
USA

Library of Congress Cataloging-in-Publication Data
Euclid.
[Optics. English & Arabic]
The Arabic version of Euclid's *optics* = Kitāb Uqlīdis fī ikhtilāf al-manāẓir/edited and translated with historical introduction and commentary Elaheh Kheirandish.
p. cm. — (Sources in the history of mathematics and physical sciences; 16)
Includes bibliographical references and index.
ISBN 0-387-98523-9 (alk. paper)
1. Optics—Early works to 1800. I. Kheirandish, Elaheh. II. Title. III. Title: Kitāb Uqlīdis fī ikhtilāf al-manāẓir. IV. Series.
QC353 .E8313 1998
535—dc21 98-17540

Printed on acid-free paper.

Production managed by Francine McNeill; manufacturing supervised by Jeffrey Taub.
Camera-ready copy prepared by the author using Microsoft Word.
Printed and bound by Braun-Brumfield, Inc., Ann Arbor, MI.
Printed in the United States of America.

9 8 7 6 5 4 3 2 1

ISBN 0-387-98523-9 Springer-Verlag New York Berlin Heidelberg SPIN 10677574

To my parents
Asadollah and Parvin-Dokht
(the "Lion of God" and the "Daughter of Pleiades")

آن قوم که راه بین فتادند وشدند
کس را بیقین خبر ندادند وشدند
آن عقده که هیچکس نتانست گشاد
هریك گرهی برآن نهادند وشدند

منسوب به خواجه نصیر الدّین طوسی
(نقل از احوال وآثار مدرس رضوی، ص ۶۱)

That group of people who, as guides, came and passed
and informed no one with certitude, as they passed
into that complex tangle, unraveled by anyone
each came to place a knot, and so they passed

Attributed to Naṣīr al-Dīn al-Ṭūsī (d. 672/1274)
(Persian poem quoted from Mudarris-i Raḍavī, *Aḥvāl va Āthār*, p. 61)

Preface

The two volumes of *The Arabic Version of Euclid's Optics* document a rare and complex history in which the translation of a single text into another language has had consequences for the transformation of the field far beyond the literal meaning implied by the title. In the fascinating account of the transition of the field from Greek *optika* to Arabic *al-manāẓir*, it was not only *what* was–or was not–transmitted that directed the course of developments; it was also *how* things were transmitted. In the case of the *Optika* of Euclid (ca. 300 B.C.), the first surviving work devoted entirely to the subject, its text was not merely transmitted to the Islamic world (and transmitted more fully and widely than the much more comprehensive work of Ptolemy composed in the 2nd century A.D. with the same title). Its initial transmission from Greek into Arabic was such as to involve textual transformations in some critical points. Not only were some central Greek terms and expressions transformed when turned into Arabic "equivalents." The passage of the most central concepts of the Greek optical tradition through a chain of flexible terminologies and interchangeable orthographies in *both* Greek and Arabic was to have notable consequences for further developments. With the publication of *The Optics* of Ibn al-Haytham (5th/11th century), itself unique in the almost miraculous share it took from the riches of Euclidean *and* Ptolemaic optics, such consequences can be readily examined in the case of what is by far the most important development of all. And, just as in the case of Ibn al-Haytham's *Kitāb al-Manāẓir*, the transmission of Euclid's *Kitāb al-Manāẓir* involved both "fortunes" and "misfortunes"–to quote A.I. Sabra's preface–from the standpoint of later developments. The transformations together with their consequences form the main core of the two volumes (Volume I containing the Arabic text and English translation, Volume II the textual and historical commentary), in the hope of capturing a reflection of the complex process of scientific development itself.

To speak of "Euclid's *Optics*" and its Arabic "version" or of "the translation of a single text into another language" would be barely meaningful with the continuing problems of textual attribution and identification in the Greek tradition, had it not been that, from the standpoint of the Arabic tradition, there *was* such a text as the "translation" of "Euclid's book on *Optics*."

From our standpoint, and perhaps that of the Greek tradition itself, "Euclid's *Optics*" has never been fully meaningful as a single point of reference. For a long time, it referred to a text with such a title in various early editions (Zamberti, 1505, Pena, 1557, David Gregory, 1703, to name but a few). About a century ago, it came to refer to *Heiberg's Euclid*, a different text based on the manuscripts of a Greek *Optika* that J.L. Heiberg edited as *Euclidis Optica*, as distinguished from the previously known set of Greek manuscripts now presented in the same edition with the title *Opticorum recensio Theonis*, as a recension by the 4th-century commentator, Theon of Alexandria (Heiberg, 1895). In the wake of the recent reexaminations of the Greek Euclidean tradition by Alexander Jones and Wilbur Knorr (in what has left less and less of Euclid in the former and more and more of it in the latter), there is now also *Euclid's Optics*, according to *their* respective standpoints.

Euclid's *Optics* of the Arabic tradition *did* represent a fixed identity, the name of Euclid being repeatedly attached to the Arabic versions passing under the title of "*Optics*"; and it was a text that, even as the product of a textual transmission involving more than one version, was passed on as a single voice. As the Arabic was *not,* in any of its close variations, exactly what was in either of the Greek versions associated with Euclid's name, and was in fact quite transformed when it came to the text's most fundamental assumptions, the "translation" was at once the Arabic version of the Euclidean text and the Arabic "version" of the Euclidean visual theory.

The challenging study of such a tradition has come to involve much more than what is to be expected even in the most straightforward textual treatments of this kind. With any study whose main sources are largely in manuscript form, various problems with extant copies, from identification to actual access, are to be reasonably anticipated. In this case, even establishing their mere existence–let alone current whereabouts–has been a problem, at times requiring persistent investigative (and occasionally detective) work. It so happens that, of the few extant manuscript copies of the main text, one has traveled apparently from Cairo to Tehran, Mashhad, New York, Paris, Basel, and on to unknown destinations, half of this in the course of the last decade; three have been housed between Istanbul and Cairo, both places with a long (and, to some, familiar) history of limited accessibility; and one is supposed to be still sitting in war-zone Kabul, with none of the request letters (even to presidents! as recommended by Afghani colleagues) bearing any fruit. The stories of the trace (or rather chase) are not as colorful in the case of the other *Manāẓir* texts, except for occasions to examine heavily guarded illuminated manuscripts in the museums of Istanbul and dust-covered codices in private libraries in Qum. But the attempts to improve the problematic documentation status of the *Taḥrīr al-Manāẓir* of Naṣīr al-Dīn al-Ṭūsī (d. 672/1274), often confused with the early Arabic version, and its short attachment *Iṣlāḥ al-Manāẓir*, occasionally taken for the *De aspectibus* of al-Kindī (d. ca. 257/870), did result in the identification of over two-thirds of their surviving manuscript copies, many being unreported (and a few misreported) in the most widely consulted of reference sources.

For the completion of the work, I feel privileged to have had timely support from many sources and in many forms. First and foremost, I greatly value the fortune of having received the inspiring guidance of Professor A.I. Sabra at a critical intellectual moment and throughout the course of a study that began life as a dissertation project and reached its current state under his genuine "supervision." An award from the National Science Foundation allowed for the initial identification and collection of manuscript sources included in the study and was particularly crucial for personal examination of those that were either difficult to obtain or, in the most critical cases, difficult to read in the best reproduced form. Subsequent appointments as a postdoctoral fellow at the Department of the History of Science at Harvard University, as a member of the Research Faculty of History and Philosophy of Science at Tehran's Institute for Humanities and Cultural Studies, and, most recently, as a Resident Fellow of the Dibner Institute for the History of Science and Technology at MIT have provided valuable opportunities and facilities for the necessary revisions and updates reflected in this publication.

The occasion also calls for grateful acknowledgment of the helpful assistance of several libraries in the Middle East and Europe that hold relevant manuscripts (itemized in Lists I and II, v. 1, pp. xxxiv–liii): the *Topkapı*, Süleymaniye, and Köprülü Libraries of Istanbul, University Library of Leiden, Bibliothèque Nationale of Paris, India Office of London, Bodleian Library of Oxford, John Rylands Library of Manchester, and Medicea-Laurenziana Library of Florence. As for largely unexplored manuscripts existing in collections throughout Iran (some in multiple copies or with the oldest transcription dates), individual acknowledgments are particularly due for the kind cooperation of the staff at Tehran University Markazī (Central) Library, Majlis-i Shawrā, Muṭahharī (Sipahsālār), and Malik Libraries in Tehran, Āstān-i Quds Library in Mashhad, Mufattiḥ (Pārs) Library in Shiraz, the Libraries of Isfahan and Shiraz University, the supplementations provided by the private libraries of Ustād ʻAlī-Naqī Munzavī in Tehran and Ayatollah Ḥasan-zādih-i Āmulī in Qum (the Library of Ayatollah Marʻashī-yi Najafī in Qum did not have facilities for women at the time of request), and, last but not least, the personal and financial support of my father, Asadollah Kheirandish, who was once my only local source of contact with these libraries. The intervention of a few other people made other important manuscripts available to me. David Partington, the former director of the Middle East Division of Widener Library, and Richard Lorch, at the University of Munich, acted as crucial links for my eventual access to a total of three manuscript copies of the main text. Professor Everett Rowson and, more recently, Skuli Sigurdsson kindly seconded my efforts in obtaining films of related manuscripts from Cairo and Krakow; and Professor Charles Burnett was instrumental in my last-minute access to the photograph of the manuscript folio included in the second volume.

Expressions of gratitude must extend to other aspects of the study. Professor John Murdoch first woke in me an appreciation for Greek and Latin intellectual history and awareness of problems associated with each. Colleagues Adam Schulman, William Newman, Dana Miller, and Tzvi Langermann offered insights from their respective fields of specialty on some complexities in the Greek and Latin texts and on the question of Syriac or Hebrew links. Professors Dimitri Gutas and Paul Kunitzsch generously shared their expert knowledge on the question of the exact name and identity of the Arabic translator and a few other specific problems. And Professors Jamil Ragep and David King took me in their own ways far beyond my expectations in improving the microscopic details of related work. Special thanks naturally goes to Professor Gerald J. Toomer for his careful and critical review of various stages of the entire work and his crucial role in having it published. It is unfortunate that the related work of Professor Roshdi Rashed remained unavailable to me until after I had submitted my manuscript for publication.

On the technical aspects of the work, the fonts for the Greek and Arabic transcriptions and English transliteration were designed (and in some cases tailor-made) by Professor Toomer, and those for Persian by Professor Wheeler Thackston. The attempted upgrades in the photographic quality of the manuscript plates carry the expert touch of Kamran Adle. And the technical support of my *truly better* half, Hormoz Goodarzy, has been indispensable throughout the work, in crisis and in calm, in substance and in form. For his caring support throughout the long course of the study itself, it is hard to find the right words.

Elaheh Kheirandish
Cambridge, Massachusetts

Contents

Volume I

Contents

Volume II

Volume I

Part I

Introduction to the Texts

1 The Arabic Versions

The corpus of sources for the study of the Arabic tradition of Euclidean optics is formed by a few genres of works. First there are the founding texts of the tradition, namely the Arabic versions of the Euclidean text, representing a major part of the original Greek as found in Heiberg's *Euclidis Optica*.[1] Then there is a set of dependent texts, which together with the Arabic versions of the text, form the tradition as it developed in the four centuries between and occasionally beyond the Arabic versions of the 3rd/9th and 7th/13th centuries. And finally there is a series of related texts, reflective of or responsive to different aspects of Euclidean optics, through which the tradition may be further studied or its chronological and geographical boundaries further extended.

The Arabic versions of the Euclidean text also survive in several forms. The earliest form is best represented by the odd-numbered set of manuscripts: m1, m3, m5 [see Description and Analysis of Manuscripts, and List of Manuscripts, hereafter Description and List, pp. xxvi–xxvii and p. xxxiv, respectively]. Characterized by the title *Ikhtilāf al-manāẓir*, this can be considered an early translation. It is a "translation" (*tarjama*) according to the colophon of one of the manuscript copies in this same set (m5), and it is early because of certain considerations including the dates of its supposed translator. The name to which the translation is attributed, Hiliyā ibn Sarjūn [reading for Hlīl ibn Sarḥūn, see Manuscript Plate V, p. 240, and The Arabic "Version," v. 2, p. xxii, n. 19], can be in fact associated with more than one early translation from Greek into Arabic: a translation, together with al-Ḥajjāj ibn Yūsuf ibn Maṭar (fl. ca. 170/786–218/833), of Ptolemy's *Almagest* (ca. 212/827–828), a text for which the *Optics*, being one of the "Intermediate Books" (*Kutub al-Mutawassiṭāt*), was to act as preparatory;[2] and with slight variation of the name, also of Cassianus Bassus's *Geoponika* [see The Arabic "Version," v. 2, p. xxii, n. 20–21]. If not the man himself, then he may be the father or the son [ibid., n. 22–23].

---

[1] Greek text ed. Heiberg, *Euclidis Optica, Opticorum recensio Theonis, Catoptrica, cum scholiis antiquis* in *Euclidis Opera Omnia*, eds. Heiberg and Menge, v. 7: *Euclidis Optica*, pp. 1–121; *Opticorum recensio Theonis*, pp. 143–247. English translation, Burton, "The Optics of Euclid"; French translation, Ver Eecke, *Euclide, l'Optique et la Catoptrique*. For a reexamination of Heiberg's position, including the attribution of the two Greek versions to Euclid and Theon of Alexandria, see Jones, "Peripatetic and Euclidean Theories," and Knorr, "Pseudo-Euclidean Reflections in Ancient Optics." See also Textual Transmission, p. lxi, and The Arabic "Version," v. 2, pp. xix–xx and n. 7.

[2] Also known as the "Middle Books" or the "Little Astronomy," this is the collective name for mathematical works to be studied intermediately between Euclid's *Elements* and Ptolemy's *Almagest*. They consist of a number of Greek texts including the *Optika*, *Data*, and *Phaenomena* by Euclid, *Spherics* of Theodosius and of Menelaus, and the *Moving Sphere* of Autolycus; see Steinschneider, "Die mittleren Bücher."

Whether or not the translation of the *Optics* was – as were the Arabic *Elements* or *Almagest*, the two masterpieces to be studied before and after it –[3] carried out during the reign of the ʿAbbasīd Caliph al-Maʾmūn (r. 198–218/813–833),[4] what circulated under the title of *Ikhtilāf al-manāẓir* must be placed among the oldest compositions of the period on independent evidence. It is with this same title that the expression "old translation" (*al-naql al-qadīm*), a common designation for the earliest phases of the 2nd/8th–3rd/9th-century translation movement, is repeatedly associated [see Summary and Commentary: Title, v. 2, p. 2], and there are references to a text with such a title starting from at least the middle of the 3rd/9th century.[5]

A slight variation to this early Arabic version is represented by the even-numbered set of manuscripts, m2 and m4, with titles representing the short form *al-Manāẓir* [see Description, pp. xxvi–xxvii and List I, p. xxxiv], and an apparently more limited circulation. The variance with the former set is most striking in the case of the very first Euclidean definition, as in the case of the two main extant Greek versions identified and edited as the respective texts of Euclid and Theon of Alexandria about a century ago, and versions A and B as reexamined and renamed recently [see Textual Transmisssion, p. lxi, and n. 51].

---

[3] The explicit statement that the "Intermediate Books" (*al-Kutub al-Mutawassiṭāt*) were to be studied after Euclid's *Elements* and before Ptolemy's *Almagest* is found in some manuscripts, for example the Arabic translation of Euclid's *Elements* [MS Oxford, Bodleian Library, Thurston 11, fol. 213a]: "*dhikr al-kutub alladhī yuḥtāju ilā qirāʾatihā qabli Kitāb al-Magisṭī wa tuʿrifu bi al-Mutawassiṭāt*" ("citation of the books needed to be read before the *Almagest* and known as the Intermediate [Work]s"). There, *Kitāb Ikhtilāf al-manāẓir li-Uqlīdis* appears as the second title in the list [ibid., fol. 213b], but the order of the "Middle Texts" within the mathematics curriculum is rather flexible for both the early version and its later recension.

[4] On the Arabic tradition of Euclid's *Elements*, see Murdoch, "Euclid," *DSB*, 4: pp. 438–443, and De Young, "The Arabic Textual Traditions of Euclid's *Elements*." For the case of Ptolemy's *Almagest*, see references to the standard literature in Toomer, "Ptolemy," *DSB*, 11, p. 203, and Kunitzsch, *Der Almagest*, pp. 60–71. On the question of dating, see The Arabic "Version," v. 2, n. 23–24.

[5] The reference "*qāla Uqlīdis fī Kitāb Ikhtilāf al-manāẓir*" ("Euclid said in the *Optics*") appears in an apparently early work [before 250/864?, according to Krause, "Stambuler Handschriften," p. 513] by Aḥmad ibn ʿĪsā, entitled *Kitāb al-Manāẓir wa al-marāyā al-muḥriqa*; see Ḥ in List II, p. xlv (ḥ1, fol. 89a); see also The Arabic Tradition, p. liv, and Summary and Commentary: Proposition 10, v. 2, pp. 47–48, n. 182–183. There is also al-Kindī's (d. ca. 257/870) *De aspectibus* [*diversitatum aspectus* (*Ikhtilāf al-manāẓir?*); see Björnbo and Vogl, *Alkindi, Tideus und Pseudo-Euklid*, pp. 1–41] based on it [see Summary and Commentary: Definition 1, v. 2, p. 12], though not preserved in Arabic for close textual comparison. Such a text by Euclid is also reported in al-Yaʿqūbī's (d. 284/897) *Taʾrīkh* (ca. 267/880), ed. Houtsma, part I, p. 139. The dates of al-Yaʿqūbī and Ibn ʿĪsā are from Krause, "Stambuler Handschriften," p. 441, n. 4, and p. 513, and Brockelmann, "al-Yaʿkūbī," *EI*[1].

The even-numbered manuscripts do have an unmistakable trace of the opening definitions of Heiberg's *Recensio Theonis* [or version B], just as the odd-numbered ones do in other parts of the text [see Summary and Commentary: Proposition 35, v. 2, p. 74]; but none have a preface or a consistent correspondence, starting from the very next definition, to be considered its direct translation.[6] If, on the other hand, an early revision of the Arabic *Optics* had ever been prepared, as in the case of other Greek scientific works,[7] then a better candidate may be the even-numbered set, representing slight and at times critical divergence, still in the name of Euclid.[8] There is, however, no evidence for the existence of such a "revision." The text *Iṣlāḥ al-Manāẓir*, attributed to al-Kindī (d. ca. 257/870) [Ṣ, see List I, p. xxxv], is, in the form in which it has reached us, a "correction" (*iṣlāḥ*) of two problematic Euclidean propositions [see Edition and Translation, pp. 226–229], not a "revision" (*taṣḥīḥ*) as such.[9] Neither is there an indication that names ordinarily associated with such "revisions" [Isḥāq ibn Ḥunayn (d. ca. 298/910), Thābit ibn Qurra (d. ca. 289/901)] or other names linked with this particular translation [Ḥunayn ibn Isḥāq (d. ca. 264/877) and Qusṭā ibn Lūqā (d. ca. 300/912)] were ever involved in such an effort [see The Arabic "Version," v. 2, p. xxii]. The existence of a revision/restoration effort, however, can be well supported, and the possibility of a prototype to which members of one or both of the above sets may be related cannot be ruled out.[10]

---

6 There is no extant text documented as the Arabic version of that particular Greek text, nor is there any reference to Theon's name in that connection. But in an unpublished study of the medieval Euclidean tradition [Theisen, *The Mediaeval Tradition*, p. 324, n. 10], it is suggested that the Vatican manuscript Urbinas 1329, with 64 propositions [and no preface], "is probably a Latin translation of an Arabic version of Theon's *Recension*."

7 On the case of the *Elements*, see Murdoch, *Euclid*, p. 443; for the *Almagest*, see Kunitzsch, *Der Almagest*, p. 67.

8 The colophon of m2 contains a reference to Euclid's name; for the details of the two variants, including stylistic differences, see Description, pp. xxvii–xxix. For discussions of authorship and of textual language, see The Arabic "Version," v. 2, p. xxv–xxviii.

9 The treatment of the two propositions in *Iṣlāḥ al-Manāẓir* does not match those in the Euclidean text [Propositions 60–61], to the numbers of which specific reference is made, such that Ṣ could be considered an excerpt from the revision of M in those parts. See respective editions, pp. 226–229 and pp. 212–215, and Summary and Commentary: Propositions 60–61, v. 2, pp. 95–97. Neither does Ṣ overlap with al-Kindī's *De aspectibus* or match Ṭūsī's *Taḥrīr* of those same propositions, where such corrections usually follow Euclidean treatments, thereby making such a sequel by Ṭūsī unnecessary.

10 The restoration assumption is supported by the missing first page of one of the manuscripts (m3), the testimony of one from the other variant set (m2) on the corrupt nature of the latter's original, and the striking manuscript variance of the opening lines. For similar arguments for the same part of the Greek, see Summary and Commentary: Definitions, v. 2, p. 6, n. 60. The prototype assumption, which is still reconcilable with restoration efforts, may be entertained in the light of the title and description in al-Ya'qūbī's *Ta'rīkh* (ca. 267/880) not being identical to any of the extant copies m1–m5; see Supplementary Passages, p. 243, and Variation Sample 1, v. 2, p. xxxviii. The

The two early variations [i.e., m1, m3, m5 and m2, m4] may be different enough [see Description, p. xxviii] to suggest a revision or restoration effort at work, but not enough to be considered as two distinct versions. They are not even different consistently enough [see Summary and Commentary: Definition 1, 2; Proposition 35, v. 2, p. 7, 17, and 74 respectively] to be respectively assigned to the traditions of the two main Greek versions. With the largely uniform content and structure of all manuscript copies, and the explicit association of members of *both* the odd- and even-numbered manuscript sets with the name of Euclid [m1, m2, m3, and m5], they have been edited here as one text [hereafter M, see Edition and Translation, pp. 2–225]. Variant readings appear in the critical apparatus below the Arabic text and in critical cases alongside the Greek separately [see Variation Samples 1–4, v. 2, pp. xxxvii–xlv]. Textual variance of all definitions and propositions with reference to Greek is treated in the corresponding Summary and Commentary, v. 2, pp. 1–99.

In addition to the "early" Arabic versions – the old translation [as early as at least ca. 212/827–828 on account of the dates of the supposed translator] and its slight variation [as early as before ca. 257/870 on account of al-Kindī's possible response to it, see Summary and Commentary: Definition 1, v. 2, p. 12] – there is a much later recension of the same 64 propositions with the title of *Taḥrīr al-Manāẓir* [T, see List I, p. xxxvii]. This can be distinguished from the earlier versions, most immediately by its extended definitions and condensed demonstrations. Composed by Naṣīr al-Dīn al-Ṭūsī (d. 672/1274), as part of his *Taḥrīr al-Mutawassiṭāt* (*Recension of the Intermediate Works*),[11] and completed in the year 651H (=1253) as reported in a number of its many manuscript copies, this late version has a well-known author and firmly established date, unlike the earlier versions. The contrast is to include the "Correction" (*Iṣlāḥ*) which though ascribed to al-Kindī in all of its identified manuscript copies, having been a companion piece to Ṭūsī's *Taḥrīr al-Manāẓir* in all but one case, has been once even labeled "among the notifications of the *Khwāja*" [namely Ṭūsī] himself.[12]

---

alternative vocabulary of the overlapping part of the Arabic text of the pseudo-Euclidean *De speculis*, suggesting an early and unsettled language [see Supplementary Passages, p. 243 and The Arabic "Version," v. 2, p. xxvi, n. 28], and comparable formulations of the same [see Supplementary Passages, pp. 244–245] may be used to support either or both assumptions [see Summary and Commentary: Definitions, v. 2, pp. 7–8, n. 61–63].

[11] Published text, Tehran lithograph edition, 1304 (=1886-87): 54–75; Hyderabad, 1358 (=1939): 2-24; *Revue de l'Institut des Manuscrits Arabes* (*RIMA*) 1383=1963, 9: 251-290.

[12] Of the extant copies (ṣ1-ṣ23), one in Isfahan (ṣ9) has the title "*Min ifādāt al-Khwāja*" ("Among the notifications of the *Khwāja*") [a common title for Ṭūsī]. All copies but the one in Paris (ṣ8), which lacks the preceding *Taḥrīr*, are cited in List I with reference to the entry under T. On the plausibility of such an early correction of the two corrupt Greek propositions, see Summary and Commentary: Propositions 60–61, v. 2, pp. 95–97.

There is at least one more text, *Tajrīd al-Manāẓir* [J in List I, p. xliv], composed by a contemporary of Ṭūsī named Ibn Abī Jarāda (ca. 676/1277),[13] that must be considered as belonging directly to this tradition. And there are a number of others in whose titles the appearance of the term *al-Manāẓir* at once reveals their close association with this tradition. These are introduced separately [see The Arabic Tradition, p. liv], and their manuscripts are listed individually [under Supplementary Texts; see List II, p. xlv].

The main sources for the study of the Arabic tradition of Euclid's *Optics* are still largely in manuscript form, and their documentation status has long remained incomplete at best, misleading at worst. For the early Arabic version, M, six extant manuscript copies are cited [four in Sezgin, *GAS V*, p. 117, two others in King, "Notes on Sources," p. 453; King, *Cairo Catalogue*, v. 2, p. 1029; and King, *Cairo Survey*, p. 24], of which at least two [from the first set] are not to be found as reported. For Ṭūsī's later recension, T, the total reported cases are close to a third of the existing ones [Sezgin, *GAS, V*, p. 117, lists 18 out of at least 58; Krause, "Stambuler Handschriften," p. 500, contains one addition; King, "Notes on Sources," p. 453; King, *Cairo Catalogue*, v. 2, p. 1029; King, *Cairo Survey*, p. 24, adds four others]; in the case of the published copies of the same work, only two out of three are usually mentioned. The earliest one, Tehran, 1304 (=1886–1887) [see t48 in List I, p. xliii] is to my knowledge cited, with one exception [Storey, *Persian Literature*, v. 2, p. 2], in only a few sources in Persian. The report of a Persian translation of *al-Manāẓir* [Dorn, *Chanykov*, p. 260; Steinschneider, "Eukild bei den Arabern," p. 101; Theisen, *The Mediaeval Tradition*, p. 10; Storey, *Persian Literature*, v. 2, part 3, p. 455: "doubtless of Ṭūsī's Arabic version"] seems to be part of a recurring confusion with a Persian *al-Masākin*.[14] In the more dramatic case of the supplementary correction Ṣ, not only is it mixed up with al-Kindī's *De aspectibus* [see repeated reference to a Latin translation: Carmody, *Arabic Astronomical Sciences*, p. 79; Sezgin, *GAS V*, p. 117] and largely overlooked even in catalogues listing its companion text T;[15] of at least 24 extant manuscript copies of it [see List I, pp. xxxv–xxxvii and p. liii], a total of three have been so far acknowledged [Brockelmann, *GALS I*, p. 374, and Sezgin, *GAS V*, p. 117, cite the same one; King, "Notes on Sources," p. 453; King, *Cairo Catalogue*, v. 2, p. 1029; and King, *Cairo Survey*, p. 24, two

---

[13] The *Tajrīd* of Ibn Abī Jarāda survives in an apparently unique manuscript copy in Cairo's Dār al-kutub, first reported by David King; see King, "Notes on Sources," p. 453; King, *Cairo Catalogue*, p. 1030; King, *Cairo Survey*, p. 24. On the author and the slight variation in the transliteration of his name, see The Arabic Tradition, p. lx.

[14] Chanykov 146, entitled *Kitāb-i Manāẓir-i Uqlīdis*, contains the text *Kitāb al-Masākin*. On other confusions between the *Mutawassiṭāt*, see The Arabic "Version," v. 2, p. xxi.

[15] The independent text ṣ8 and two of the attached copies (ṣ14, ṣ17) are usually the only ones cited; see List I, pp. xxxv–xxxvi. I am grateful to Jamil Ragep for supplying the requested information on manuscripts ṣ12 and ṣ13 while in Istanbul and attempting the same for the Atif Collection (1712, 1716), which remained inaccessible through his stay.

additional ones]. As for other texts falling within this tradition, the apparently unique manuscript copy of *Tajrīd al-Manāẓir* is known through a single source [King, ibid.]. And in the case of a shorter and much later version in Persian, *Dar 'Ilm-i manāẓir* (*On the Science of Optics*) [D, see List II, p. lii], with a revealing list of few otherwise unknown titles, including a *Sharḥ-i Manāẓir* (*Commentary on the Optics*) [see The Arabic Tradition, p. lx], no direct references are made to any of its known copies.

The citations of those manuscript copies already identified are also defective in many ways. In the most extreme case, the early and late Arabic versions of the Euclidean text are mixed up. Nor is it merely the case that the much more widely available recension of Ṭūsī is reported – and the report often duplicated – as the Arabic translation [Assemani, *Bibliotheca Medicea*, p. 392; Uri, *Bibliotheca Bodleiana*, p. 189; Nicoll and Pusey, *Bibliotheca Bodleiana*, p. 260; Wenrich, *De Auctorum Graecorum*, p. 182; Winter, "British Isles," p. 76; King, "Notes on Sources," p. 453 referring to the latter]; the opposite is also the case [Dār al-kutub's cover page of one of the two manuscripts there (m4) labels the earlier version as if composed by Ṭūsī].

Besides the false identity of manuscript copies, there is also the problem of their current locations. A copy of the main text reported to be in Kabul (m6) [de Beaurecueil, *Manuscrits d'Afghanistan*, p. 294; Sezgin, *GAS V*, p. 117] has yet to be located. Another copy, recorded as part of the collection of the H. P. Kraus Library in New York (m5) [Sezgin, *GAS V*, p. 117], is not there now; nor is it a "13. Jh. H." copy [Sezgin, ibid.], though it is 13th century A.D., as suggested by the name of its copyist [see Description, p. xxvii]. Two copies of the text's later recension, cited as being in Berlin (t51, t52) [Ahlwardt, *Königlichen Bibliothek*, v. 5, p. 354, no. 6017; see List I, p. xliv] were moved to Krakow in the early 1940s. Another copy of the same, supposed to be in Istanbul (t8) [Krause, "Stambuler Handscriften," p. 500; Sezgin, *GAS V*, p. 117, see List I, p. xxxviii], continues to be of uncertain status.

Numerous other forms of incorrect citations range from codex references and textual entries to folio numbers and copy dates [detected errors are underlined in Lists I–II, pp. xxxiv–liii]. Unverifiable reports of versions with various lengths and assignments of the best known of the translators have also long permeated the literature [see The Arabic "Version," v. 2, pp. xxi–xxii, and n. 12–18]. In most such cases, the problem seems to be the result of the confusion of one of the *Mutawassiṭāt* works with the text of *Manāẓir*, which is the first among them to be published in the form of a critical edition, English translation and historical introduction and commentary, as presented in these two volumes.[16]

Volume I is primarily devoted to the edition and translation of *Kitāb Uqlīdis fī Ikhtilāf al-manāẓir*, with a critical apparatus based on its five available manuscripts and with the English translation on facing pages [see Editorial Procedures and Apparatus, p. xxix]. This is supplemented by a similar treatment

---

16 The initial form of the study was my Ph.D. dissertation, *The Medieval Arabic Tradition of Euclid's Optika*, Harvard University, 1991.

of *Iṣlāḥ al-Manāẓir*, the alleged "correction" of two propositions, attributed to "Abū Yūsuf Ya'qūb ibn Isḥāq al-Kindī" in the opening line of all its 24 identified manuscripts. A full textual treatment of Ṭūsī's recension, itself extant in nearly two and a half times as many manuscript copies as its attached *Iṣlāḥ*, is not included in Volume I, as the Arabic text of *Taḥrīr al-Manāẓir* has been published more than once [Tehran, 1304 (=1886–1887); Hyderabad, 1358 (=1939); *Revue de l'Institut des Manuscrits Arabes* (*RIMA*), 1383=1963, 9: 251–290]. Since the most glaring defect of the existing editions, or rather transcriptions, is the publication as part of *Taḥrīr* what is in fact the first diagram of the companion text *Iṣlāḥ* [*RIMA*, p. 290], Volume I does include a critical edition and English translation of the full text of the latter with its two diagrams [for the 64 diagrams of the former reproduced from a manuscript folio that uniquely presents them all on the same page, see v. 2, pp. viii–ix]. As an initial step toward the publication of a truly "critical" edition of *Taḥrīr al-Manāẓir*, a list of all its identified manuscript copies [many hitherto unknown, including a few transcribed during Ṭūsī's lifetime or shortly after], is also provided, along with those of the other main and supplementary texts in both Arabic and Persian [see Lists I–II, pp. xxxiv–liii; see also The Arabic Tradition, pp. liv–lxi]. Textual transmission, as far as it can be determined at present, is represented in the section with this title [from p. lxi] through two charts [Greek-Arabic-Latin Transmission; Internal Transmission, pp. lxiv–lxv].

Volume II is about the nature of transmission and the conceptual transformations often tied to the very act of translation or transcription. Introduction to the Commentary contains sections on the transformed text [see The Arabic "Version," v. 2, p. xix], its structural correspondence to Greek and Latin versions [see Correspondence Tables 1 and 2, v. 2, pp. xxxiii–xxxiv], and the textual variance of critical Greek and Arabic passages [Variation Samples 1–4, v. 2, pp. xxxvii–xlv]. In the section Summary and Commentary, the Summary provides a synopsis of all Arabic definitions and propositions and the division of propositions. The Commentary consists of a textual treatment [starting with Textual Correspondence] of the relationship of each definition and proposition first to the corresponding ones in the two main Greek versions O and R [see Textual Transmission, p. lxi and The Arabic "Version," v. 2, p. xix] and then to the later Arabic version T, with historical treatments of important cases [starting with Historical Context]. The textual and historical commentaries cover, in the case of all transmitted definitions and a select number of propositions, the initial occurrence and eventual consequence of a "transformed" Arabic version. An Index of Arabic Terms is included to provide a sense of the style of the early edition/translation and of the many problems inevitably presented by this edition and translation. The Outline of Key Points is meant to highlight a range of conclusions scattered throughout the two volumes and, it is hoped, deter any possible forms of misinterpretation.

2 Description and Analysis of Manuscripts

There are at least five extant copies representing the manuscript tradition of the early Arabic version. The following description and analysis is based on the available copies [five of the six reported ones], to be supplemented with a reproduction of opening and closing pages [see Manuscript Plates, pp. 232–241].

The copy with the earliest known date (600H) is located in Cairo's Dār al-kutub: DR 260, 2: fols. 59a–92a [see List I, p. xxxiv and Manuscript Plate III, pp. 236–237]. It lacks a separate title page, and its first page lacks the definitions. The title appears as *Manāẓir* at the top of its first page and as *Kitāb Uqlīdis fī Ikhtilāf al-manāẓir* in the colophon, where the name of Euclid is attached to the longer title. The proposition numbers, which are transcribed in *abjad* (alphabetical) form, are unique in having been placed at the end rather than the beginning of the propositions, and they appear unclear in the first two cases. The figures are mostly sideways. The text begins with the enunciation of the first proposition: "*Laysa shay'un min al-mubṣarāt yubṣaru jamī'an ma'an.*"

The copy dated 625H [date in codex: fols. 115b, 170a, 188a, and 222b], is located in Istanbul, Aḥmad Thālith (III) or New Library at Topkapi Palace: A3464, 4: fols. 59b-74a [see List I, p. xxxiv, and Manuscript Plate I, pp. 232–233]. The title appears as *Kitāb Uqlīdis fī Ikhtilāf al-manāẓir wa al-shu'ā'āt* on both its first and last pages, a form closest to that in al-Ya'qūbī's *Ta'rīkh* [see The Arabic Versions, p. xx]. The text is part of a codex contained in a middle-size brown cover, with small and rather illegible naskh script and marginal notes in a different hand, and with an apparently wide circulation. The collection is prefaced by a statement in Persian: "*Kitāb-i Mutawassiṭāt ghiyr az Taḥrīrāt-i Khwājih, az avval yik ṣafḥih kam*" ("The Book of Intermediate [Work]s, other than the Recensions of *Khwāja* [i.e., Ṭūsī], a page missing from the beginning"). The figures are mostly upright. The opening line is "*Al-shu'ā' yakhruju min al-'ayn 'alā khuṭūṭin mustaqīmatin. . . ,*" preceded in the margin by "*Za'ama Uqlīdisu*" ("Euclid maintained") and filled with other marginal additions. About other unique features, see Editorial Procedures and Apparatus, p. xxx, and Summary and Commentary, Definitions 1, Propositions 3, 9, 10, v. 2, p. 7–48.

The copy dated 692H [date in codex in the same hand: p. 28, 65, 111, no folio numbers or title], is located in Leiden, Universiteitsbibliotheek: Or 133, 5: pp. 81-105 [see List I, p. xxxiv, and Manuscript Plate II, pp. 234–235]. The title on the cover page is *al-Manāẓir li-Uqlīdis* [with two owner's marks dated 697H and 756H] and in the colophon is "*Kitāb al-Manāẓir li-Uqlīdis al-Ṣūrī,*" where the name of the copyist appears as Ibn al-Ghulām al-Qūnawī and the place of copy as Balkh, along with the revealing statement "*kāna aṣluhu saqīman, mughlaqan* [or *muqlaqan*], *batran* [or *jiddan*]" ("its original was poor, obscure, and cut off") [or "its original was really poor and mixed-up:" alternative readings kindly suggested by Mr. Hossein Masoumi Hamedani]. The figures are mostly upright. The opening line is "*Al-manāẓir min al-'ayn hiya anna al-shu'ā' yakhruju min al-'ayn 'alā khuṭūṭin mustaqīmatin mukhtalifat al-ab'ād wa al-maqādīr . . . .*"

The copy dated 751H [date in the colophon] is located in the form of photos of a manuscript in Damietta Institute at Cairo's Dār al-kutub: DR 720: 2–77 [see List I, p. xxxiv, and Manuscript Plate IV, pp. 238–239]. The head title is *Al-Manāẓir min al-'ayn* [on title page, followed by "*ta'līf Khaṭīr al-qadr . . . ?min al-afāḍil al-akrām*" (Ṭūsī?)], but the colophon states that "this is called *Kitāb al-Manāẓir*." Propositions 4 and 37 are transcribed twice. The figures are mostly upright. The opening line is "*Inna al-shu'ā' yakhruju min al-'ayn 'alā khuṭūṭin mustaqīmatin mukhtalifat al-ab'ād wa al-maqādīr . . . .*"

The copy to be dated before 1049H [earliest stamp of ownership] and most likely ca. 680H [suggested by the name of the copyist] is no longer in the H. P. Kraus Collection in New York [see List I, p. xxxiv], as the codex containing it has since been sold [according to letter from Kraus Library]. I have had access to a microfilm copy labeled Universitätsbibliothek Basel, fols. 20b–32b [see Manuscript Plate V, pp. 240–241], the place where the film was made, *not* the current site of the original codex or its microfilm, which remains unknown. The stamps of two one-time owners of the collection, dated 1049H Ṭihrān and 1269H Āstān Ḥaḍrat al-Raḍawī (Mashhad), indicate that it has not been copied as late as "13th-century H" [Sezgin, *GAS, V*, p. 117; *H. P. Kraus Catalogue 137*, no. 29; caption on p. 67 has only 13th century], or that it has remained in Egypt, the place where it has most likely been transcribed. The copyist named on the title page is "al-Shaykh Abī 'Alī al-Mashhūr, the composer of *Kitāb al-Mabādī' wa al-ghāyāt*," a description which matches closely Abū 'Alī al-Ḥasan ibn 'Alī al-Marrākushī (ca. 680/1281-82), the "famous" author of *Kitāb Jāmi' al-mabādī' wa al-ghāyāt fī 'ilm al-mīqāt* [see King, "al-Marrākushī," EI^2, VI, p. 598, where it is stated that Sédillot (J. J.) has "misdated him to 660/1262"]. The Maghribī style of the script employed in the so-called "Egyptian" copy [Sezgin, GAS V, p. 117; *H. P. Kraus Catalogue 137*, no. 29] accords well with the copyist being the Maghribī astronomer of Cairo, and the title "al-Mashhūr" (well-known) is well suited to the author of a work of such high caliber [King, "al-Marrākushī," EI^2, VI, p. 598]. In that case, the copy, whose exceptionally clear text and figures could very well be the product of a specialist's touch, would have to be considered as ca. 680H.

The title of the collection itself is not "*Maqūsṭāt*," derived from Qusṭā! [i.e., Ibn Lūqā: *H. P. Kraus Catalogue 137*, no. 29; caption, p. 67], but *Mutawassiṭāt* (Intermediate [Work]s), as transcribed in the table of contents, where the title of Euclid's *Optics* appears as *Kitāb Ikhtilāf al-manāẓir li-Uqlīdis*, distinct from *Kitāb al-Manāẓir li-Uqlīdis* in the colophon. Here, the explicit attribution of the original composition to Euclid is followed by the unique and important reference to the name of the translator: "translated by (*tarjamahu*) Hlīl ibn Sarḥūn" [i.e., Hiliyā ibn Sarjūn, see Manuscript Plate V, p. 240, and The Arabic "Version," v. 2, n. 19–23]. Particularly notable among the unique features of this copy is the lack of any corresponding part to the last Euclidean definition on clarity of appearance [see Mdef. 4-3 (17), Edition and Translation, pp. 2–3]. The text begins with "*Al-shu'ā' yakhruju min al-'ayn 'alā khuṭūṭin mustaqīmatin . . . .*"

The five available copies represent at least two variations, one having the longer expression *Ikhtilāf al-manāẓir* as the title and the term *al-shu'ā'* as the opening term [those dated 600H, 625H, ca. 680H], the other containing the short form *al-Manāẓir* as part of the title and the expressions *al-manāẓir min al-'ayn* (visual rays) and *mukhtalifat al-ab'ād wa al-maqādīr* (unequal in distance and magnitude) as distinctive features of the opening lines [those dated 692H and 751H]. In the absence of any other identified manuscript variants [the status of the one reported as *Kitāb al-Manāẓir li-Uqlīdis*, ca. 999/1000H – see m6 in List I, p. xxxv – remains uncertain], the slight variation between the two sets may be interpreted in at least two ways: either as two distinct compositions, where one is a revised version of the other, or as a single composition, where one is the restored version of the other. The "revision" hypothesis is in accord with many other such cases of the period, as well as such stylistic differences between variants as *fa-l-yakun* vs. *mithāluhu fa-l-yakun* or *dhālika mā aradnā an-nubayyina* vs. *dhālika mā aradnā bayānahu,* respectively, in the *mithāl* (illustration) and *khātima* (conclusion) parts of the propositions. The "restoration" hypothesis, which still accounts for such stylistic differences, is supported by the evidence of the missing opening part of the earliest extant manuscript copy [dated 600H], the testimony of one variant [dated 692H] regarding the corrupt nature of its original, and above all, the variance among manuscripts, itself being most striking in the very same opening page.

As a more exact measure for determining the relationship between the five available manuscript copies, a quantitatively based comparison between them was conducted in the following manner: 10 columns were drawn up to record the similarities between each one of the manuscript copies with the others: the earliest one dated 600H, with the four dated 625H, 692H, 751H, and ca. 680H [4 columns]; the one dated 625H with the ones dated 692H, 751H, and ca. 680H [3 columns]; the one dated 692H with those dated 751H and ca. 680H [2 columns]; and the one dated 751H with that to be dated ca. 680H [1 column]. Through a line-by-line comparison of the text, identical cases were marked under the appropriate column, the total marks added for each case, and the sum of all 10 columns compared, the range from the largest to smallest numbers representing the highest and lowest degrees of closeness between each two copies compared. According to this quantitative measure, the copies dated 692H and 751H [844 marks] are closest to each other, while the pairs dated 600H, 625H [693 marks] and 692H, 751H [636 marks] are the next closest, and the copies dated 625H and ca. 680H [376 marks] the least close. This strongly supports the assumption of the pair dated 600H and 625H being in one category and the ones dated 692H and 751H in another. But the case of the copy to be dated ca. 680H cannot be as decisive, despite some advantages of this numerical version of a stemma. Closer to the second pair on purely quantitative measures, this copy may be better grouped with the first pair on qualitative grounds, both for its agreement with the latter [having *Ikhtilāf al-manāẓir* rather than *al-Manāẓir* as title and *shu'ā'* rather than *al-manāẓir* as the visual ray vocabulary] and its key differences with the former [lacking *mukhtalifat al-ab'ād wa al-maqādīr* in the first definition].

For purposes of reference, the five manuscript copies are designated in the form of numbers following the letter m [m1-m5, m standing for *al-manāẓir* in the titles of all]. The joint letter-number designation is meant to reflect the distinct variance between the main two sets [m1, m3, m5 vs. m2, m4] and the measure of the closeness of members in each set to the original, which in this case is largely chronological from lower to higher numbers.

The oldest copy [dated 600H] being nonextant in the portion most critical for identification purposes, the manuscript copy denoted as m1 [dated 625H] and closest to it among all copies [see p. xxviii] seems to be the one closest to the original composition on many counts [see Editorial Procedure and Apparatus, pp. xxix–xxx]. The numbering of the other manuscript copies follows according to the odd-even principle and in chronological order, such that the copies dated 600H and ca. 680H are marked as m3 and m5 and the two dated 692H and 751H as m2 and m4 as the respective members of the odd-numbered and even-numbered variant sets.

The number-based denotation further reflects the nature of accessibility to manuscript copies. In the case of m1 and m2, located in Istanbul and Leiden respectively [see List I, p. xxxiv], I have personally examined their original texts, while for m3 and m4 in Cairo [the second filmed from a manuscript in Damietta Institute], and m5, filmed at Basel from the manuscripts once recorded in H. P. Kraus Library, New York [see List I, p. xxxiv] access was limited to reproduced copies alone. One copy, m6, cited to be in Kabul [List I, p. xxxv], has remained inaccessible after years of effort through a series of independent channels to verify its reported identity and locality.

3 Editorial Procedures and Apparatus

The critical edition of *Kitāb Uqlīdis fī Ikhtilāf al-manāẓir* in volume I has been prepared from all five available manuscript copies, denoted here as m1–m5 [see Description, pp. xxvi–xvii, and List I, p. xxxiv]. The establishment of the text (M) is primarily based on m1 [dated 625H], a copy likely to be closest to the original composition, as suggested by some identifiable features including the title [the variants of the other four manuscripts, m2–m5, dated 692H, 600H, 751H, and ca. 680H respectively, are presented in the critical apparatus]. The odd-numbered manuscripts, m1, m3 and m5, would all qualify for such a selection. Not only do their titles as *Kitāb Uqlīdis fī Ikhtilāf al-manāẓir* refer to the name of Euclid, as do those with the short form *al-Manāẓir*, but the association with "an old translation" (*al-naql al-qadīm*) is more explicit in their case [see Summary and Commentary: Title, v. 2, p. 2, n. 43]. The curious description of the visual rays as "unequal in distances and magnitudes" (*mukhtalifat al-ab'ād wa al-maqādīr*) in the set m2, m4 [see Description, pp. xxvi–xxvii] and especially the statement in one of them (m2) regarding the "poor" (*saqīm*), "obscure" (*mughlaq*) and "cut off" (*batr*) nature of its "original" (*aṣl*) [including the alternative readings in Description, p. xxvi] further point to members of the odd set (m1, m3, m5) as more authentic and complete texts.

From among the odd-numbered manuscript copies, m1 was selected as the primary source for the critical edition, largely on account of elements indicating the closeness of this copy to the original composition. A few rarely encountered terms, expressions, and constructs unique to this copy may be considered reflections of the unsettled language of a composition as early as ca. 212/827–828. One is the term *al-manẓar* [in the singular form] in the curious expression *fī al-manẓar* [see Summary and Commentary: Proposition 9, v. 2, p. 41]; another is its cognate in the joint singular-plural form, "*nukhriju al-shu'ā' shu'ā'āt . . .*" [see Summary and Commentary: Propositions 1–7, v. 2, pp. 23–40]; another is the equally uncommon construct *makhraj al-shu'ā'* [see Summary and Commentary: Proposition 4, v. 2, p. 37]; and yet another is the ending *wa al-shu'ā'āt* to the *Ikhtilāf al-manāẓir* of the title. The last two cases have correspondences in some early accounts [in the forms *makhraj al-naẓar* and *wa al-shu'ā'* in al-Ya'qūbī's *Ta'rīkh* (ca. 267/880), and as *makhraj al-nūr min al-baṣar* in Ibn 'Īsā's (ca. 250/864?) *al-Manāẓir*: see Supplementary Passages, pp. 243–244]. Copy m1 is to be further noted as the only variant in which an explicit association with the name of Euclid appears within the text, not just in the title or the colophon [The definition section begins with the opening words "Euclid maintained that" ("*za'ama Uqlīdisu*") in the top margin]. The copy also notably contains marginal notes, many with frequently encountered vocabulary to suggest wide circulation. The very small, at times illegible, script of both the text and the marginal notes, together with missing letters and lines in the diagrams, do provide disadvantages from the standpoint of preparing a critical edition [see Manuscript Plate I, pp. 230–231]. I hope to have minimized problems arising from the poor quality of the copy by personally examining the original manuscript in Istanbul's Ahmet III (New) Library at Topkapi Palace.

In the case of the supplementary text Ṣ, the preparation of the critical edition involved consultation of more manuscript copies, in accordance with the greater number of extant manuscripts and the nature of access to them. The text known as *Iṣlāḥ al-Manāẓir*, which is attributed to al-Kindī and which follows Naṣīr al-Dīn al-Ṭūsī's *Taḥrīr al-Manāẓir li Uqlīdis* in several of its manuscript copies, survives in at least 23 such companion pieces (ṣ1–ṣ7, ṣ9–ṣ24) and only one independent copy (ṣ8) [see List I, pp. xxxv–xxxvii and p. liii]. Of the copies so far identified, all but the independent excerpt (ṣ8) lack a title; 15 (ṣ1, ṣ2, ṣ4, ṣ6, ṣ7, ṣ10, ṣ12, ṣ13, ṣ17–ṣ22, ṣ24) have instead of *Iṣlāḥ* (*Correction*), the supposed title of the text, "*iṣlāḥihi*" (his correction) [al-Kindī's] as part of its introduction; 8 (ṣ3, ṣ8, ṣ9, ṣ18-ṣ21, ṣ23) lack the two accompanying figures; and three (ṣ4, ṣ16, ṣ22) lack clear lines or letters. Among the otherwise almost identical copies, close to half (ṣ1–ṣ9) with the earliest copy dates [ṣ1 (671H), ṣ2 (676H), ṣ3 (701H), ṣ4 (720H), ṣ5 (725H), ṣ6 (784H), ṣ7 (845H), ṣ8 (10th-century H), and ṣ9 (1092H)] were used for the critical edition. The first of these (ṣ1), the copy that has, in addition to the earliest date of transcription, the clearest figures, is used as a primary copy, and the rest, about half of which are being held in restrictive manuscript collections, are reflected in the critical apparatus.

The establishment of the two texts, M and Ṣ, their translations into English, and the representation of their figures have been carried out so as to include a reflection of the textual problems involved. Especially in the case of M, where such problems seem to have been carried over to other texts, relaying a sense of the textual problems, some quite complex, is crucial to the understanding of a tradition developed through some of those very problems. With such priorities, when variations among manuscripts existed, agreement with Heiberg's edition of the Greek has not in itself served as the basis for the reading. Accordingly, the difficult task of translating the contents of a text of this nature into modern English has been quite a sensitive undertaking. The task has been made more difficult by the particular style of the Arabic text: sometimes different terms are employed for one and the same concept [e.g., *mubṣar*, *manẓūr ilayh*, and *shay'* for the object of vision, here translated as visible object, viewed object, and object, respectively, and *munkhafiḍ*, *taḥt*, *dūn*, and *suflā*, translated as lower, under, beneath, and underneath, to reflect the original distinctions; exceptions are the translation of both *miqdār* and *'iẓam* as magnitude and *a'lā* and *arfa'* as higher; see Index of Arabic Terms, v. 2, pp. 127-152]; other times the same term is employed for different concepts [e.g., *baṣar* for both eye and sight], in some cases becoming particularly ambiguous [e.g., *al-shu'ā'* for both the singular and collective sense of the term, here translated as the ray, the ray[s], or the [cone of] ray[s], accordingly; even more problematic, *kathīr* [not to mention *kathra* or *akthar* from the same verbal root] is employed in at least three senses – great in number, size, and clarity, here translated as many, much, and more [clear] as consistently as possible; see Index of Arabic Terms, p. 147]. Such cases have at times produced what may seem too literal a form of translation, presented as such in the hope of reflecting the nature of the problems as closely as possible.

Presentation of Text and Translation: The Arabic text and its corresponding English translation are presented on the right and left side of the same spread, with the corresponding definitions and propositions aligned line by line [numbered every third line]. The guiding principle behind this arrangement has been the breakdown of the texts into a series of corresponding premises, rather than paragraphs, for purposes of easy reference.

Critical Apparatus: Beneath the Arabic text on each right-hand page, the corresponding critical apparatus is placed in smaller type. While the upper text is supplied with vocalization [when needed] and *hamza*, *tashdīd*, *tanwīn,* and bars over letters [throughout, including the lower text behind the dividing brackets], variant readings are presented without any such marks. In a few critical cases, Arabic variants are presented without dots following the manuscript copies to reflect the possibility of multiple reading depending on the positions of the dots.

Figures and Letters: Beneath the corresponding English translation on each left-hand page, the figures are presented with English letters and are accompanied by an English-Arabic letterkey containing the original lettering. For propositions of more than one page, figures are repeated on successive English pages, and for those with more than one figure [40, 41, 45, 46, 47, 63], the figures are in some cases [46, 47, 63] presented on the same page, and in other cases [40, 41, 45] presented on different pages, depending on the nature of textual correspondence.

Numbers and Titles: The transmitted definitions [all but two of the seven Euclidean definitions], not numbered in the original, are presented here as three one-part definitions [1-3 as in the Greek] and one four-part definition [numbered 4.1-4.4 (4 and 7 in Heiberg's edition)], the numbers inserted [e.g., 4.3] next to the translation of each definition. The 64 propositions, originally numbered in *abjad* letters next to enunciations, are numbered at the top middle of the first page of each proposition [for those more than one page, also in the top margins of all consecutive pages] and supplied with titles and subtitles [not in the original text] in the commentary. The 72 figures [70 in M and 2 in Ṣ] are numbered according to the order of propositions. When figures of propositions with more than one figure are on different pages, the proposition number is divided accordingly [40.1, 40.2; 41.1, 41.2; 45.1, 45.2]; when on the same page, the same number is given to both such figures [46, 47, and 63].

Dates: Dates are presented in Hijra/Western years simultaneously [according to table in Bacharach, *A Near East Studies Handbook*]. Dates of composition and of historical figures or periods are presented with a slash between the respective years [e.g., composed 651/1253, d. 672/1274, fl. ca. 3rd/9th century]. For modern sources with Hijra publication dates, the equivalent Western year is included [e.g., 1340=1962] and placed in parentheses if not given in the original [e.g., 1340 (=1962)]. Copy dates of manuscripts are given as one date and depending on their place of copy, either with a letter H for Hijra [e.g., 671H or 671H (*abjad*)], or letters A.D. next to the Western year [e.g., 1269 A.D.].

Transliteration: The system of Arabic and Persian transliteration is adopted from the Library of Congress, with few exceptions: the form ḍ is used for both Arabic and Persian [Library of Congress has two dots under z for the corresponding Persian letter]. Persian names and terms are transliterated according to their native pronounciation [e.g., Raḍavī, Ṭūsī, *nuskhih* (= Riḍawī, al-Ṭūsī, *nuskhah* in Arabic)] unless current in a different form [e.g., Tehran (Ṭihrān only for manuscript inscriptions]. Transliteration of Greek terms is given next to the textual form [e.g., ὄψεις/*opseis*] to serve both classicists and nonspecialists.

Indices: In addition to the General Index, an Index of Arabic Terms is included to transmit a sense of the linguistic style of the texts presented in part II of this volume. The terms index provides references to every occurrence of a given term in an effort to record its relative use and distribution. In some critical cases it also provides references to different senses of the same term [e.g., *al-shuʿāʿ* (the ray) vs. *al-shuʿāʿ* ([collection] of ray[s]); *kathīr* (many) vs. *kathīr* ([more] clearly); see Index of Arabic Terms, pp. 139-140 and p. 147, respectively].

References: The Bibliography contains full references: author, title, volume, and place and date of publication. Works by the same author are entered chronologically, and those by more than one author according to their order in the published volumes. Abbreviated titles precede frequently cited articles and books [e.g., King, "Notes on Sources" = King, David, A., "Notes on the Sources for the History of Islamic Mathematics," *Journal of the American Oriental Society*; King, *Cairo Catalogue* = King, David A., *A Catalogue of the Scientific Manuscripts in the Egyptian National Library*]. Primary sources accompanied by studies are listed in the bibliography by cross-references to modern editions/translations [e.g., Ibn al-Haytham, *Kitāb al-Manāẓir*: see Sabra, *Kitāb al-Manāẓir: Books I-II-III* and Sabra, *The Optics of Ibn al-Haytham*, 2 vols.]. Footnotes and references within the text are given in abbreviated form, with reference to modern edition/translation and volume number when applicable [e.g., Ibn al-Nadīm, *Kitāb al-Fihrist,* ed. Flügel, v. 1, p. 129; tr. Dodge, v. 2, p. 665]. Publication dates are provided in the case of error duplication. Cross-references within the two volumes are given, in the case of the texts, in this format: abbreviated letter, definition or proposition number, line number [e.g., Mdef. 1(3,5), M2(8–10)], otherwise as the section title followed by volume and page number [e.g., The Arabic "Version," v. 2, p. xix]. No volume numbers are given for cross-references within the same volume.

Use of Brackets: Square brackets are employed for editorial notes and additions in English and cross-references within the two volumes, parentheses for presenting dates or equivalent forms, and angled brackets for conjectural additions to the Arabic text and corresponding English translation. Other conventions for the critical edition are as follows:

Apparatus

Text above the line	فس
Text below the line	تس
Text in the margins	هش
Reference to lines in text reading	(،)
Reference to consecutive lines in text reading	(–)
Separation of text and variant reading	[
Addition of variant reading to the text	+ [
Absence of text from manuscript X	X – [
Separation of variant reading from manuscript X	X :
Indication of other variant readings	=
Conjectural additions to the text and translation	< . . . >

4 List of Manuscripts

I Main Texts

كتاب أُقليدس في اختلاف المناظر

Kitāb Uqlīdis fī Ikhtilāf al-manāẓir M/م

كتاب اقليدس في اختلاف المناظر والشعاعات

Kitāb Uqlīdis fī Ikhtilāf al-manāẓir wa al-shuʻāʻāt

Istanbul: Topkapı Sarayı, Ahmet III 3464, 4: 59b–74a, 625H. m1/م١

Date: codex. Some marginal notes.
[Krause, "Stambuler Handschriften," p. 441; Sezgin, *GAS V*, p. 117].

كتاب المناظر لاوقليدس الصورى

Kitāb al-Manāẓir li-Ūqlīdis al-Ṣūrī

Leiden: Universiteitsbibliotheek, Or. 133, 5: 81–105, 692H. m2/م٢

Title, copyist (Ibn al-Ghulām al-Qūnawī, Balkh): colophon. Date: codex. [Vander, *Biblioteca Publica Universitatis Lugduno-Batavæ*, p. 453; Wenrich, *De Auctorum Graecorum*, p. 182; de Jong and de Goeje, *Bibliotheca Academiae Lugduno Batavae*, v. 3, p. 43; Steinschneider, "Euklid bei den Arabern," p. 100; Klamroth, "Schriftstellern bei al-Jaʻqūbī," p. 9; Suter, "Die Mathematiker," p. 40; Brockelmann, *GALS I*, p. 368; Voorhoeve, *Handlist*, p. 184; Dānishpazhūh, *Mikrufilm'hā*, v. 1, p. 557; Sezgin, *GAS V*, p. 117].

كتاب اوقليدس في اختلاف المناظر

Kitāb Ūqlīdis fī Ikhtilāf al-manāẓir

Cairo: Dār al-Kutub al-Miṣriyya, Dr 260, 2. 59a–92a, 600H. m3/م٣

Title: colophon. Lacks page with Euclidean Definitions.
[King, "Notes on Sources," p. 453; King, *Cairo Catalogue*, v. 2, p. 1029; King, *Cairo Survey*, p. 24: 600H].

المناظر من العين

Al-Manāẓir min al-ʻayn

Cairo: Dār al-Kutub al-Miṣriyya, Dr 720: 2–77, 751H. m4/م٤

Title on title page: *Kitāb al-Munāẓira min al-ʻayn*. Date: colophon.
[King, "Notes on Sources," p. 453; King, *Cairo Catalogue*, v. 2, p. 1029; King, *Cairo Survey*, p. 24: "photos of a MS in Damietta Institute," 751H].

كتاب اوقليدس في اختلاف المناظر

Kitāb Ūqlīdis fī Ikhtilāf al-manāẓir

Basel: Universitätsbibliothek Microfilm, 2: 20b–32b, ca. 680H. m5/م٥

Copyist (al-Shaykh Abī ʻAlī al-Mashhūr) and owner stamps (1049H, Ṭihrān, and 1269H, Mashhad): title page.
[Sezgin, *GAS V*, p. 117: 13. Jh. H.; *H. P. Kraus Catalogue 137*, no. 29].

– Entries are presented as they occur in MSs. All identified errors are underlined.

كتاب المناظر لاقليدس

Kitāb al-Manāẓir li-Uqlīdis

Kabul: Wizārat Maṭbū‘āt 45, 55: 255–262, ca. 999–1000H? m6/م٦

[de Beaurecueil, *Manuscrits d'Afghanistan*, p. 294; p. 290: 999–1000H); Sezgin, *GAS V*, p. 117].

إصلاح المناظر

Iṣlāḥ al-Manāẓir Ṣ/ص

Attributed to Abū Yūsuf Ya‘qūb ibn Isḥāq al-Kindī (d. ca. 257/870)

Tehran: Muṭahharī (Sipahsālār) 4727, 8: 101b (194), 671H. ṣ1/ص١

Copyist and date: codex [see t1/ت١].

[Dānishpazhūh and Munzavī, *Sipahsālār*, v. 3, p. 350: "by Kindī on correcting himself" (*dar iṣlāḥ-i khudash*)].

Istanbul: Topkapı Sarayı, Ahmet III 3453, 9: 119b, 676H. ṣ2/ص٢

Date: codex [see t3/ت٣]. The 2nd figure has unclear letters.

First figure published as part of T3: *RIMA*, 1963, 9: 290.

Tehran: Muṭahharī (Sipahsālār) 559, 16: 47a (115), 701H. ṣ3/ص٣

Copyist and date: codex. [see t6/ت٦]. No figures.

[Dānishpazhūh and Munzavī, *Sipahsālār*, v. 3, p. 349: *Fāyidih-i riāḍī*].

Istanbul: Topkapı Sarayı, Ahmet III 3456, 8: 43a, 720H. ṣ4/ص٤

Date: codex [see t10/ت١٠]. Both figures unclear.

Istanbul: Köprülü 931, 7: 81b, 725H. ṣ5/ص٥

Date: codex [see t12/ت١٢].

Tehran: Muṭahharī (Sipahsālār) 597, 8: 78a, 784H. ṣ6/ص٦

Copyist and date: codex [see t13/ت١٣].

[Dānishpazhūh and Munzavī, *Sipahsālār*, v. 3, p. 349: copyist, date].

Istanbul: Süleymaniye, Ayasofya 2760, 10: 87b, 845H. ṣ7/ص٧

Date: colophon [see t14/ت١٤].

من المناظر

Paris: Bibliothèque Nationale 2467, 2: 56b, ca. 900 H. ṣ8/ص٨

Date: codex. No figures.
[de Slane, *Bibliothèque Nationale*, p. 436: XVIe siécle; Brockelmann, *GALS I*, p. 374; Sezgin, *GAS V*, p. 117: 10 Jh. H, ff. 56–58, refers to Latin *De aspectibus*; King, *Cairo Survey*, p. 34].

Isfahan: University, Dānishkadih-i Adabiyyāt 285 (16955), ca. 1000H ṣ9/ص ٩

Tehran University Microfilm no. 4537 [see t21/ت ٢١]. No figures.

[Dānishpazhūh, *Mikrufilm'hā*, v. 3, p. 69; Afshār and Dānishpazhūh, *Iṣfahān*, p. 886, ca. 1000H, refer to Microfilm no. 4537].

Tehran: Majlis-i Shawrā 658, 3: 125–126, ca. 1000H. ṣ10/ص ١٠

[Ḥā'irī, *Majlis*, v. 22, p. 349: fols. 63b–64a, date, see t22/ت ٢٢].

Istanbul: Atif 1712, 5, ca. 1100H. [see t28/ت ٢٨]. ṣ11/ص ١١

Istanbul: Süleymaniye, Carullah 1455, 40b, 1105H [see t30/ت ٣٠]. ṣ12/ص ١٢

Istanbul: Süleymaniye, Beşir Ağa 440, 100b, 1134H [see t31/ت ٣١]. ṣ13/ص ١٣

Cairo: Dār al-kutub al-Miṣriyya, MR40, 7: fol. 88a, 1146H. ṣ14/ص ١٤

Copyist: codex. Date: colophon [see t32/ت ٣٢].

[King, "Notes on Sources," p. 453; King, *Cairo Catalogue*, v. 2, p. 1029: MR 40, 6; King, *Cairo Survey*, p. 34: MR40, 2, fol. 88r].

Princeton: Princeton University 4855 (Yehud. 1553): 17b, 1179H. ṣ15/ص ١٥

Copyist, date: colophon [see t33/ت ٣٣; Mach, *Garrett Collection*, p. 419].

Istanbul: Topkapı Sarayı, Hazine 603, 3, 46a, 1194H. ṣ16/ص ١٦

Copyist: colophon. Date: codex [see t35/ت ٣٥]. Both figures unclear.

Manchester: John Rylands 350 (381), 5: 79b–80a, 1196H. ṣ17/ص ١٧

Copy: colophon. Date: codex [see t36/ت ٣٦].

[Mingana, *John Rylands*, p. 546: late 18th century; King, "Notes on Sources," p. 453: 1150H; King, *Cairo Survey*, p. 34: 1196H].

Tehran: Malik 3078, 3: 101b, 1206H. ṣ18/ص ١٨

Copyist and date: colophon [see t40/ت ٤٠]. Margin of t40. No figures.

Tehran: Majlis-i Shawrā 208, 3: 16b, 1251H. ṣ19/ص ١٩

Date: codex [see t45/ت ٤٥]. No figures.

Mashhad: Āstān-i Quds 399 (5428), 3: 99 (69), 1267H. ṣ20/ص ٢٠

Copyist and date: codex [see t46/ت ٤٦]. No figures.

[Gulchīn Ma'ānī, *Āstān-i Quds*, p. 360: cites last line as part of T].

Mashhad: Āstān-i Quds 597 (12124), 8: 28, 1309H. ṣ21/ص ٢١

Copyist and date: colophon [see t49/ت ٤٩]. No figures.

['Irfāniān, *Āstān-i Quds*, v. 10, p. 155: cites last line as part of T].

Krakow: Jagiellonska, Or. Qu. 559, 3: 117a–118b, ?H. ṣ22/ص ٢٢

[See t51/ت ٥١]. Both figures unclear.

Krakow: Jagiellonska Or. 258, 14: 163b, ?H. ṣ23/ص ٢٣

[See t52/ت ٥٢]. No figures. See also Addendum, p. liii.

[Ahlwardt, *Königlichen Bibliothek*, v. 5, p. 354: lists ṣ22–23 as Berlin 6017].

تحرير المناظر لأقليدس

Taḥrīr al-Manāẓir li-Uqlīdis T/ت

by Naṣīr al-Dīn al-Ṭūsī (d. 672/1274)

تحرير كتاب المناظر لاقليدس

Taḥrīr Kitāb al-Manāẓir li-Uqlīdis

Tehran: Muṭahharī (Sipahsālār) 4727, 7: 97b–101b (186–194), 671H. t1/ت ١

Copyist (Muḥammad ibn Aḥmad Muḥammad ibn Rabī' al-Zarkashī): codex; date [671 H (abjad)]: colophon.
[Dānishpazhūh and Munzavī, *Sipahsālār*, v. 3, p. 349: 96b–101b (p. 330: codex entry); Mudarris-i Raḍavī, *Ahvāl va āthār*, p. 208 (p. 209: codex date, no MS nos.); Mudarrisī, *Sarguzasht*, p. 112 (no MS nos.)].

تحرير كتاب المناظر لاقليدس

Taḥrīr Kitāb al-Manāẓir li-Uqlīdis

Istanbul: Selim Ağa 743, 16: 263b–271a, 671H. t2/ت ٢

Süleymaniye Microfilm no. 2531. Date: codex. Many marginal notes.
[Krause,"Stambuler Handschriften," p. 500: 743, 6; Sezgin, *GAS V*, p. 117].

تحرير المناظر لاقليدس

Taḥrīr al-Manāẓir li-Uqlīdis

Istanbul: Topkapı Sarayı, Ahmet III 3453, 8: 115a–119b, 676H. t3/ت ٣

Date: codex. Many marginal notes.
[Dimirdash (ed.), *Revue de l'Institut des Manuscrits Arabes* (*RIMA*), 1963, 9: 251–290: includes 1st figure of Ṣ as part of T; *Topkapı Sarayı* 15, p. 735; Sezgin, *GAS V*, p. 117: 3453/7:115a–126b, 677H].

تحرير كتاب المناظر لاقليدس

Taḥrīr Kitāb al-Manāẓir li-Uqlīdis

Leiden: Universiteitsbibliotheek, Or. 133, 3: 93b–99a, 692H. t4/ت ٤

Copyist (Ibn al-Ghulām al-Qūnawī), place (Balkh) and date: codex.
[Wenrich, *De Auctorum Graecorum*, p. 182; de Jong and de Goeje, *Bibliotheca Academiae Lugduno Batavae*, v. 3, p. 43; Steinschneider, "Euklid bei den Arabern," p.100; Steinschneider, "Die arabischen Uebersetzungen," p. 171: Leyden 977; Suter, "Die Mathematiker," p. 40; Brockelmann, *GALS I*, p. 368, p. 930: Leid. 977; Voorhoeve, *Handlist*, p. 184: CCO 977; Dānishpazhūh, *Mikrufīlm'hā*, v. 1, p. 557; Sezgin, *GAS V*, p. 117].

تحرير المناظر لاوقليدس

Taḥrīr al-Manāẓir li-Ūqlīdis

Tehran: Tehran University, Markazī 2432, 5: 97a–103b, 695H. t5/ت ٥

Date: codex. Many marginal notes.
[Dānishpazhūh, *Markazī*, v. 9, p. 1099–1100; v. 15, p. 214].

تحرير المناظر لاوقليدس

Taḥrīr al-Manāẓir li-Ūqlīdis

Tehran: Muṭahharī (Sipahsālār) 559, 16: 44b–47a (111–115), 701H. t6/ت ٦

Copyist (ʻAbd al-Qādir ibn ʻAbd al-Rashīd Abarqūhī),
Place (Tabriz) and date: codex.
[Dānishpazhūh and Munzavī, *Sipahsālār*, v. 3, p. 349 (ibid., p. 333: cites p. 99r of codex for date and Mullā Muḥammad Narāqī as owner)].

تحرير المناظر لاقليدس

Kītāb al-Manāẓir li-Uqlīdis

Rampur, 410?, 709H. t7/ت ٧

Date and place of copy (Tabriz): colophon.
Published in *Majmūʻ al-rasāʼil* , Hyderabad, 1358 (=1939): 2– 24.
[Sezgin, *GAS V*, p. 117; Brockelmann, *GALS I*, p. 930: Rampur 410].

Istanbul: ʻAskarī Müze, 716H? t8/ت ٨

Unable to locate (*Taḥrīrāt* Codex: 723H, did not include *Manāẓir*).
[Krause, "Stambuler Handschriften," p. 500: Armeemus. 769, 12: 181–187, 716H; Sezgin, *GAS V*, P. 117: 769, 5:109–121, 716H; Şeşen, *Maktabāt Turkiyya*, v. 1, p. 33, no. 83: 108–120: 715H].

تجريد المناظر لاوقليدس

Tajrīd al-Manāẓir li-Ūqlīdis

Oxford: Bodleian 875, 6: 86b–96a, 719H. t9/ت ٩

Copyist (al-Shaykh Zayn al-Dīn al-Abharī) and date: codex. Marginal notes.
[Uri, *Bibliotheca Bodleiana*, p. 189; Wenrich, *De Auctorum Graecorum*, p. 182; Steinschneider, "Euklid bei den Arabern," p. 100; Klamroth, "Schriftstellern bei al-Jaʻqūbī," p. 9; Suter, "Die Mathematiker," p. 40; Brockelmann, *GALS I*, p. 930 (Bodl. 895 cited, lacks *Taḥrīr al-Manāẓir*, see Uri, *Bibliotheca Bodleiana*, p. 194); Winter, "British Isles," p. 76: Seld A 45: 719H; King, "Notes on Sources," p. 453: refers to Winter].

تحرير المناظر لاقليدس

Taḥrīr al-Manāẓir li-Uqlīdis

Istanbul: Topkapı Sarayı, Ahmet III 3456, 7: 40a–43a, 720H. t10/ت ١٠

Date: codex. Many marginal notes.
[Krause, "Stambuler Handschriften," p. 500: 3456, 8, 720 h; Sezgin, *GAS V*, p. 117: 3456, 8: 40a–43a; Karatay, *Topkapı Sarayı* 15, p. 736].

تحرير المناظر لاوقليدس

Taḥrīr al-Manāẓir li-Ūqlīdis

Paris: Bibliothèque Nationale 5974, 7: 73b–79b, 722H. t11/ ت١١

Date: colophon.
[Derenbourg, *Bibliothèque Nationale*, p. 47: "copié sur un manuscrit copié en 700 de hégire sur le manuscrit original de Nasir ad Din al-Tousi"; Blochet, *Bibliothèque Nationale*, p. 153: "copié en 722 (1322) sur un exemplaire de 700 (1300) exécuté d'apres le manuscrit original"; Sezgin, *GAS V*, p. 117: refers to Vajda, *Bibliothèque Nationale*, p. 449].

تحرير المناظر لاقليدس

Taḥrīr al-Manāẓir li-Uqlīdis

Istanbul. Köprülü 931, 6: 75a–81b, 725H. t12/ ت١٢

Date: codex.
[Krause, "Stambuler Handschriften," p. 500: 75–82, 725 H; Sezgin, *GAS V*, p. 117: 75–82; Şeşen, *Köprülü*, p. 469: 75–82].

تحرير المناظر لاقليدس

Taḥrīr al-Manāẓir li-Uqlīdis

Tehran: Muṭahharī (Sipahsālār) 597, 7: 71b–78a, 784H. t13/ ت١٣

Copyist (Ḥusayn ibn Muṭahhar) and date: codex.
[Dānishpazhūh and Munzavī, *Sipahsālār*, v. 3, p. 349: naskh, Ḥusayn ibn Muṭahhar, 12th Muḥarram 784H, microfim at Markazī Library, Tehran].

تحرير المناظر لاقليدس

Taḥrīr al-Manāẓir li-Uqlīdis

Istanbul: Süleymaniye, Ayasofya 2760, 9: 80b–87b, 845H. t14/ ت١٤

Date: codex.
[Krause, "Stambuler Handschriften," p. 500: 11, 134b–141b, 845H; Sezgin, *GAS V*, p. 117: 11, 134b–141b, 845H].

تحرير المناظر لاوقليدس

Taḥrīr al-Manāẓir li-Ūqlīdis

Florence: Med.-Laurenziana Or. 164, 6: 112b–124a, 872H. t15/ ت١٥

Date: codex.
[Assemani, *Bibliotheca Medicea*, , p. 381: no. 271; Suter, "Die Mathematiker," p. 40; Brockelmann, *GALS I*, p. 930].

تحرير كتاب المناظر لاقليدس

Taḥrīr Kitāb al-Manāẓir li-Uqlīdis

Tehran, Malik 687, 3: 26b–55a, 881H. t16/ ت١٦

Copyist (Manṣūr ibn Mu'īn), and place (Harāt): colophon
[Afshār and Dānishpazhūh, *Malik*, v. 5, pp. 146–147, no. 117: cites copyist and date and marginal signature as N. Badr].

تحرير المناظر الاقليدس

Taḥrīr al-Manāẓir al-Uqlīdis

Qum: Mar'ashī-yi Najafī 4521, 6: 32a–38a, 880–881H. t17/ت ١٧

Copyist (Ḥusayn ibn Ḥasan, Shahanshāh-i Simnānī): catalogue. Few notes. [Ḥusaynī and Mar'ashī, *Mar'ashī*, v. 12, p. 99: refers to no. 4009 (t23/ت ٢٣); ibid., p. 103: cites copyist and date].

Cairo: Dār al-Kutub DR 897, 3: 93b–99a, ca. 900H. t18/ت ١٨

[King, *Cairo Survey*, p. 24: photos of MS Leiden 897, 3].

Tehran: Muṭahharī (Sipahsālār) 689, 4: 96a–106a, ca. 900H. t19/ت ١٩

Owner stamps (1262H and 1297H): title page. [Dānishpazhūh and Munzavī, *Sipahsālār*, v. 3, p. 350 (ibid., p. 343: ca. 900H)].

Tehran: Muṭahharī (Sipahsālār) 688, 4: 57a–63a, ca. 1000H. t20/ت ٢٠

[Dānishpazhūh and Munzavī, *Sipahsālār*, v. 3, p. 350 (p. 332: ca. 1000H)].

تحرير المناظر

Taḥrīr al-Manāẓir

Isfahan: Dānishkadih-i Adabiyyāt 285 (16955), ca. 1000H. t21/ت ٢١

Tehran: Tehran University Markazī. Microfilm no. 4537. Marginal notes. [Dānishpazhūh, *Mikrufilm'hā*, v. 3, p. 69; Afshār and Dānishpazhūh, *Iṣfahān*, p. 886, ca. 1000H, refers to Microfilm no. 4537].

تحرير كتاب المناظر لاوقليدس الصورى

Taḥrīr Kitāb al-Manāẓir li-Ūqlīdis al-Ṣūrī

Tehran: Majlis-i Shawrā 658, 2: 89–124, ca. 1000H. t22/ت ٢٢

[Ḥā'irī, *Majlis*, v. 22, p. 349: fols. 45b–63a, date].

تحرير المناظر لاقليدس

Taḥrīr al-Manāẓir li-Uqlīdis

Qum: Mar'ashī-yi Najafī 4009, 4: 111–123, 1009H. t23/ت ٢٣

Copyist ('Abd al-Riḍā ibn 'abd al-Majīd al-Ḥusaynī): codex. [Ḥusaynī and Mar'ashī, *Mar'ashī*, v. 11, p. 10: "recension of a text translated into Arabic during Ma'mūn's reign"; cites copyist and date].

Cairo: Dār al-Kutub al-Miṣriyya DR 704, 1: 1b–15b, 1011H. t24/ت ٢٤

[King, "Notes on Sources," p. 453; King, *Cairo Catalogue* v. 2, p. 1029; King, *Cairo Survey*, p. 24: "with additional notes by the editor not in published text"].

تحرير المناظر لاقليدس

Taḥrīr al-Manāẓir li-Uqlīdis

Berlin: 6016 (pet. 677): 62b–70b, 1064H. t25/ت ٢٥

[Ahlwardt, *Königlichen Bibliothek,* v. 5, p. 354: no date [in colophon]
Steinschneider, "Die arabischen Uebersetzungen," p. 171: Ahlw. 6017 (3MSs); Brockelmann, *GAL I*, p. 674; Sezgin, *GAS V*, p. 117].

تحرير المناظر لاوقليدس

Taḥrīr al-Manāẓir li-Ūqlīdis

Tehran: Tehran University Markazī 1246, 1: 2b–12a, 1084H. t26/ت ٢٦

Date ("compared with original"): colophon. Figures in Red. Few notes.
[Dānishpazhūh, *Mishkāt*, v. 3, n. 533, p. 2623: 1084 or 984: "damaged"].

تحرير المناظر لاقليدس

Taḥrīr al-Manāẓir li-Uqlīdis

London: India Office 1249 (743), 2: 36–56, ca. 1100H. t27/ت ٢٧

[Steinschneider, "Euklid bei den Arabern," p. 100; Steinschneider, "Die arabischen Uebersetzungen," p. 171; Suter, "Die Mathematiker," p. 40; Brockelmann, *GALS* I, p. 930; Winter, "British Isles," p. 86; Sezgin, *GAS V*, p. 117: India Off. 1249/2 (ff. 36–56, 12 Jh. H., s. Kat. No. 743)].

Istanbul: Atif 1712, 4: 30a–36b, ca. 1100H. t28/ت ٢٨

[Sezgin , *GAS V*, p. 117: 12. Jh. H.].

Istanbul: Atif 1716, 8: 13ff., 1103H. t29/ت ٢٩

[Sezgin , *GAS V*, p. 117: copy date].

Istanbul: Süleymaniye, Carullah 1455, 4, 36b–40b, 1105H. t30/ت ٣٠

[Sezgin, *GAS V*, p. 117: 5ff. , copy date].

Istanbul: Süleymaniye, Beşir Ağa 440, 6, fols. 91b–100a, 1134H. t31/ت ٣١

[Sezgin, *GAS V*, p. 117: 10ff., copy date].

تحرير كتاب المناظر لاقليدس

Taḥrīr Kitāb al-Manāẓir li-Uqlīdis

Cairo: Dār al-Kutub al-Miṣriyya, MR40, 6: 80b–88a, 1146H. t32/ت ٣٢

Copyist (Muṣṭafā Ṣidqī) and date: codex.
[King, "Notes on Sources," p. 453: MR 40, 2; King, *Cairo Catalogue*, v. 2, p. 1029: MR40, 6; King, *Cairo Survey*, p. 24].

تحرير المناظر لاقليدس

Taḥrīr al-Manāẓir li-Uqlīdis

Princeton: Princeton University 4855 (Yehud. 1553): 1b–17a, 1179H. t33/ت ٣٣

Copyist (Fayḍullāh Shakar-zadah) and date: codex.
[Mach, *Garrett Collection*, p. 419].

تحرير كتاب المناظر لاقليدس

Taḥrīr Kitāb al-Manāẓir li-Uqlīdis

Mashhad: Āstān-i Quds 402 (5450), 1183H. t34/ت ٣٤

Copyist (Muḥammad Bāqir ibn Sheykh Ḥusayn Gīlānī), and date: colophon.
[Gulchīn Ma'ānī, *Āstān-i Quds*, v. 8, pp. 365–366: copyist and date].

تحرير المناظر

Taḥrīr al-Manāẓir

Istanbul: Topkapı Sarayı, Hazine 603, 2: 32b–45b, 1194H. t35/ت ٣٥

Copyist: 'Abd al-Ḍa'īf al-Ḥāj al-Ḥāfiẓ al-Aḥmad. Date: codex.
[Sezgin, *GAS V*, p. 117].

تحرير كتاب المناظر لاوقليدس الصورى

Taḥrīr Kitāb al-Manāẓir li-Ūqlīdis al-Ṣūrī

Manchester: John Rylands 350 (381), 4: 60b–79b, 1196H. t36/ت ٣٦

"Copied from the copy of a copy in the hand of Quṭb al-Dīn al-Shīrāzī": colophon. Date: codex.
[Mingana, *John Rylands*, p. 546; Winter, "British Isles," p. 85: Crawford (381): 73 figures; King, "Notes on Sources," p. 453: 1150H].

Cairo: Dār al-Kutub ṬR 102, 5: 115–126a, ca. 1200H. t37/ت ٣٧

[King, "Notes on Sources," p. 453 1130H; King, *Cairo Catalogue*, v. 2, p. 1029: ca. 1200H; King, *Cairo Survey*, p. 24].

مناظر اقليدس

Manāẓir-i Uqlīdis

Tehran: Majlis-i Shawrā 209, 4: 62–94,. ca. 1200H. t38/ت ٣٨

Date suggested by Mr. 'Abd al-Ḥusayn Hā'irī (Librarian). Copy incomplete and in two hands. A marginal note (1st page) is signed Bīrjandī.
[I'tiṣāmī, *Majlis*, p. 118: cites title with *Ṣūrī*) and different beginning].

Mashhad: Āstān-i Quds 395 (5456): 397–400, ca. 1200H [fragment] t39/ت ٣٩

[Gulchīn Ma'ānī, *Āstān-i Quds*, v. 8, p. 353: "lacks 1st part and figures" (ibid., p.355: ca. 1200H)].

كتاب المناظر لاقليدس الصور

Kitāb al-Manāẓir li-Uqlīdis al-Ṣūrr

Tehran: Malik 3078, 2: 79b–102a, 1206H. t40/ت ٤٠

Copyist (Ḥusayn-'Alī ibn Luṭf-'Alī) and date: codex.
[Afshār and Dānishpazhūh, *Malik*, v. 6, p. 201: no. 732].

Rampur: Riāsat 63, 1214H. t41/ ت ٤١

Copyist ('Abd al-Raḥmān) and date: catalogue.
[Bahādur, *Riāsat*, p. 418: codex 944 fols., "*manqūl az nuskhih-i* (communicated from the copy of) *Maḥmūd ibn Mas'ūd al-Shīrāzī*"].

رسالة فى علم المناظر والمرايا

Risāla fī 'Ilm al-manāẓir wa-al-marāyā

St. Petersbourg: M. E. Saltykov-Shchedrin State Public Library.
Chanykov 144: 281a–287a, 1232H. t42/ ت ٤٢

[Dorn, *Chanykov*, p. 258: lists this as 11th entry, with date from colophon (p. 259: lists as the 13th entry, a *Kitāb al-Manāẓir* for *Kitāb al-Masākin*?)].

تحرير المناظر لاقليدس

Taḥrīr al-Manāẓir li-Uqlīdis

Tehran: Majlis-i Shawrā (former Sinā) 1189, 12: 145–166, ca. 1237H. t43/ ت ٤٣

Copyist (Aḥmad Bābī) and date: catalogue. Includes notes in Persian.
[Dānishpazhūh and 'Ilmī-yi Anvārī, *Majlis*, p. 164: copyist and date].

تحرير المناظر لاقليدس

Taḥrīr al-Manāẓir li-Uqlīdis

Mashhad: Āstān-i Quds 594 (12169), 3, 47 folios, 1237H. t44/ ت ٤٤

Copyist (Muḥammad Khabūshānī) and date: colophon. No folio numbers. Few notes. Ref. to "Muḥammad Bāqir al-Yazdī": codex.
['Irfāniān, *Āstān-i Quds*, v. 10, p. 149 (p. 150: copyist, date)].

Tehran. Majlis-i Shawrā 208, 2, fols. 6a–16b, 1251. t45/ ت ٤٥

Date: codex.
[I'tiṣāmī, *Majlis*, p. 117: Manāẓir Uqlīdis, 1252 H].

Mashhad: Āstān-i Quds 399 (5428), 2: 81–99 (51–69), 1267H. t46/ ت ٤٦

Copyist ('Abdullāh ibn Muḥammad): codex. Marginal notes.
[Gulchīn Ma'ānī, *Āstān-i Quds*, 8, p. 360: cites last line from Ṣ].

تحرير كتاب المناظر لاقليدس

Rampur: Riāsat , 163?, 3: 451–470, 1281H. t47/ ت ٤٧

[Bahādur, *Riāsat*, 3, p. 2: "*Ākhar nā tamām*" (incomplete ending)].

كتاب المناظرة (فى كيفية الابصار)

Kitāb al-Munāẓira (*Fī Kayfīyat al-ibṣār*)

Tehran: Lithograph, 1304 (=1886–1887), 2: 54–75, 1304H. t48/ ت ٤٨

Copyist ('Abd al-Karīm al-Sharīf al-Shīrāzī): p. 53. Date: colophon.
[Ṭihrānī, *Al-Dharī'a*, v. 3, p. 393; Mudarris-i Raḍavī, *Ahvāl va āthār*, p. 208; Dānishpazhūh and Munzavī, *Sipahsālār*, v. 3, p. 349; Mushār, *Mu'allifīn*, p. 736; Mushār, *Chāppī-yi 'arabī*, p. 163; Storey, *Persian Literature*, v. 2, p. 2 (referring to Ṭihrānī)].

كتاب المناظر والمرايا لاقليدس الصورى

Kitāb al-Manāẓir wa-al-marāyā li-Uqlīdis al-Ṣūrī

Mashhad: Āstān-i Quds 597 (12124), 7: 1–28, 1309H. t49/ت ٤٩

Copyist (Asadullāh Munajjim-i Māzandarānī), and date: codex.
['Irfāniān, *Āstān-i Quds*, p. 155: cites last line from Ṣ].

تحرير المناظر لاوقليدس

Taḥrīr al-Manāẓir li-Ūqlīdis

Florence: Med.-Laurenziana, Or. 15, 5: 312a–344, ?H. t50/ت ٥٠

[Assemani, *Bibliotheca Medicea*, p. 392: no. 286; Suter, "Die Mathematiker," p. 40; Brockelmann, *GALS I*, p. 930].

تحرير كتاب المناظر الاوقليدس الصورى

Taḥrīr Kitāb al-Manāẓir al-Ūqlīdis al-Ṣūrī

Krakow: Jagiellonska Qu. 559: 100a–117a,?H. Marginal notes. t51/ت ٥١

"Copied from a copy, itself copied from a manuscript in the hand of Quṭb a-Dīn al-Shīrāzī": colophon.
[Steinschneider, "Euklid bei den Arabern," p. 100; Ahlwardt, *Königlichen Bibliothek*, v. 5, p. 354: listed under Berlin 6017(2); Sezgin, *GAS V*, p. 117].

تحرير المناظر لاقليدس

Taḥrīr al-Manāẓir li-Uqlīdis

Krakow: Jagiellonska Or. 258, 13: 154a–163b, ?H. t52/ت ٥٢

[Ahlwardt, *Königlichen Bibliothek*, v. 5, p. 354: listed under Berlin 6017(1); Sezgin, *GAS V*, p. 117].

Ankara: Saib 4186: 1a–14a, ?H t53/ت ٥٣

[Sezgin, *GAS V*, p. 117].

Istanbul: 'Askarī Müse, 82: 221a–230b, ?H t54/ت ٥٤

[Şeşen, *Maktabāt Turkiyya*, v. 1, p. 33: "*katabat li khazānat al-Sulṭān Muḥammad Fātiḥ*"].

Berlin Qu. 1867, 6: 91b–100b, ?H. t55/ت ٥٥

[Sezgin, *GAS V*, p. 117]. See also Addendum, p. liii.

تجريد أقليدس فى المناظر

Tajrīd Uqlīdis fī al-Manāẓir J /ج

by Jamāl al-Dīn Muḥammad ibn al-Ṣaḥib Kamāl al-Dīn 'Umar ibn Aḥmad ibn Hibatallāh (Ibn Abī Jarāda): colophon.
Cairo: Dār al-Kutub, DR 638: 12 folios, ca. 1100H, unique?
[King, "Notes on Sources," p. 453: 2 fols. ca. 1100H; King, *Cairo Catalogue*, v. 2, p. 1030: 12 folio, ca. 1100H; King, *Cairo Survey*, p. 24: "12 folio, ca. 1100H, unique (?)"].

II Supplementary Texts

كتاب المناظر والمرايا المحرقه

Kitāb al-Manāẓir wa al-marāyā al-muḥriqa Ḥ/ح

by Aḥmad ibn ʿĪsā (ca. 250/864?)

[Krause,"Stambuler Handschriften," p. 513: "vielleicht schon vor 250h verfasst"].

كتاب فيه المناظر والمرايا المحرقه تاليف احمد بن عيسى على مذهب اقليدس فى علل البصر

Kitāb fīhi al-Manāẓir wa al-marāyā al-muḥriqa ta' līf Aḥmad ibn ʿĪsā ʿalā madhhab Uqlīdis fī ʿilal al-baṣar

Istanbul: Süleymaniye, Ragip Paşa 934, 141 fols., undated (ca. 600 H?). ḥ1/ح١

[Krause, "Stambuler Handschriften," p. 514; Sabra, *The Optics of Ibn al-Haytham*, v. 2, p. xxxvii, n. 39: "?thirteenth century A.D."].

كتاب المناظر والمرايا المحرقه من تاليف احمد بن عيسى على مذهب اقليدس من علل البصر

Kitāb al-Manāẓir wa al-marāyā al-muḥriqa min ta' līf Aḥmad ibn ʿĪsā ʿalā madhhab Uqlīdis min ʿilal al-baṣar

Istanbul: Süleymaniye, Laleli 2759, 2: 21b–151b, before 870H. ḥ2/ح٢

[Krause, "Stambuler Handschriften," p. 514: "vor 870h geschrieben"; Şeşen, *Maktabāt Turkiyya*, v. 1, p. 318: "Kitāb al-Manāẓir wa al-marāyā al-muḥriqa ʿalā madhhab Uqlīdis min ʿilal al-baṣar", 7th century, 21–150b; Sabra, *The Optics of Ibn al-Haytham*, v. 2, p. xxxvii, n. 39: 2752, "incomplete"; ownership date: "870/1465–66"].

رسالة فى ايضاح وجدان أبعاد ما بين الناظر ومراكز أعمدة الجبال

Risāla fī Īḍāḥ wijdān abʿād mā bayn al-nāẓir W/و

wa marākiz aʿmadat al-jibāl

by Yaʿqūb al-Kindī (d. ca. 257/870)

رسالة يعقوب بن اسحق الكندى فى ايضاح وجدان ابعاد ما بين الناظر ومراكز اعمدة الجبال وعلواعمدتها وعلم عمق الابار وعروض الانهار وغيرذلك وهو تسمى موريسطس

Risāla Yaʿqūb ibn Isḥāq al-Kindī fī Īḍāḥ wijdān abʿād mā bayn al-nāẓir wa marākiz aʿmadat al-jibāl wa ʿulūw aʿmidatihā wa ʿilm ʿumq al-ābār wa ʿurūḍ al-anhār wa ghayr dhālika wa huwa tussamā Mūrīsṭus.

Istanbul: Süleymaniye, Ayasofya 4830, 13, 626H. w1/و١

[Krause, "Stambuler Handschriften," p. 449: 204b–210b: date; Brocklemann, *GALS I*, p. 374].

رسالة الكندى فى ايضاح وجدان ابعاد ما بين الناظر
ومراكز اعمدة الجبال وعلو اعمدة الجبال

Risālat al-Kindī fī Īḍāḥ wijdān ab'ād mā bayn al-nāẓir wa marākiz a'madat al-jibāl wa 'ulūw a'midat al-jibāl

Istanbul: Süleymaniye, Ayasofya 4832, 28: 66b–70a (218b–222a), 568H. w2/و٢

[Krause, "Stambuler Handschriften," p. 449: 66b–70a, 5 s. h.; Ritter and Plessner, "Stambuler Bibliotheken," p. 370: 66b–70a, 568H; Brocklemann, *GALS I*, p. 374].

كتاب فى علل ما يعرض فى المرايا من اختلاف المناظر

Kitāb fī 'Ilal mā ya'riḍu fī al-marāyā min ikhtilāf al-manāẓir L/ل

by Qusṭā ibn Lūqā al-Yūnānī (d. 287/912)

Mashhad: Āstān-i Quds 392 (5593), 1: 1–31, 867H.

[Gulchīn Ma'ānī, *Āstān-i Quds*, v. 8, p. 344: page nos., date; Toomer, *Diocles on Burning Mirrors*, pp. 26–28: "unique"?; refers to Gulchīn Ma'ānī].

المساحات المناظريه

al-Misāḥāt al-manāẓiriyya Ẓ/ظ

by Sinān ibn al-Fatḥ (ca. 4th/10th century)

Cairo: Dār al-Kutub Dr 260, 3: 92b–94a, ca. 600H.

[King, *Cairo Catalogue*, v. 2, p. 1030; King, *Cairo Survey*, p. 39: "*Qawl fī l-misāḥāt al-manāẓirīya*": unique; Sezgin, *GAS VII*, p. 406].

القول فى الإبصار والمبصر

al-Qawl fī al-Ibṣār wa al-mubṣar IM/ام

by Abū al-Ḥasan al-'Āmirī (d. 382/992)

Cairo: Dār al-kutub, Taymūr Ḥikma 98, no. folio nos., unique?

[Khalīfāt (ed.), *'Āmirī, Rasā'il*, pp. 335–363].

فى معرفة ارتفاع الأشخاص القائمة وأعمدة الجبال وارتفاع الغيوم

Fī Ma'rifat irtifā' al-ashkhāṣ al-qā'ima KH/خ

wa a'midat al- jibāl wa irtifā' al-ghuyūm

by al-Shaykh Abī 'Alī ibn al-Haytham (d. ca. 432/1040)

Leiden: Universiteitsbibliotheek, Or. 14, 8: 236–237, ca. 500H KH1/ خ١

[de Jong and de Goeje, *Bibliotheca Academiae Lugduno Batavae*, v. 3, p. 60; Voorhoeve, *Handlist*, p. 195; Sezgin, *GAS V*, p. 370: 6 Jh., lists other MSs in Oxford and Tehran].

في استخراج أعمدة الجبال

Fī Istikhrāj a'midat al-jibāl S/س

by al-Ḥasan ibn al-Ḥasan ibn al-Haytham (d. ca. 432/1040)
Oxford: Bodleian Library Seld A32: 187a–188b.

شعاع

Shu'ā' [in Persian] SH/ش

by Naṣīr al-Dīn al-Ṭūsī (d. 672/1274)
[Not in Storey, *Persian Literature*].

Istanbul: Topkapı Sarayı, Ahmet III 3455, 9: 72a, 663H. sh1/ش ١

[Krause, "Stambuler Handschriften," p. 498: cites no., but no title].

Istanbul: Süleymaniye, Ayasofya 2587, 5: 128a, 725H. sh2/ش ٢

[Not included in Krause, "Stambuler Handschriften"].

Shiraz: Shiraz University, Dānishkadih-i Adabiyyāt, 925H. sh3/ش ٣

[Munzavī, *Persian Manuscripts*, v. 1, p. 182: cites date].

قوس وقزح

Qaws va Qizaḥ

Tehran: Malik 4681, 24: 50a-53a , ca. 1000H sh4/ش ٤

[Afshār and Dānishpazhūh, *Malik*, v. 8, p. 27; Munzavī, *Persian Manuscripts*, v. 1, p. 182: cites copyist (Muḥammad Qāsim, farzand-i Bihzād), and date (11th century)].

Tehran: Tehran University Markazī 4100, 3: 45–50, 1003H. sh5/ش ٥

[Munzavī, *Persian Manuscripts*, v. 1, p. 182: cites date].

Tehran: Malik 6192, 8: 71–74, 1085H. sh6/ش ٦

[Afshār and Dānishpazhūh, *Malik*, v. 9, p. 231; Munzavī, *Persian Manuscripts*, v. 1, p. 182: cites) copyist (Muḥammad, farzand-i Ḥusayn 'Aqīlī-yi Istirābādī) and date (1085H)].

Tehran: Tehran University Markazī 5911, 5: 75–78, 1243H. sh7/ش ٧

[Not in Munzavī, *Persian Manuscripts*].

Tehran: Tehran University Markazī 487, 1: 1b–3a, 1324 H. sh8/ش ٨

[Munzavī, *Persian Manuscripts*, v. 1, p. 182].

Isfahan: Isfahan University 286, 2, ?H sh9/ش ٩

Tehran University Markazī microfilm no. 4545.
[Dānishpazhūh, *Mikrufīlm'hā*, v. 3, p. 70: *Risālih-i Shu'ā'*].

رسالة فى انعكاس الشعاعات وانعطافها

Risāla fī In'ikās al-shu'ā'āt wa in'iṭāfihā I/الف

by Naṣīr al-Dīn al-Ṭūsī (d. 672/1274)

رسالة فى انعكاس الشعاعات وانعطافها

Risālah fī In'ikās al-shu'ā'āt wa in'iṭāfihā

[Krause, "Stambuler Handschriften," p. 498: Seray 3453, 2 Bl (673h) codex does not contain this treatise].

Istanbul: Süleymaniye, Ayasofya 2587, 4: 127a–127b, 725H. i1/الف ١

[Krause, "Stambuler Handschriften," p. 498: 2587, 7, 125a–125b, 725h].

Istanbul: Süleymaniye, Fatih 3387, 4: 193b–194a, 823H. i2/الف ٢

[Krause, "Stambuler Handschriften," p. 498: cites date].

Istanbul: Köprülü 927, 2: 113a–114a, 842H. i3/الف ٣

[Krause, "Stambuler Handschriften," p. 498: 839h; Şeşen, *Köprülü*, p. 461].

رسالة فى انعكاس الشعاع وانعطافه

Risāla fī In'ikās al-shu'ā' wa in'iṭāfihi

Tehran: Tehran University Markazī 8591, 3: 155–157, 887H. i4/الف ٤

[Dānishpazhūh, *Markazī*, v. 18: 8591, 5].

رسالة فى انعكاس الشعاعات وانعطافها

Risālah fī In'ikās al-shu'ā'āt wa in'iṭāfihā

Cairo: Dār al-kutub al-Miṣriyya 3957, 2: 152a–155b, 900H i5/الف ٥

[King, *Cairo Catalogue*, v. 2, p. 1032; King, *Survey*, p. 152].

Tehran: Majlis-i Shawrā 658, 4, fols. 64b–66b, ca. 1000H. i6/الف ٦

[Ḥā'irī, *Majlis*, v. 22, p. 349: cites folios, title (*In'iṭāf va in'ikās*), and date].

Tehran: Tehran University Markazī 2136, 1: 1b–4a, 1024H. i7/الف ٧

[Dānishpazhūh, *Markazī*, v. 9, p. 777].

رسالة فى انعكاس الشعاع وانعطافه

Risāla fī Ini'ikās al-shu'ā' wa in'iṭāfihi

Berlin: Deutsche Staatsbibliothek 6020 (671): 83b–84a, 1064H. i8/الف ٨

[Ahlwardt, *Königlichen Bibliothek*, p. 355: no date; Brockelmann, *GAL I*, p. 674: Berl. 1091; *GALS I*, p. 930].

رسالة فى انعكاس الشعاع وانعطافه

Risāla fī Ini'ikās al-shu'ā' wa in'iṭāfihi

Manchester: John Rylands 350 (381), 5: 80b–82b, 1096. i9/الف ٩

[Mingana, *John Rylands*, p. 546: "*'Akkās al-shu'ā'*;" Brockelmann, *GALS I*, p. 930; Winter, "British Isles," pp. 84–85].

Mashhad: Āstān-i Quds 594 (12169), 4, 2 fols., 1237H. i10/الف ١٠

Copyist (Muḥammad Khabūshānī) and date: codex.
['Irfāniān, *Āstān-i Quds*, v. 10, p. 149 (ibid., p. 150: copyist and date)].

Mashhad: Āstān-i Quds 575 (12225), 3: 1306H. i11/الف ١١

Copyist (Asadullāh al-Munajjim, Ṭihrān) and date (1306H): codex.
['Irfāniān, *Āstān-i Quds*, v. 10, p. 110; (ibid., p. 112: "copied from the copy of Mu'īn al-Dīn Munajjim-i Kāshī, 826H")].

Mashhad: Āstān-i Quds 597 (12124), 9: 54, 1309H. i12/الف ١٢

Copyist (Asadullāh Munajjim-i Māzandarānī, Ṭihrān): codex.
['Irfāniān, *Āstān-i Quds*, v. 10, p. 156].

رسالة فى انعكاس الشعاع وانعطافه

Risāla fī In'ikās al-shu'ā' wa in'iṭāfihi

Manchester: John Rylands 348 (447), 1: 1b–2b, ca. 1100–1200H i13/الف ١٣

[Mingana, *John Rylands*, p. 538: "*Risāla 'Akkās al-shu'ā' wa al-'iṭāfuhu*;" Brockelmann, *GALS I*, p. 930; Winter, "British Isles," p. 84: Crawford 447].

رسالة فى انعكاس الشعاع وانعطافه

Risāla fī In'ikās al-shu'ā' wa in'iṭāfihi

Krakow: Jagiellonska Or. 258,14: 163–164b, ?H. i14/الف ١٤

[Ahlwardt, *Königlichen Bibliothek*, p. 355: Berlin 6020 (no date)].

Tehran: Tehran University Markazī 821, 6: 12b–13b, ?H. i15/الف ١٥

[Dānishpazhūh, *Mishkāt*, v. 3, p. 835]. See also Addendum, p. liii.

كتاب البصائر فى العلم المناظر

Kitāb al-Baṣā'ir fī al-'ilm al-manāẓir B/ب

by Kamāl al-Dīn al-Fārisī (d. ca. 718/ 1318)

Tehran: Muṭahharī (Sipahsālār) 554, 731H. b1/ب ١

[Dānishpazhūh and Munzavī, *Sipahsālār*, v. 3, pp. 255–256: "copied from a copy, itself copied from one in author's hand"; refer to the date 718H in the colophon as the year of Fārisī's death].

Istanbul: Süleymaniye Ayasofya 2451, 837H. b2/ب ٢

[Krause, "Stambuler Handschriften," p. 509].

Tehran: Muṭahharī (Sipahsālār) 2911/27: 98–114, 1092H. b3/ب ٣

[Dānishpazhūh and Munzavī, *Sipahsālār*, v. 3, pp. 254–255: title cited as *al-Basā'ir fī Ikhtiṣār Tanqīḥ al-Manāẓir*].

Mashhad: Āstān-i Quds 5434, ca. 1000. b4/ب ٤

[Gulchīn Ma'ānī, *Āstān-i Quds*, v. 8, p. 47: "composed in 708H"].

Istanbul: Esat 2006, ?H b5/ب ٥

[Krause, "Stambuler Handschriften," p. 509].

ارتفاع

Irtifā' [in Persian] E/ا

by Badr al-Dīn al-Ṭabarī (ca. 824/1420)
Tehran: Tehran University: Markazī 1246, 4: 60b–65b, 1094 or 984H? e1/١ ا

[Munzavī, *Persian Manuscripts*, v. 1, p. 132: 1246/1, 57b–63a, 1094 or 984].

مرآتيه

Mir'ātiyya

Tehran: Majlis-i Shawrā 6594/31: 312–319, 1069H. e2/٢ ا

[Munzavī, *Persian Manuscripts*, v. 1, p. 132: "uncatalogued," cites date and page numbers].

مرآتيه

Mir'ātiyya

Tehran: Muṭahharī (Sipahsālār) 2911/18: 73–75, 1092H. e3/٣ ا

[Munzavī, *Persian Manuscripts*, v. 1, p. 132: cites date and page number; Dānishpazhūh and Munzavī, *Sipahsālār*, v. 3, p. 100].

Mashhad: Āstān-i Quds 5878/9: 47–51, ?H. e4/٤ ا

[Munzavī, *Persian Manuscripts*, v. 1, p. 132: "undated," cites page numbers and refers to Tehran University Markazī microfilm no: 2253].

فى تحقيق الإبصار

Fī Taḥqīq al-ibṣār Q/ ق

Anonymous [on the views of 'Alī al-Qūshjī (d. 879/1424)]
Tehran: Tehran University Markazī 6616, 20: 67b–69b, unique?

[Dānishpazhūh, *Markazī*, v. 16, p. 310: cites filio nos.].

رساله فى سبب رؤية الأشياء وبيان المذاهب فيه
وبيان رؤية الأشياء فى المرايا المصنوعة

*Risāla fī Sabab ru'yat al-ashyā' wa bayān al-madhāhib fīhi wa bayān ru'yat al-ashyā' fī al-marāyā al-maṣnū'*a R/ر

Anonymous (ca. 918/1512–927/1520: dedicated to Sultan Selim I)

Istanbul: Topkapı Sarayı Hazin, Hazine 603/5, unique?

[Karatay, *Topkapı Saray*ı 12, p. 321: no. 919, lists as 3rd text in codex; not in Krause,"Stambuler Handschriften," or Storey, *Persian Literature*].

إبصار

Ibṣār IB/اب

by Ghiyāth al-Dīn Manṣūr al-Dashtakī (d. ca. 948/1541)
Shiraz: Kitābkhanih-i Mufattiḥ (Millī-yi Pārs) 595/1, 934H, unique?
[Afshār, *Pārs*: *Risalih dar Ibṣār*, p. 254; Munzavī, *Persian Manuscripts*, v. 1, p. 132: Kitābkhanih-i Millī-yi Pārīs (Paris!); copied by Abul-khayr Muḥammad Taqī Fārisī, Qazvīn, 934H; Bihrūzī and Faqīrī, *Fārs*, v. 2, p. 184: *Risalih dar Ibṣār*, includes Arabic opening and closing lines; the damaged manuscript does not allow verification of the transcription of title or author].

كتاب نور حدقة الإبصار ونور حديقة الأنظار

Kitāb Nūr ḥadaqat al-ibṣār wa nūr ḥadīqat al-anẓār N/ن

by Taqī al-Dīn ibn Ma'rūf (ca. 982/1574–1003/1595)

كتاب نور حدقة الابصار ونور حديقة الانظار

Kitāb Nūr ḥadaqat al-ibṣār wa nūr ḥadīqat al-anẓār
Oxford: Bodleian Marsh 119, 982H. n1/ن ١

[Uri, *Bibliotheca Bodleiana*, pp. 201–202; Nicoll and Pusey, *Bibliotheca Bodleiana*, pp. 248–249; Sabra, *The Optics of Ibn al-Haytham*, p. lxxii, n. 116: "the author's autograph, completed in A.H. 982/1574–75. The book was dedicated to Sultan Murād III"].

نور حدقة الابصار ونور حديقة الانظار

Nūr ḥadaqat al-ibṣār wa nūr ḥadīqat al-anẓār
Cairo: Dār al-kutub al-Miṣriyya 893, 52 folios, 1004H. n2/ن ٢

[King, *Cairo Catalogue*, v. 2, pp. 1033–1034].

مناظر ومرايا

Manāẓir va Marāyā [in Persian] Y/ی

by Qāsim-'Alī al-Qāyinī (ca. 1070/1659).
[Not in Storey, *Persian Literature*].
Tehran: Tehran University Markazī 3530/5: 149–234, 1097H, unique?
"Qāsim al-arzāq Qāsim-'Alī al-Qāyinī": p. 149 (no folio numbers); "completed" at Madrasih-i Chahār Bāgh, Mashhad 1097H: colophon.
[Dānishpazhūh, *Markazī*, v. 12, p. 2551; Munzavī, *Persian Manuscripts*, v. 1, p. 457].

در علم مناظر (درجامع بهادرخانی)

Dar 'Ilm-i manāẓir [in Persian] D/د

in *Jāmi'-i Bahādur-Khānī* [or *Mafātīḥ/Miftāḥ al-raṣad*],
2nd Khazīna: *Manāẓir va Marāyā*
by Qāḍī Ḥusayn al-Jawnpūrī (d. 1279/1867)

Calcutta: Lithograph Publication, 1835 (=1250H). d1/د١

[Ṭihrānī, *Al-Dharī'a*, v. 4, p. 43; Mushār, *Chāppī-yi fārsī*, v. 2, p. 1488; Dānishpazhūh and Afshār, *Nuskhihā-yi khaṭṭī*, v. 4, p. 38; Storey, *Persian Literature*, v. 2, p. 20; Mushār, *Mu'allifīn*, v. 4, p. 684; Munzavī, *Persian Manuscripts*, v, 1, p. 155].

Bengal: Asiatic Society, Curzon 580, 2, 1250H. d2/د٢

[Ivanow, *Curzon Collection*, pp. 402–403: completed 1248/1833, copied in the middle of xiii/xix century, refers to extract in R1038; Storey, *Persian Literature*, v. 2, p. 20].

Tehran: Muṭahharī (Sipahsālār) 553, 1: 7–96, 1272H. d3/د٣

"Contained in a book owned by . . . 'Alī-Naqī Mīrzā"?, 1272H: title page. [Dānishpazhūh and Munzavī, *Sipahsālār*, v. 5, p. 653: *Manāẓir va marāyā*, 13th. century].

Tehran: Malik 3255: 1–78, before1300H? d4/د٤

Contains only the 2nd Khazīna: *Manāẓir va Marāyā*.
[Afshār and Dānishpazhūh, *Malik*, v. 2, pp. 174–175: 2nd Khazīna, ca. 13th century; Munzavī, *Persian Manuscripts*, v. 1, p. 155].

Mashhad: Āstān-i Quds 597, 6, 1309H. d5/د٥

Contains only the 2nd Khazīna: *Manāẓir va Marāyā*
Copyist Asadullāh Munajjim, from copy of Mīrzā 'Abdullāh?, itself copy of original: colophon. Marginal notes by Seyyed Jalāl Ṭihrānī, owner .
['Irfāniān, *Āstān-i Quds*, p. 155: *Dar Manāẓir va marāya*, 2nd Khazīna].

ضياء البصاير

Ḍiyā' al-Baṣā'ir Ḍ/ض

by Karīm ibn Ibrāhīm al-Kirmānī (d. ca. 1268/1852)
Los Angeles: UCLA, Bio. Med. Ar. 31: 1–144, ca. 1268H, unique?
[Brockelmann, *GALS II*, p. 901; Iskandar, *University of California*, p. 40].

فايده در كيفية الإبصار

Fāyidih dar Kayfiyat al-ibṣār F/ف

by Muḥammad-Ismā'īl Muḥammad-'Alī al-Qāyinī (d. 1305/1887)
[on the views of Naṣīr al-Dīn Ṭūsī? (d. 672/1274), not in Storey, *Persian Literature*].

فايده در كيفية الابصار

Fāyidih dar Kayfiyat al-ibṣār

Mashhad: Āstān-i Quds 370 (5526), 3: 118, 1286H [in author's hand?] f1/ف١

[Gulchīn Ma'ānī, *Āstān-i Quds*, v. 8, p. 310: "Mīrzā Muḥammad 'Alī ibn Muḥammad-Ismā'īl Qāyinī"; cites title and date above].

برهان لطيف على انّ الابصار ليس بخروج الشعاع

Burhān laṭīf 'alā anna al-ibṣār laysa bi khurūj al-shu'ā'

Mashhad: Āstān-i Quds 597 (12124), 6: 1309H. f2/ف٢

['Irfāniān, *Āstān-i Quds*, v. 10, p. 155: "on light reflection, apparently by Mīr Abū l-Qāsim Findiriskī"; cites date of codex as 1308–1312H].

رؤية الشجر فى الماء

Ru'yat al-shajar fī al-mā' G/ج

Anonymous.
Tehran: Malik 2534, 14: 157b, ca. 1200–1300H, unique?
[Afshār and Dānishpazhūh, *Malik*, v. 6, p. 36: no. 592, cites date].

Addendum (MSs identifed since the compilation of Lists I–II):

تحرير كتاب المناظر لاوقليدس الصورى

Taḥrīr Kitāb al-Manāẓir li-Ūqlīdis al-Ṣūrī (by Naṣīr al-Dīn al-Ṭūsī)

Tehran: Malik 3179 [772], 6: 68b-80a, 986H. t56/ت٥٦

Copyist: 'Alī ibn Naṣr-allāh al-Dihlī (going back to handcopy of Shīrāzī).
Copy date: fol. 95a; codex includes *Iṣlāḥ al-Manāẓir* (fols. 80a–80b) ṣ24/ص٢٤
and Ṭūsī's *Risāla fī In'ikās al-shu'ā' wa in'iṭāfihi* (fols. 81a–82a). i16/الف١٦

تحرير المناظر لاقليدس

Taḥrīr al-Manāẓir li-Uqlīdis (by Naṣīr al-Dīn al-Ṭūsī)

Tehran: Majlis-i Shawrā 772, 4: 157–178, ca. 1020H. t57/ت٥٧

[Ḥā'irī, *Majlis*, v. 23, p. 75: fols. 79b–90a, copyist (Mullā Mūsā Farmalī), date].
Tehran: Majlis-i Shawrā 809, 4: 102–112, ca. 1268–1269H. t58/ت٥٨

[Ḥā'irī, *Majlis*, v. 23, p. 99: fols. 52a–57a, copyist ('Alī Muḥammad), date].

5 The Arabic Tradition

The sources for the study of the Arabic tradition of Euclidean optics include, in addition to the Arabic versions already introduced [see The Arabic Versions, pp. xix–xxv, and List I, pp. xxxiv–xliv)], a set of supplementary texts, which contain besides optics proper (*al-manāẓir*) material related to neighboring disciplines, from catoptrics (*al-marāyā*) to surveying (*al-misāḥa*) [see List II, pp. xlv–liii].[17] These texts, which span the long chronological period of close to a millennium and a wide geographical area from Baghdad to Calcutta,[18] date from a period shortly after the Arabic translation of Euclid's *Optics* (ca. 212/827–828), a text upon which many are directly dependent.

Among the earliest compositions is a long treatise by a certain Aḥmad ibn 'Īsā (ca. 250/864?), entitled *Kitāb al-Manāẓir wa al-marāyā al-muḥriqa* [Ḥ in List II, p. xlv][19] and explicitly placed in the "tradition of Euclid" (*'alā madhhab Uqlīdis*) in its full title. This combined treatment of optics and burning mirrors contains a thorough coverage of most Euclidean definitions and propositions of the *Optics*, mainly under the section "*Kalām fī Ikhtilāf al-manāẓir*" ("Discourse on Optics"), with relevant discussions scattered with a range of other subjects throughout the rest of the text. Another dependent text by Qusṭā ibn Lūqā (d. 287/912), this time bringing together the more commonly coupled topics of

---

17 Texts primarily involving direct vision – vision through air (*manāẓir*) – are M, Ṣ, T, J in List I, pp. xxxiv–xliv; those involving indirect vision (*marāyā*) are SH, I, G in List II, pp. xlvii–xlix; those involving a combined treatment are Ḥ, L, IM, B, Q, R, IB, Y, N, D, Ḍ, and F; those applying vision to surveying (*misāḥa*) are W, Ẓ, KH, S, and E in List II, pp. xlv–liii.

18 See Kheirandish, "Optics: Highlights from Islamic Lands", *Science in Islam*, UNESCO, and Kheirandish, "The 'Manāẓir' Tradition through Persian Sources," *La science dans le monde iranien*, Tehran: Institut Français de Recherche en Iran (IFRI), forthcoming. The disciplinary, theoretical and methodological developments are discussed in Kheirandish, "Optics in the Islamic World," *Encyclopaedia of the History of Science, Technology, and Medicine in Non-Western Cultures*," pp. 40–44.

19 *The Book on Optics and Burning Mirrors*. The two extant manuscripts of this text in Istanbul [see List II, p. xlv: ḥ1 and ḥ2] have *'alā madhhab Uqlīdis fī 'ilal al-baṣar* (in the Tradition of Euclid on the Reasons for Vision) as part of the title. Krause, "Stambuler Handschriften," pp. 513–514, includes chapter titles with German translations; Sabra, *The Optics of Ibn al-Haytham*, v. 2, p. xxxii, n. 31, pp. xxxvi–xxxvii, and n. 39, provides more details of the text's contents and manuscript copies, including excerpts in Hebrew characters. Sabra, referring to Krause's "conjecture" of an early date of composition (before 250H), makes note of textual features which support such an assumption. On the vocabulary of the Euclidean definitions, see Summary and Commentary: Defintions 1–2, v. 2, pp. 10–11, n. 69 and n. 72), and p. 17, n. 89; see also Kheirandish, "The Arabic 'Version' of Euclidean Optics: Transformations as Linguistic Problems in Transmission," *Tradition, Transmission, Transformation*, Ragep and Ragep (eds.), p. 232, n. 17.

manāẓir and *marāyā*[20] under the title *Kitāb fī 'Ilal mā ya'riḍu fī al-marāyā min ikhtilāf al-manāẓir* [L in List II, p. xlvi],[21] contains a discussion of all the transmitted Euclidean definitions and only a limited selection of propositions. A few works by al-Kindī (d. ca. 257/870), composed during the same early period, cover much of the Euclidean text, though in varying forms and proportions. Among al-Kindī's extant works, an elaborate one on the subject, which survives as a Latin translation known as *De aspectibus*,[22] includes a range of Euclidean discussions, while in the shorter texts the coverage is much more specific: an *Iṣlāḥ al-Manāẓir* attributed to al-Kindī [Ṣ in List I, pp. xxxv–xxxvi], and at times confused with his *De aspectibus*,[23] is a "correction" of two problematic propositions on appearances involving movements, and a *Risāla fī Īḍāḥ wijdān ab'ād mā bayn al-nāẓir wa marākiz a'madat al-jibāl* [W in List II, pp. xlv–xlvi], cited among his works even less frequently,[24] is the application of four

---

20 As in the apparently lost text by al-Kindī reported by Ibn al-Nadīm as *Kitāb risālatuhu fī Ikhtilāf al-manāẓir al-mir'āt;* see *Kitāb al-Fihrist*, ed. Flügel, p. 258; tr. Dodge, p. 619.

21 *The Book on the Reasons for What Occurs in Mirrors on Account of the Difference of Aspects.* On the manuscript of this text in Mashhad; see List II, p. xlvi. The text is not listed among Qusṭā's works as reported in the *Kitāb al-Fihrist*, ed. Flügel, p. 295; tr. Dodge, p. 695, nor is it reported by Ibn Abī Uṣaybi'a, as pointed out by Toomer, *Diocles on Burning Mirrors*, pp. 27–28, the first source in European languages to refer to the text and its unique manuscript copy; see also Kheirandish, "The Arabic 'Version' of Euclidean Optics: Transformations as Linguistic Problems in Transmission", p. 232, n. 17.

22 See Björnbo and Vogl, *Alkindi, Tideus und Pseudo-Euklid*, pp. 1–41. The full title, as reconstructed from the Latin translation, *De causis diversitatum aspectus et dandis demonstrationibus geometricis super eas*, may have been *Fī 'Ilal ikhtilāf al-manāẓir wa al-barāhīn al-handasiyya 'alayhā.*

23 Carmody, *Arabic Astronomical Sciences*, p. 79, cites "*Iṣlāḥ al-Manāẓir*, 23 chapters" under *De aspectibus*, a title that in the edition of Björnbo and Vogl, *Alkindi, Tideus und Pseudo-Euklid*, pp. 4–41, has 24 sections. In Sezgin *GAS V*, p. 117, the citation of *Iṣlāḥ al-Manāẓir* is followed by a reference to Carmody's citation of Gerard of Cremona's Latin translation of *De aspectibus*, while specific reference is also made to Björnbo and Vogl, *Alkindi, Tideus und Pseudo-Euklid*, p. 148, where this work is differentiated from *De aspectibus*.

24 *Treatise on Clarification of Finding Distances between an Observer and the Centers of Mountain Heights*. Of the two extant manuscripts in Istanbul [see List II, w1, w2, pp. xlv–xlvi], the first has the following addition to the title: *wa 'ulūw a'midatihā wa 'ilm 'umq al-ābār wa 'urūḍ al-anhār wa ghayr dhālika wa huwa tussamā Mūrīsṭus (and the Elevation of its Height, and the Science of the Depth of Wells and the Width of Rivers and other Things, and he? is called Mūrīsṭus* [there is an entry under this Greek author in *EI*²]. The short Arabic treatise is not among the list of al-Kindī's works reported by Ibn al-Nadīm; see *Kitāb al-Fihrist*, ed. Flügel, pp. 257–261; tr. Dodge, pp. 618–620. Atiyeh, *Al-Kindi: The Philosopher of the Arabs*, 1966, p. 200, lists it as no. 230, where reference is made to Brockelmann, *GALS I*, p. 374; it is also cited in Ritter and Plessner, "Stambuler Bibliotheken," p. 370.

Euclidean propositions involving the determination of unknown heights, widths, and depths of natural objects [see Summary and Commentary: Propositions 60–61 and 19–22, v. 2, pp. 95–97 and pp. 54–61 respectively], belonging more properly to the "*al-misāḥa*" core of the "*al-manāẓir*" tradition.

Of the variety of works within that core, the only one that has these expressions as part of its title is *al-Misāḥāt al-manāẓiriyya* of Sinān ibn al-Fatḥ (ca. 4th/10th cent.) [Ẓ in List II, p. xlvi].[25] This is a text with similar methods to the surveying propositions of Euclid and al-Kindī in calculating unknown values from three known ones (*ma'lūm*), presumably to be determined through measurements.[26] Comparable works of a slightly later period include two short tracts carrying the name of Ibn al-Haytham (d. ca. 432/1040),[27] a name long known for the important and influential *Kitāb al-Manāẓir*[28] and one to be most directly associated with the Euclidean tradition through an apparently lost summary of the respective works of Euclid and Ptolemy.[29] One of the surveying works, bearing the long title *Maqāla fī Ma'rifat irtifā' al-ashkhāṣ al-qā'ima wa a'midat al- jibāl wa irtifā' al-ghuyūm* [KH in List II, p. xlvi], is attributed to Shaykh Abī 'Alī Ibn al-Haytham; while the other, entitled *Qawl fī Istikhrāj a'midat al-jibāl* [S in List II, p. xlvii], carries the more commonly encountered

---

25 The apparently unique manuscript copy is in Dār al-kutub of Cairo; see King, *Cairo Catalogue*, v. 2, p. 1030; *Cairo Survey*, p. 39. For references to the author, see *Kitāb al-Fihrist*, ed. Flügel, p. 281, tr. Dodge, p. 665; Sezgin, *GAS V*, p. 301; *GAS VI*, p. 207, Sezgin, *GAS VII*, p. 406.

26 The last section, on the determination of the width of a large river, which involves conceiving of the width as the side of the square and taking square roots, is an exception.

27 See Sabra's article on Ibn al-Haytham in *DSB*, 6: 189–210.

28 See Sabra, *The Optics of Ibn al-Haytham*, Books I–III: On Direct Vision, 2 vols., [includes an English translation and commentary based on an earlier critical edition; see Bibliography. A similar study of books IV–VII is forthcoming].

29 The long list of "Ibn al-Haytham's Writings on Light and Vision, " Sabra, *The Optics of Ibn al-Haytham*, v. 2, pp. xxxii–xxxiv, begins with references to the *Talkhīṣ Kitābay Uqlīdis wa Baṭlamyūs fī al-Manāẓir*, based on the report of Ibn Abī Uṣaybi'a (d. 669/1270); see *'Uyūn al-anbā' fī ṭabaqāt al-aṭibbā'*, ed. Müller, p. 94. The *Optics* of Ibn al-Haytham is rather unique in belonging more properly to the tradition of Ptolemaic *Optics*; the only other case is an earlier [lost] work by Abū Sa'd Ibn Sahl (ca. 374/984), a supposed review of the fifth book of Ptolemy's *Optics*, entitled *al-Burhān 'alā anna al-falak laysa huwa fī ghāyat al-ṣafā', istakhrajahu Abū Sa'd al-'Alā' ibn Sahl 'inda taṣaffuḥihi Kitāb Baṭlamyūs fī al-Manāẓir wa arāda an uḍamminahu jumlat al-taṣaffuḥ li-l-maqalā al-khāmisa min hādhā al-kitāb*; see Sabra, *The Optics of Ibn al-Haytham*, pp. lii–liii and pp. lviii–lx. On this and other works by Ibn Sahl, see Rashed, *Géometrie et Dioptrique au Xe siécle: Ibn Sahl, al-Qūhī et Ibn al-Haytham*, especially pp. 53–56, and Rashed, "A Pioneer in Anaclastics: Ibn Sahl on Burning Mirrors and Lenses," reprinted in *Optique et mathématique*.

name al-Ḥasan ibn al-Ḥasan ibn al-Haytham.[30] A much later surveying treatise simply called *Irtifā‘* [E in List II, p. l], by a Persian commentator on Ṭūsī by the name of Badr al-Dīn al-Ṭabarī (ca. 824/1421),[31] treats height determination of lower objects, once by means of a "rod" (*‘amūd*), and again by a plane mirror [hence its alternative title, *Mir'ātiyya*: e2, e3, ibid.], as in the corresponding Euclidean propositions [see Summary and Commentary: Propositions 19 and 20, v. 2, pp. 54–59]. The respective methods of "how to know" *(ma‘rifat)* or "extract" (*istikhrāj*) the magnitude of a height, whether of mountains and clouds or of lower objects, distinguish these texts from other treatments involving elevations, from the appearance of the stars in Ibn al-Haytham's own *Risāla fī Ru'yat al-kawākib*[32] to the visibility of trees in the water (*Ru'yat al-shajar fī al-mā'*) in an anonymous and probably later text [G in List II, p. liii].

In contrast to the traditions of *al-manāẓir* or direct vision [i.e., appearances through air], and *al-misāḥa*, its application in surveying, the subject of *al-marāyā* or indirect vision [i.e., appearances through a medium] is rather under-represented in the Arabic Euclidean tradition. The subject, appearing in Euclid's *Optics* only in the form of a one-line statement of the principle of reflection from a plane mirror in a single proposition [Greek Proposition 19/Arabic Proposition 20], has still created its own specific genre of literature. Included here and in this specific context are two short treatises by Naṣīr al-Dīn al-Ṭūsī, one in Arabic, known as *Risāla fī In‘ikās al-shu‘ā‘āt wa in‘iṭāfihā,* the other in Persian, called *Shu‘ā‘* [I and SH respectively in List II, pp. xlvii–xlix].[33] These are discussed alongside Ṭūsī's *Taḥrīr al-Manāẓir,* the recension of the Euclidean text, in the context of the related treatments of Ṭūsī's student, Quṭb al-Dīn al-Shīrāzī (d. 711/1311), and the latter's student, Kamāl al-Dīn al-Fārisī (d. ca. 718/1318) [for

[30] *Treatise on the Knowledge of the Height of Upright Figures and Elevation of Mountains and Height of Clouds* and *Statement on the Extraction of the Height of Mountains*, respectively, see List II, pp. xlvi–xlvii. The former seems more directly in the Arabic Euclidean tradition for including references to the eye (*baṣar*) and its rays (*shu‘ā‘*) in the course of height determinations.

[31] *Height.* The author is reported as a commentator on two works, the Persian *Sī Faṣl* (*Thirty Chapters*) attributed to Ṭūsī and Euclid's *Elements*; see Munzavī, *Persian Manuscripts*, v. 1, p. 132.

[32] *On Seeing the Stars*, alternatively entitled *Risāla fī Manāẓir al-kawākib*; see Sabra and Heinen, "On Seeing the Stars, Edition and Translation of Ibn al-Haytham's *Risāla fī Ru'yat al-kawākib.*" See also Sabra, "Psychology vs. Mathematics: Ptolemy and Alhazen on the Moon Illusion," where the work's composition date is discussed with respect to Ibn al-Haytham's related writings.

[33] *On the Reflection and Deflection of the Rays* (I) is better known than *Ray* (SH). For a brief description and English translation of it, see Winter and ‘Arafat, "A Statement on Optical Reflection and ‘Refraction’ Attributed to Naṣīr ud-Dīn aṭ-Ṭūsī." See also Wiedemann, "Über die Reflexion und Umbiegung des Lichtes von Naṣīr al Dīn al Ṭūsī," and Meyerhof, "Die Optik der Araber," pp. 52–54.

this dating, see b1 in List II, p. xlix], best known for his important commentary on Ibn al-Haytham's *Optics*, with the modest title of *Tanqīḥ al-Manāẓir* [see Summary and Commentary: Proposition 20, v. 2, pp. 55–59].

There also exists a range of optical works, whose titles contain the verbal root B^{a}Ṣ^{a}R^{a} (to see) as opposed to N^{a}Ẓ^{a}R^{a} [as in *manāẓir*] or R$^{a'\bar{a}}$ [as in *ru'ya*]. There is a *Kitāb al-Baṣar wa al-baṣīra*, once attributed to Thābit ibn Qurra (d. 289/901), and a *Kayfiyat al-ibṣār*, attributed to Abū Bakr Zakariyā al-Rāzī (d. 311/923),[34] as well as the *al-Qawl fī al-Ibṣār wa al-mubṣar* of Abū al-Ḥasan al-'Āmirī (d. 382/992) [IM in List II, p. xlvi], *Kitāb al-Bāṣā'ir fī 'ilm al-manāẓir* of Kamāl al-Dīn al-Fārisī [B in List II, p. xlix], *Fī Taḥqīq al-ibṣār* on the views of 'Alī al-Qūshjī (d. 879/1474) [Q in List II, p. l], *Ḍiyā' al-Baṣā'ir* of Karīm ibn Ibrāhīm al-Kirmānī (d. ca. 1268/1852) [Ḍ in List II, p. lii], *Rafī' al-baṣar* of Rafī' al-Dīn Khān (d. 1294/1877),[35] and *Fāyidih dar Kayfiyat al-ibṣār* of a Mīrzā Muḥammad-'Alī al-Qāyinī (d. 1305/1887) [F in List II, p. liii].

To the last few titles must be added a number of others in other forms and lengths, which represent the nature of the field after the classical period, some of much historical interest for the reconstruction of the earlier optical traditions.[36] Dating from the early 10th/16th century is a long anonymous Persian treatise with a long Arabic title, *Risāla fī Sabab ru'yat al-ashyā' wa bayān al-madhāhib fīhi wa bayān ru'yat al-ashyā' fī al-marāyā al-maṣnū'a* [R in List II, p. l].[37] Composed for another Ottoman Sultan, this time in Arabic, is *Nūr ḥadaqat al-Ibṣār wa nūr ḥadīqat al-anẓār* [N in List II, p. li] by Taqī al-Dīn ibn Ma'rūf (d. 994/1585), with three main sections on direct vision, vision by reflection, and refraction (*Ru'yat al-istiqāma, Ru'yat al-in'ikās* and *Ru'yat al-in'iṭāf* (respectively).[38] From Islamic Persia comes a series of late optical texts, starting

---

[34] The latter is reported as *Kayfiyāt al-ibṣār* in Ibn al-Nadīm's *Kitāb al-Fihrist*, ed. Flügel, p. 299, and Ibn al-Qifṭī's *Ta'rīkh al-ḥukamā'*, ed. Lippert, p. 273; the former is cited by Sabra, *The Optics of Ibn al-Haytham*, v. 2, p. lxvi, as a post-Rāzī compilation.

[35] Hyderabad edition, 1250/1834-45; see Storey, *Persian Literature*, v. 2, part 1, and p. 21, p. 100, and part 3, pp. 455–456, where another late Persian optical text is reported with the alternative titles *Manāẓir-i Turābiyya* and *Tanwīr al-'uyūn*, by a M. Turāb-'Alī ibn Ghulām-'Alī Bulandshahrī, with no dates or further references.

[36] See Kheirandish, "The 'Manāẓir' Tradition through Persian Sources."

[37] *Treatise on the Cause of the Visibility of Objects, and the Explanation of Methods in it, and Vision of Objects in Constructed Mirrors.* The composition date of the anonymous text is known only through its dedication to Sultan Selim I (r. 918/1512–927/1520). On the apparently unique manuscript copy of this text in Istanbul and the state of its documentation, see List II, p. l, and Kheirandish, "The 'Manāẓir' Tradition through Persian Sources."

[38] "Composed for the Ottomon Sultan Murād III" (r. 982/1574–1003/1595), see Sabra, "Optics, Islamic," p. 246; see also Sabra, *The Optics of Ibn al-Haytham*, v. 2, p. lxxii, n. 116. For the manuscripts of this text in Oxford and Cairo, see List II, p. li.

with one entitled *Ibṣār* [IB in List II, p. li][39] by Ghiyāth al-Dīn Manṣūr al-Dashtakī (d. ca. 948/1541), after whom the *Manṣūriyya* school in Shiraz is named. A related work by Ghiyāth al-Dīn's son, Muḥammad ibn Manṣūr, is particularly notable for references to "*madhhab-i manṣūr*" and "*madhhab-i mashhūr*" as two distinct traditions in "'*ilm-i manāẓir*".[40] Of a few late Persian optical works with the combined title *Manāẓir va marāyā*, one is composed by Qāsim-'Alī al-Qāyinī (ca 1072/1661) [Y in List II, p. li], an author whose few surviving works in unique and largely unknown Persian manuscripts add to our historical knowledge much beyond optics.[41] Another, no less important in terms of historical interest, is a much later text by Qāḍī Ḥusayn al-Jawnpūrī (d. 1279/1867), a text forming the second khazīna (treasure) of his better known *Jāmi'-i Bahādur-Khānī*.[42] Jawnpūrī's *Manāẓir va marāyā* contains, in addition to the independent chapter *Dar 'Īlm-i manāẓir* (on the science of optics) [D in List II, p. lii], a historical table that reveals the sources (*ma'khadh*) for the enunciations (*da'vā*) and proofs (*burhān*) of its 45 propositions in optics and 14 propositions in catoptrics. Included alongside the author's own treatments [under

---

39 *Treatise on Vision*. On the apparently unique manuscript copy of this text in Shiraz, see List II, p. xlviii. In Munzavī, *Persian Manuscripts*, v. 1, Tehran, 1348 =1969, p. 132, the manuscript is cited as being at Kitābkhānih-i Millī-yi "Pārīs" [i.e., Paris!], rather than "Pārs" [i.e., Pārs Library in Shiraz (now called Mufattiḥ), see Afshār, *Pārs*, p. 254, also Bihrūzī and Faqīrī, *Fārs*, v. 2, p. 184]. For references to the author and sources for the list of his works, see Kheirandish, "The 'Manāẓir' Tradition through Persian Sources."

40 The expressions "*madhhab-i manṣūr*" and "*madhhab-i mashhūr*" introduced in the course of the introductory sections of his text on the rainbow and lunar halo [respectively, *kamān-i Rustam* and *kharman-i māh* in Persian] seem to be a play on words to mean, respectively, "the triumphant" [i.e., more justifiable] for the author's family name and "the well-known" for the more established traditions being challenged. The text is extant in more than one manuscript; see Munzavī, *Persian Manuscripts*, v.1, p. 337, where it is listed under *Qaws quzaḥ* (rainbow), as in the case of Ṭihrānī, *al-Dharī'a*, v. 17, p. 207. I have consulted Tehran University (Markazī) MS 5911: 100–106; see also Kheirandish, "The 'Manāẓir' Tradition through Persian Sources."

41 The text, which contains references to the *Optics* of Euclid and al-Kindī, Ibn al-Haytham and of Abū Rayḥān! [i.e., Bīrūnī], is reported to have been completed (*bi itmām risīd*) in the Chahār Bāgh School in Mashhad. In the opening folio of the codex containing the text, Qāyinī is mentioned as being "among the outstanding students" ("*az nukhbih talāmudh*") of "Mullā Muḥammad Ḥusayn ibn Muḥammad-Bāqir-i Yazdī" (ca. 1047/1637). On the apparently unique manuscript copies of the astronomical works of Qāyinī and his teacher, Muḥammad Ḥusayn (ca. 1072/1661), see Kheirandish, "The 'Manāẓir' Tradition through Persian Sources."

42 The author's name is sometimes written as Chawnpūrī and sometimes followed by Karbalā'ī; the title is written in the alternative forms *Mafātīḥ* and *Miftāḥ al-raṣad*. For a description of the work, see Tytler, "Analysis and Specimens of a Persian Work on Mathematics and Astronomy," which makes no reference to the section on optics or to the historical table.

mu'allif] and those by Euclid [under *Manāẓir-i Uqlīdis*: 16 propositions in optics and one in catoptrics] is a list of otherwise unknown related works, presumably written mostly in Arabic during the classical period, in this order: *Risālih-i* Hāmānī [i.e., Māhānī: 7 propositions in *manāẓir*]; *Ḍiyā'* by Abī Manṣūr [possibly Muḥammad ibn Manṣūr, already mentioned, reading Ibn for Abī: 7 propositions in *manāẓir*];[43] *Lama'āt* by Abī Rayḥān [i.e., Bīrūnī: 4 propositions in *manāẓir*, 2 involving mediums]; *Ta'līqāt-i Manāẓir* by Muḥaqqiq-i Ṭūsī [i.e., Naṣīr al-Dīn: 3 propositions in *manāẓir*, 6 in *marāyā*, different from Ṭūsī's treatments in *Taḥrīr*, *In'ikās* and *Shu'ā'*]; *Sharḥ-i Manāẓir* by Abī Ja'far Khāzin [1 proposition in *manāẓir*, 3 in *marāyā*]; and Vajīzi-yi Banū Mūsā [2 propositions in *manāẓir*, and 1 in *marāyā*].[44] Within the text there is also a reference to a *Sharḥ-i Manāẓir-i Uqlīdis* by "Kamāl al-Dīn Yūnus," the supposed mathematics teacher of Ṭūsī [i.e., in the form Kamāl al-Dīn ibn Yūnus], yet none to the *Tajrīd al-Manāẓir* of Ṭūsī's contemporary, Ibn Abī Jarāda.[45] Neither is there a single reference to the most outstanding of all the *Manāẓir* texts, namely *Kitāb al-Manāẓir* of Ibn al-Haytham, or its prototype by Ptolemy with the same title;[46] and unless the reference to "Kamāl al-Dīn

---

[43] A marginal note in d5 [see List II, p. lii] suggests Abī Manṣūr to be Yaḥyā ibn Manṣūr or Abū Manṣūr Baghdādī; see manuscript folio in Kheirandish, "The 'Manāẓir' Tradition through Persian Sources."

[44] On Banū Mūsā, Khāzin, Māhānī, Bīrūnī, and Ṭūsī, see the articles under these names in *DSB*. The critical issue of the reliability of the historical report is verifiable in the case of the correspondence of the propositions with the only survivng text of this list, namely Euclid's *Manāẓir*, though some present a clearly modern touch. For more details, see Kheirandish, "The 'Manāẓir' Tradition through Persian Sources."

[45] On Kamāl al-Dīn ibn Yūnus (551–639/1156–1242) as the mathematics teacher of Ṭūsī, see Ragep, *Ṭūsī's Memoir on Astronomy*, v. 1, pp. 6–8. On Ibn Abī Jarāda [the colophon name is Jamāl al-Dīn Muḥammad ibn al-Ṣaḥib Jamāl 'Umar ibn Hibatullāh], see King, *Cairo Survey*, pp. 37 and 154, reporting a *Taḥrīr* (recension) by him of Thābit ibn Qurra's *Kitāb Quṭū' al-usṭuwāna wa-basīṭuhā*, compiled in 691 (=1291). Krause, "Stambuler Handschriften," p. 440, cites and dates as 676H (=1277) a handcopy of an Arabic manuscript of the *On the Sphere and Cylinder* of Archimedes [Istanbul, Fatih 3414] by "Ibn Abī Garrāda;" King, *Cairo Survey*, p. 154, has it "between 676 and 684H." An Ibn Abī Jarrāda [again with two r's] is also mentioned in connection with the ownership and annotation of Ibn al-Haytham's handcopy of a manuscript of the *Conics* of Apollonius; see Toomer, *Apollonius*, v. 1, p. lxxxviii [reference kindly provided by the author].

[46] Lejeune, *L'Optique de Claude Ptolémée*, new edition with French translation, 1989; for a recent English translation and study, see Smith, "*Ptolemy's Theory of Visual Perception*;" on the arguments for the limited accessibility of Ptolemy's and Ibn al-Haytham's works in the Islamic world derived from a thorough survey of related literature, see Sabra, *The Optics of Ibn al-Haytham*, v. 2, pp. lviii–lix and pp. lxv–lxviii, respectively.

Yūnus" is meant to be directed to Kamāl al-Dīn al-Fārisī, the same can be said of his *Tanqīḥ al-Manāẓir*,[47] the commentary with internal circulation as late as its supposed "criticism," under the title *Ḍiyā' al-Baṣā'ir* [Ḍ in List II, p. lii].[48]

From this long list of sources, what may or may not belong to "the Arabic tradition of Euclidean optics" lies in the qualification of such a collective appellation. Here, "Arabic" is to be qualified as the dominant textual language and, as it turns out, the theoretical turning point, "Euclidean" as the methodological mode of geometrical over experimental reasoning,[49] and "optics" as a textual rather than disciplinary catagory.

6 Textual Transmission

In this final part of the introductory sections, the main core of the *al-Manāẓir* texts are presented within the context of the Greco-Arabic and Arabo-Latin transmissions [see Chart 1: Greek-Arabic-Latin Transmission and Chart 2: Internal Transmission, pp. lxiv–lxv].[50]

The Greek textual tradition consists of the following: two extant Euclidean texts edited by Heiberg as *Euclidis Optica* and *Opticorum recensio Theonis* [as represented respectively by Vienna codex Vind. xxxi 13, 12th cent. and Vatican codex, gr. 204, 10th cent.], renamed versions A and B by Jones and Knorr in a recent reexamination of Heiberg's attributions;[51] Pappus's *Collection* (C) – certain parts of its sixth book falling within that tradition [see note 58 and Summary and Commentary: Propositions 36–42, v. 2 pp. 75–81]; and the *Optica* of Ptolemy (P), which survives only in a Latin translation made from the Arabic [see Lejeune, *L'Optique de Claude Ptolémée*].

---

47 For an exhaustive treatment of the unknown status of Ibn al-Haytham's *Kitāb al-Manāẓir* in various traditions in the Islamic world, as well as an updated treatment of Fārisī's *Tanqīḥ al-Manāẓir li-dhawī al-abṣār wa al-baṣā'ir* [2 vols., Hyderabad, 1347–1348 (=1928–1930)], see Sabra, *The Optics of Ibn al-Haytham*, v. 2, pp. lxiv–lxxiii.

48 On the text and such a description of it, see Iskandar, *University of California*, p. 40.

49 After Euclid's own *Elements* and *Optics*, in contrast to Ptolemy's *Optics*. So works bearing "*al-Manāẓir*" in their titles but no Euclidean demonstrations or titles containing *ru'ya* or *ibṣār* in any form or subjects beyond optics proper altogether [e.g., burning mirrors (*marāyā al-muḥriqa*)], cannot be fully assigned to it.

50 The titles, authors, and dates are given in Arabic/Latin script and, according to Hijra/Western calendars depending on the place of composition, with solid or broken connecting lines representing the definite or indefinite nature of textual transmission.

51 See Heiberg, *Euclidis Optica and Opticorum recensio Theonis*, *Euclidis Opera Omnia*, v. 7, pp. 1–121, and pp. 143–247, respectively; on versions A and B, see Jones, "Peripatetic and Euclidean Theories," and Knorr, "Pseudo-Euclidean Reflections in Ancient Optics;" on the designation O and R here, see The Arabic "Version," v. 2, p. xix.

The Arabic textual tradition consists of both Arabic translations of the Greek Euclidean texts and compositions made originally in Arabic based on these and other texts. Of the Arabic translations, the only one still extant is Euclid's *Kitāb al-Manāẓir*, or *Ikhtilāf al-manāẓir*, depending on the manuscipt copy [dating from 600/1203 to at least 750/1349, see Description, pp. xxvi–xxvii]. In the case of Ptolemy's work, while no Arabic translation of it has survived, it is known to have existed at some point, for the extant Latin version of Admiral Eugenius of Sicily in the 12th century is a translation from the Arabic; that its title was *Kitāb al-Manāẓir* is also known through a lost summary and reconstruction by Ibn al-Haytham,[52] before he composed his own voluminous and influential work based on it, and bearing the same title. As for the other two Greek texts, it cannot be determined whether an Arabic translation was ever made, though in the case of the Euclidean text long known as Theon's Recension, its availability in some form is suggested by both the Arabic and Latin traditions.[53]

Among Arabic compositions based on these translations, al-Kindī's *De aspectibus* [*Ikhtilāf al-manāẓir?*] and Ibn al-Haytham's *Kitāb al-Manāẓir* stand out for being not just outstanding contributions but also important sources for further transmission. The Latin tradition, consisting of both translations and compositions, is represented here through the following texts: the Latin versions of the *Optica* of Euclid and Ptolemy, the former in four versions variously linked to the Arabic textual tradition,[54] as well as the *De aspectibus* of al-Kindī and Ibn al-Haytham, the latter with a prominent place in the transmission by virtue of its further circulation via an Italian translation and a printed edition.[55]

---

[52] *Talkhīṣ Kitābay Uqlīdis wa Baṭlamyūs fī al-Manāẓir;* see The Arabic Tradition, p. lvi.

[53] See The Arabic Versions, p. xxi, n. 6, and The Arabic "Version," v. 2, p. xix.

[54] Theisen, *The Mediaeval Tradition,* includes an edition, translation, and commentary on the "Greco-Latin" version, *De visu*, now published; see Theisen, "Liber de visu." Also included are editions of the "Arabo-Latin" versions, *Liber de aspectibus, Liber de radiis visualibus* [in 38 and 54 propositions respectively] – see pp. 336–384 and pp. 403–422 – and citation of another: *Euclidis de aspectuum diversitate* [in 4 definitions and 64 propositions], still in manuscript form, p. 324, n. 10. On the overlaps of M with the first of these, see The Arabic "Version," v. 2, p. xxix and n. 32; for the case of the others, see Summary and Commentary: Propositions 1, p. 23, and Propositions 16–17, pp. 52–53.

[55] On the Latin translation, *Perspectiva or De aspectibus,* of the late 12th or early 13th century, the Italian translation, *Prospettiva* of the 14th century, and the printed edition, as part of Risner's *Opticae Thesaurus* of 1572, see Sabra, *The Optics of Ibn al-Haytham,* v. 2, pp. lxxiv–lxxv. Lindberg, *Roger Bacon and the Origins of Perspectiva in the Middle Ages*, pp. xcvi-xcvii, gives the number of Latin manuscript copies of the *Optics* of Euclid (*De visu, De aspectibus*), Ptolemy, al-Kindī, and Ibn al-Haytham (*De aspectibus*) [the latter referred to as Alhacen], as 44, 15, 14, and 20, respectively.

That the Greek texts did not have the same fate in Islamic languages either, can be reasonably gathered from these charts. The main Greek Euclidean texts, Heiberg's *Euclidis Optica* [O or version A] and the one deviating from it in few but critical points, namely *Recensio Theonis* [R or version B], were transmitted most fully;[56] Ptolemy's *Optica* [P], richer both in content and method, was transmitted marginally, while most fruitfully,[57] and Pappus's *Collection* [C], uncertain in its place within the Arabic textual tradition, was transmitted only possibly or indirectly.[58] What is not transparent from these or any other charts is that in the case of the more widely circulated *Optics* of Euclid, the Arabic text, which is, no surprise, the only one of the Arabic translations still extant, the process of the Greco-Arabic transmission had involved a transformation of the Greek visual models as contained in the Greek originals.[59] The Arabic version of Euclid's *Optics*, forming the subject of the second volume, was indeed literally such: a "version" in Arabic, and one from which dependent "versions" were subsequently drawn.

---

56 Despite the unclear nature of latter's exact routes of transmission. On the alternative designations and their sources, see The Arabic "Version," v. 2, p. xix.

57 On the limited circulation of Ptolemy's *Optics* before the time of Ibn al-Haytham, with the exception of the supposed "Review of Ptolemy's *Optics*" by Ibn Sahl, see Sabra, *The Optics of Ibn al-Haytham*, v. 2, pp. lii–liii, lix–lx, and lviii–lx. See also The Arabic Tradition, p. lvi, n. 29.

58 On Pappus, see Hultsch, *Pappus Alexandrinus Collectio*, and Ver Eecke, *Pappus: La Collection*. For the Greek context of the sixth book of Pappus's *Collection*, see Knorr, "When Circles Don't Look Like Circles: An Optical Theorem in Euclid and Pappus." There is no reference in the Arabic tradition to the name of Pappus or his *Collection* in this connection, though in a few cases the Arabic text is closer to it than to the main Greek versions O and R; see Summary and Commentary: Propositions 37–38, v. 2, pp. 77–79.

59 See Kheirandish, "The Arabic 'Version' of Euclidean Optics: Transformations as Linguistic Problems in Transmission," *Tradition, Transmission, Transformation*, Ragep and Ragep (eds.), pp. 227–247.

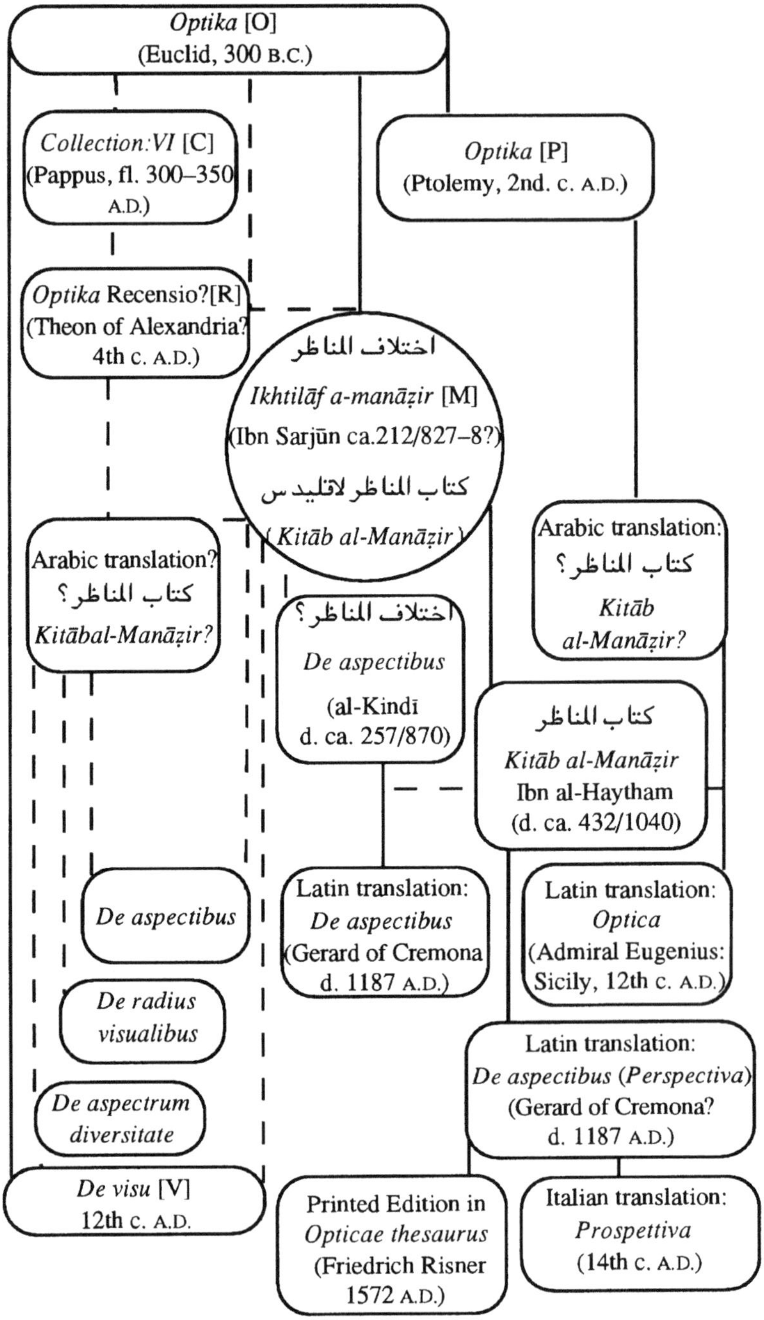

Chart 1 Greek-Arabic-LatinTransmission

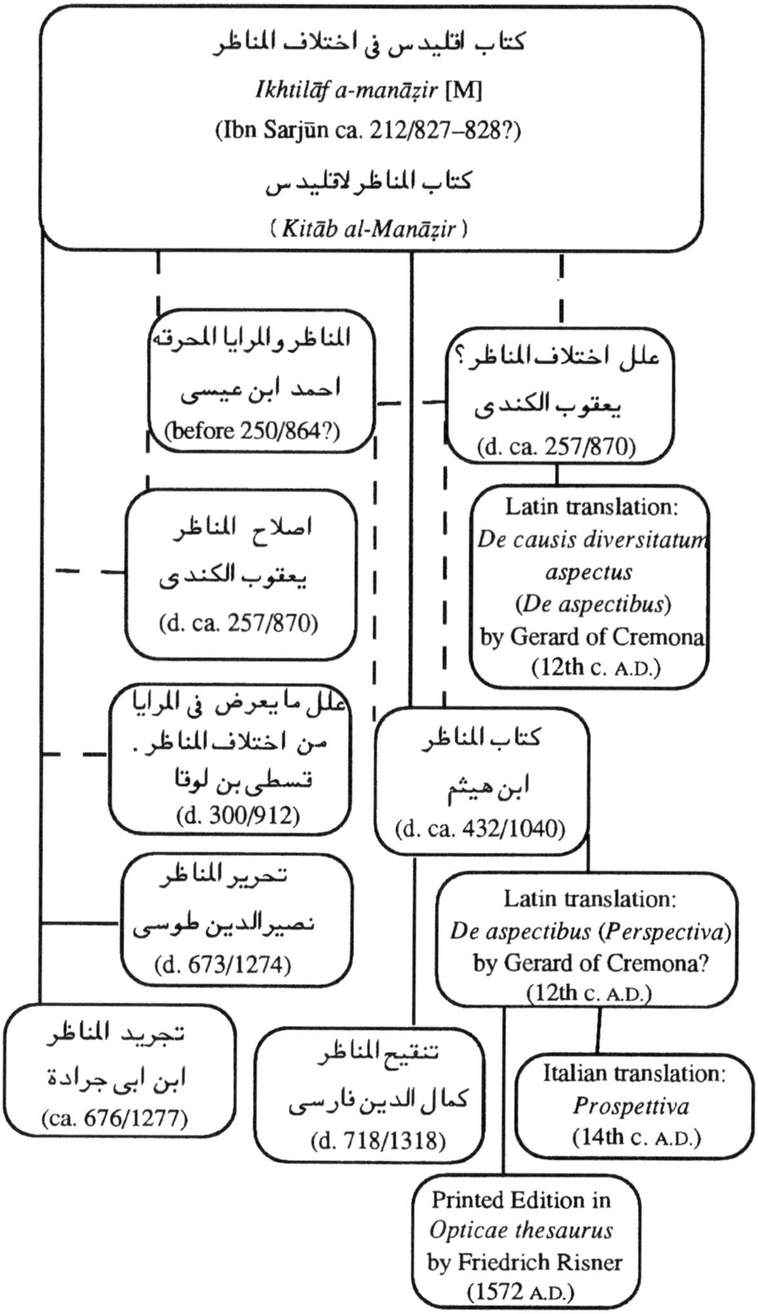

Chart 2 Internal Transmission

Part II

Edition and Translation

In the Name of God, the Compassionate, the Merciful.

The Book of Euclid on the Difference of Aspects

[1] The ray issues from the eye in straight lines while producing straight paths of infinite multitude.

[2] The figure enclosed by the ray is a cone whose apex is next to the eye and whose base is next to the extremity of the visible object.

[3] Objects upon which the ray falls are seen, and those upon which it does not fall are not seen.

[4-1] That which is seen through a large angle appears large;

[4-2] that which is seen through a small angle appears small;

[4-3] that which is seen through many angles appears more [clearly];

[4-4] that which is seen through equal angles appear equal.

بسم الله الرّحمن الرّحيم
كتاب أقليدس فى اختلاف المناظر

الشعاع يخرُج من العين على خطوَط مستقيمة
ومعه يحدُث سموتٌ مستقيمةٌ لا نهاَية لكثرتها ٥°
والشكل الذى يحيط به الشعاع مخروطٌ رأسه يلى العين
وقاعدته تلى نهاية المبصر ٥°
والأشياء التى يقَع عليها الشعاع تُبصَر
والتى لا يقع عليها لا تُبصَر ٥°
وما أُبصر من زاوية عظيمة ظهَر عظيماً
وما أُبَصر من زاوية صغيرة ظهَر صغيراً
وما أُبصِر من زوايا كثيرة ظهَر كثيراً
والأشياء التى تُبصَر من زوايا متساوية تظهَر متساويةً ٥°

----

(١) بسم الله الرّحمن الرّحيم] + وعليه التوكّل: م١ = + وما توفيقى الا بالله: م٢ = رب وفق برحمتك والله؟ ولى التوفيق: م٤ = وصلى الله على محمد وعلى آله: م٥ .

(٢) كتاب أقليدس فى اختلاف المناظر] + والشعاعات: م١ = مناظر: م٣ = المناظر من العين: م٤ = كتاب اوقليدس فى اختلاف المناظر: م٥ ، — م٢ .

(٣-٤) الشعاع يخرج ...ويحدث] زعم اقليدس ان الشعاع نحرج من العين على خطوط مستقيمة ومعه نحدث: م١ = المناظر من العين هى ان الشعاع نخرج من العين على خطوط مستقيمة مختلفة الابعاد والمقادير ويحدث به: م٢ = ان الشعاع يخرج من العين على خطوط مستقيمة مختلفة الابعاد والمقادير وتحدث به: م٤ = الشعاع نخرج من العين على خطوط مستقيمة وىحدث بعده: م٥ .

(٣) الشعاع] + المؤدى: فس م١ . (٤) يحدث] + بعد: فس م١ .

(٦) المبصر] + اليه: فس م١ . (٦) يلى ...تلى] ىلى ...ىلى: م١ = + اليه: فس م١ = يلى ...يلى: م٢ = على ...يلى: م٥ .

(٧) يقع عليها الشعاع تبصر] ىقع عليها: م١ = + اشعاع ىبصر: هش م١ = ىقع عليها الشعاع تبصر : م٢ . (٨) والتى لا يقع عليها لا تبصر] لا ىبصر: م١ = والذى لا تقع عليها لا ىبصر: هش م١ = والأشياء التى لا يقع عليها لا تبصر: م٤ = والتى لا ىقع الشعاع لا تبصر: م٥ .

(٩، ١٠، ١٢) ظهر...ظهر...تبصر] يرى ...يرى ...تظهر: م٥ .

(١١) وما أبصِر من زوايا كثيرة ظهر كثيراً] هش م١ = وما ابصر من راوىه كىسره طهر كىسرا: م١ = وما ابصر من زوايا كىسرة ظهر كثيراً: م٢ ، — م٥ .

1

Nothing among visible objects is seen all at once.
For let line AB be a visible object.
Then I say that it is not possible that line AB is seen all at once.
To demonstrate this we set the eye at mark G and draw the ray
as rays GA, GD, GE, GW, GB.
Now ray GE falls before ray GD, and GD before GA,
and GE before GW, and GW before GB;
consequently, magnitude ED is seen before magnitude DA;
and in the same way EW <is seen> before magnitude BW,
[which] is seen before AB.
Therefore nothing among visible objects is seen all at once,
though the object may be imagined to be seen all at once
because of the speed of sight's glance.
And that is what we wished to demonstrate.

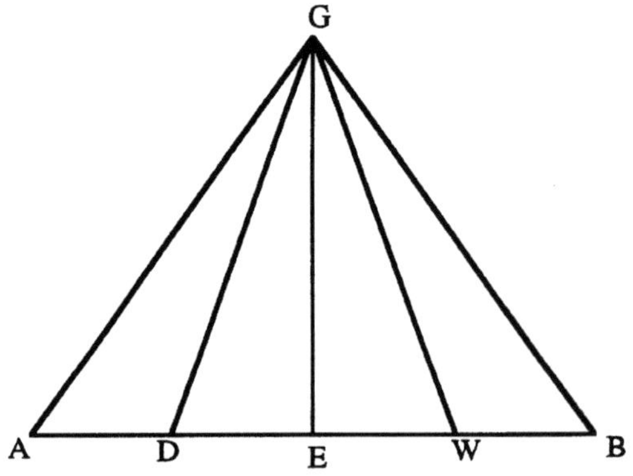

FIGURE 1

A	B	G	D	E	W	Z	H	T	I	K	L	M	N	S	O	F
ا	ب	ج	د	ه	و	ز	ح	ط	ى	ك	ل	م	ن	س	ع	ف

آ

ليس شئٌ من المبصرات يُبصَر جميعاً معاً
فليكن خطّ اب مبصراً
فأقول إنّ اب لا يمكن أنْ يُبصَر جميعاً معاً
برهانه أنْ نجعَل العين علامة جـ ونخرِج الشعاع
شعاعات جا جد جه جو جب
فشعاع جه يقع قبل شعاع جد وجد قبل جا
وَ جه قبل جو وجو قبل جب
فمقدار ه د يُبصَر قبل مقدار د ا
وكذلك مقدار ه و ‹يُبصَر› قبل مقدار ب و
يُبصَر قبل اب
فليس شئٌ من المبصرات يُبصَر جميعاً معاً
وقد يُتوهّم الشيء مبصراً جميعاً معاً
لسرعة لَمْح البصر
وذلك ما اردنا أنْ نبيّن ٥٥

----

(٣) فليكن] مثاله فليكن: م٢ ، م٤ .
(٤) إنّ اب] إنّ خط اب : م٥ .
(٥) برهانه أن] برهانه انا : م٥ .
(٥) العين] هش م٢ .
(٥-٦) الشعاع شعاعات] شعاع: م٢ ، م٣ ، م٤ ، م٥ .
(٧) شعاع جد] + لانه عمود: هش م١ .
(٧-٩) فشعاع جه يقع ... مقدار دا] فشعاع جا يقع قبل شعاع جد وشعاع جد يقع قبل جه وجه قبل جو وجو قبل جب فمقدار اد قبل مقدار ده: م٢= فشعاع جا يقع قبل شعاع جد وجد قبل جه و جه قبل جو وَجو قبل جب فمقدار اد قبل مقدار ده: م٤= فشعاع جا ...يبصر قبل مقدار ده: م٣ ، م٥ .
(٨) و جه] + ايضاً: فس م١ . (٨) قبل جو] + قبل جو: م٤ .
(١٠) وكذلك ... اب] _ م٢ ، م٣ ، م٤ ، م٥ .
(١٣) قد يتوهّم الشيء مبصراً جميعا معاً] قد يتوهَم الشيء جميعا مبصرا معا: م١= قد يتوهم يبصر جميعا: م٢= قد يتوهم الشيء جميعا معا: م٤= قد نتوهم الشيء يبصر جميعا معا: م٥، _ م٣ .
(١٥) أن نبيّن] بيانه: م٥ .

2

[Of] equal magnitudes unequally distant [from the eye],

those closer [to the eye] are seen more accurately.

For let two magnitudes AB, GD be equal, and [let] the eye be E,

and [let] GD be closer to E.

Then I say that GD is seen more accurately than AB.

To demonstrate this we draw the ray as rays EG, ED, EB, EA.

Now more of the ray falls on GD than on AB;

and that upon which more of the ray falls is seen more accurately;

therefore GD is seen more accurately than AB.

And that is what we wished to demonstrate.

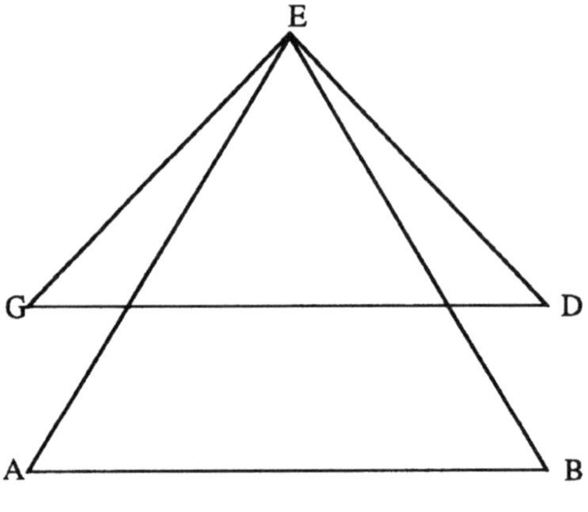

FIGURE 2

A	B	G	D	E	W	Z	H	T	I	K	L	M	N	S	O	F
ا	ب	ج	د	ه	و	ز	ح	ط	ى	ك	ل	م	ن	س	ع	ف

ب

المقادير المتساوية المختلفة الأبعاد
أقربها أصدقها رؤيةً
فليكن مقدارا اب جد متساويين والعين ه
و جد أقرب إلى ه
فأقول إنّ رؤية جد أصدق من رؤية اب
برهانه أنْ نخرج الشعاع شعاعات هج هد هب ها
فالواقع على جد من الشعاع أكثر من الواقع على اب
وما وقع عليه الشعاع أكثر فرؤيته أصدق
فرؤية جد أصدق من رؤية اب
وذلك ما أردنا أنْ نبيّن ە°ە

----

(٢) المقادير] المقدار: م٥ .
(٢) الأبعاد] + من البصر: هش م١ .
(٣) أقربها] + انه: فس م٣= + من البصر: م٥ .
(٤) فليكن] مثاله فليكن: م٢ ، م٤ .
(٤) اب جد] ا ب وجد: م٥ .
(٤) العين ه] العين نقطة ه: م٣ .
(٥) أقرب إلى ه] الاقرب: م٢= الاقرب من ه= م٤ ، م٥ .
(٦) رؤية ... رؤية] زاوية ... زاوية: م١ .
(٧) شعاعات] – م٢ ، م٣ ، م٤ ، م٥ .
(٧) هب ها] ها هب: م٢ ، م٣ ، م٤=
اه هب: م٥ .
(٩) الشعاع أكثر] من الشعاع اكثر: م٣ .
(٩) رؤيته] زاويته: م١ .
(١٠) رؤية اب] + فمقدار جد يرى اكثر من
مقدار اب بقدر جزحد: هش م١ .
(١٠) رؤية ... رؤية] زاوية ... زاوية: م١ .
(١١) وذلك ما أردنا أن نبيّن] – م٢= وذلك ما اردنا بيانه: م٥ .

3

Every visible object has a limit of distance that once crossed is not seen.
For let the visible object be line AB, and [let] the eye be [point] G.
Then I say that if AB becomes distant from the eye a certain extent, it is not seen.
To demonstrate this we draw the ray [as rays] GA, GB,
so that magnitude AB is seen;
then we move magnitude AB until it crosses ZH and we draw DE upon it.
Now because the ray falls upon AB, it is seen;
and because the ray does not fall upon DE, it is not seen;
but DE is the same as AB;
therefore, when magnitude AB becomes distant a certain extent, it is not seen.
And that is what we wished to demonstrate.

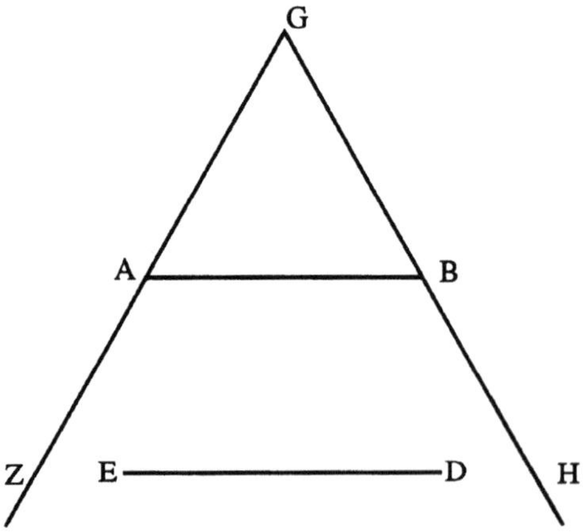

FIGURE 3

A	B	G	D	E	W	Z	H	T	I	K	L	M	N	S	O	F
ا	ب	ج	د	ه	و	ز	ح	ط	ى	ك	ل	م	ن	س	ع	ف

جـ

كلّ مبصرٍ فله غايةٌ من البُعد إذا جازها لم يُبصَر
فليكن المبصَر خطّ اب والعين جـ
فأقول إنّ اب إذا بَعُد من العين حداًّ لم يُبصَر
برهانه أنْ نخرِج الشعاع جا جب
فمقدار اب يبصر
وننقُل مقدار اب حتّى يجوز زَح ونرسُم عليه د ه
فلأنّ اب يقع عليه الشعاع يُبصَر
ولأنّ د ه لا يقع عليه الشعاع لا يُبصَر
ولكنّ د ه هو مثل اب
فمقدار اب إذاً بَعُد حداًّ لم يُبصَر
وذلك ما أردنا أنْ نبيّن ٭٭

---

(٢) من البعد] فس م١ .
(٢) جازها] جاوزها : م٢ .
(٣) فليكن] مثاله فليكن: م٢ ، م٤ .
(٣) والعين جـ] والموضع العين نقطة ج: م٣ .
(٤) إذا] كلما : م١ .
(٤) حداًّ] جداً : م٤ ، م٥ ، ـ م٢ .
(٤) بعد من] + العين صغر لان الزاوية التى يبصر اليه منها تصغر حتى اذا بعد الى غاية لم يبصر اليه وحتى زاوية غاب فلم يبصر: هش م١ .
(٥) الشعاع] شعاع : م٢ ، م٣ ، م٤ ، م٥ .
(٥) جا جب] جا ز جـ ب ح: م٢ ، م٣ ، م٤ ، م٥ .
(٦) فمقدار اب يبصر] + من زاوية ج ثم يصيره اب د فيرا اصغر منه اولاً وتضيق زاوية ج فيصير اصغر من د ه و تضيق زاوية ج ثم يصير ط ى فيرا ج فيرى اصغر من زح و تضيق زاوية ج ولا يرى كذلك كلما بعد حتى لا يبصر وذلك ما اردنا ان نبين: هش م١ = ـ م٢ ، م٣ ، م٤ ، م٥ .
(٧) د ه] ه د : م١ .
(٨-٩) يبصر ولان د ه لا يقع عليه الشعاع] ـ م٣ .
(١٠) هو مثل اب] د ه مثل اب: م٣ .
(١١) فمقدار اب] ف اب جـ : م٢ = ومقدار اب : م٣ = ف اب : م٤ ، م٥ .
(١١) حداًّ لم يبصر] حداً الم يبصر: م١ = جداً لم يبصر: م٢ ، م٣ ، م٤ ، م٥ .
(١٢) وذلك ما أردنا أن نبيّن] ـ م٢ = وذلك ما اردنا بيانه: م٥ .

4

When equal magnitudes are [placed] on the same line, the one toward which the path of the ray is the longest is seen as the smallest. For let AB, BG, GD be equal [magnitudes] on line AD, and [let] the eye be E, and [let] EA be perpendicular to AD. Then I say that AB is seen as greater than BG, and BG is seen as greater than GD. To demonstrate this we draw the ray as rays EA, EB, EG, ED, and from point B in triangle AGE we draw a line parallel to line GE [and] it cuts line AE at point W. Now [the ratio of] magnitude AB to BG is the same as that of AW to WE, so AW is the same as WE; but BW is greater than WE, and [so] angle WEB is greater than angle WBE;

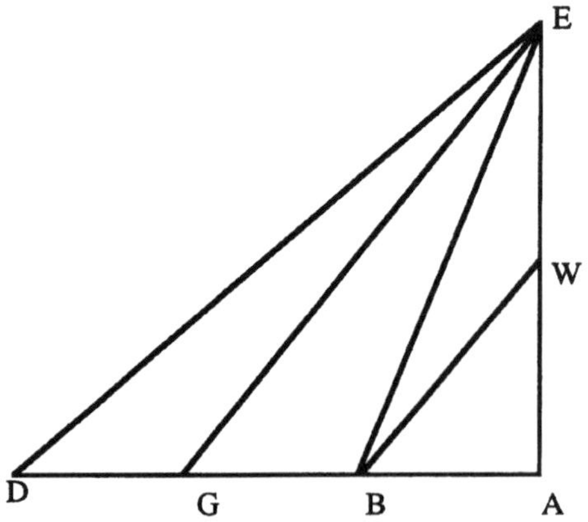

FIGURE 4

A	B	G	D	E	W	Z	H	T	I	K	L	M	N	S	O	F
ا	ب	ج	د	ه	و	ز	ح	ط	ى	ك	ل	م	ن	س	ع	ف

د

إذا كانت مقاديرٌ متساويةٌ على خطّ واحد
فإنّ الذى سمت الشعاع إليه أطولً يرى ًأصغر
فليكن اب بج جد متساويةً على خطّ اد
والعين ه وَ ها عمود على اد
فأقول إنّ اب يرى أعظم من بج
وَ بج يرى أعظم من جد
برهانه أنْ نخرجِ الشعاع شعاعات ها هب هج هد
ونخرِج من نقطة ب فى مثلث اجه خطّاً موازياً لخطّ جه
يقطَعَ خطّ اه على نقطة و
فقدر اب عند بج كقدر او عند وه
فَ او مثل وه
ولكنّ بو أعظم من وه وزاوية وهب أعظم من الزاوية وبه

---

(٣) إليه] فس م١ .

(٤) فليكن] مثاله فليكن: م٢ ، م٤ .

(٤) اب بج جد متساوية] اب جد: م٢ = اب جد متساوية: م٥ .

(٥) والعين ه] وموضع العين نقطة ه: م٣ .

(٧) وَ بج يرى] وان بج يرى: م٢ ، م٤ ، م٥= وان بج يرا: م٣ .

(٨) الشعاع شعاعات] شعاع: م٢ ، م٣ ، م٤ ، م٥ .

(٩) نقطة] _ م٢ ، م٥ .

(١٠) نقطة] _ م١ .

(١١) وه] هو: م١ .

(١١) فقدر] فمقدار= م٥ .

(١٢) مثل وه] + لان اب مثل بج: هش م١ .

(١٣) لكنّ بو] لكن ب: م٣ .

(١٣) وهب] اهب: م١ ، م٣ .

(١٣) من الزاوية] من زاوية: م٥ .

but the two angles WBE, BEG are equal, therefore angle WEB is greater than angle BEG; so the opening of the [cone of] ray[s] to AB is greater than its opening to BG, so magnitude AB is seen as greater than magnitude BG; and in the same way we demonstrate that BG is seen as greater than GD. And that is what we wished to demonstrate.

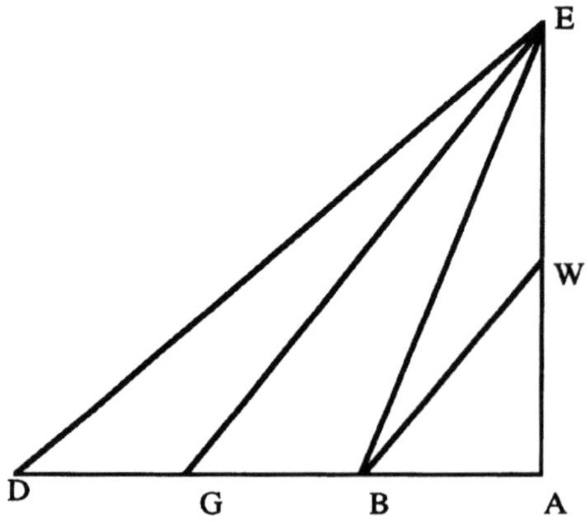

FIGURE 4

A	B	G	D	E	W	Z	H	T	I	K	L	M	N	S	O	F
ا	ب	ج	د	ه	و	ز	ح	ط	ی	ك	ل	م	ن	س	ع	ف

لكنّ زاويتا وب ه ب ه ج متساويتان
١٥ فزاوية وه ب أعظم من زاوية ب ه ج
فمخرج الشعاع إلى ا ب أعظم من مخرجه إلى ب ج
فمقدار ا ب يرى أعظم من مقدار ب ج
١٨ وكذلك نبيّن أنّ ب ج يرى أعظم من ج د
وذلك ما أردنا أنْ نبيّن ٥°

----

(١٤) لكنّ زاويتا] فزاويتى: م٢ ، م٤ ، م٥= ولكن زاويتا: م٣ .
(١٤) متساويتان] + لانهما متبادلتان: هش م١ .
(١٥) ب ه ج] ب ه د: م٤ .
(١٦) فمخرج] ومخرج: م٢ ، م٤ ، م٥ .
(١٧-١٨) مقدار ب ج وكذلك نبيّن أنّ أعظم من] ب ج وكذلك نبيّن أن اعظم من: هش م١ .
(١٩) أردنا أن نبيّن] اردناه: م٢= اردنا بيانه: م٥ .

5

Equal magnitudes unequally distant [from the eye] are seen as unequal, and the one closer to the eye is seen as the greatest. For let the two magnitudes AB, GD be equal, and [let] the eye be at E; and [let] AB be the closer of the two to E. Then I say that AB is seen as greater than GD. To demonstrate this we draw the ray as rays EA, EB, EG, ED. Now angle AEB is larger than angle GED; and what is seen through a large angle is seen as greater than what is seen through a small angle; therefore, magnitude AB is seen as greater than magnitude GD. And that is what we wished to demonstrate.

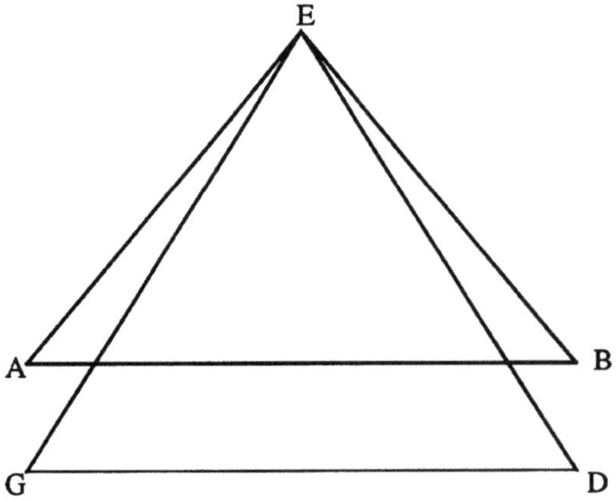

FIGURE 5

A	B	G	D	E	W	Z	H	T	I	K	L	M	N	S	O	F
ا	ب	ج	د	ه	و	ز	ح	ط	ى	ك	ل	م	ن	س	ع	ف

ه

المقادير المتساوية المختلفة الأبعاد تُرى مختلفةً
وأقربها من العين يرى أعظم
فليكن مقدارا اب جد متساويين والعين ه
واب أقربهما إلى ه
فأقول إنّ اب يرى أعظم من جد
برهانه أنْ نخرج الشعاع شعاعات ها هب هج هد
فزاوية اهب أعظم من زاوية جهد
وما أُبصر من زاوية عظيمة يرى أعظم
ممّا أُبصر من زاوية صغيرة
فمقدار اب يرى أعظم من مقدار جد
وذلك ما أردنا أنْ نبيّن ٥٥

---

(٣) من العين] الى العين: م٢ ، م٤ .

(٣) يرى] + بعد: فس م١ .

(٣) أعظم] اعظمها: م٢ ، م٣ ، م٤ ، م٥ .

(٤) فليكن] مثاله فليكن: م٢ ، م٤ = فيكون: م٥ .

(٤) مقدارا] مقدار: م٣ ، م٥ .

(٤) والعين ه] وموضع العين نقطة ه: م٣ .

(٥) إلى ه] الى نقطة ه: م٣ .

(٥) أقربهما] اقربها: م٢ ، م٣ ، م٤ = اقرب: م٥ .

(٧) برهانه أن] برهانه انا: م٥ .

(٧) الشعاع شعاعات] شعاع: م٢ ، م٣ ، م٤ ، م٥ .

(١٠) ممّا ابصر من زاوية صغيرة] وما ابصر من زاوية عظيمة صغيرة: م٥ .

(١١) أن نبيّن] بيانه: م٥ .

6

Parallel lines are seen from a distance as [being] unequal in width. For let AB [and] GD be parallel, and [let] the eye be E. Then I say that AB, GD are seen [such that] the [separating] lines of width between them are unequal, and the line closest to the eye is seen as greatest. To demonstrate this we draw the ray as rays EB, EW, EZ, ED, EH, ET; and we join the [separating] lines of width BD, WH, ZT. Now angle BED is larger than angle WEH, so line BD is seen as greater than line WH, and in the same way line WH is seen as greater than line ZT; therefore, the two lines AB, GD are seen in such a way that the [separating] lines of width between them are unequal. And that is what we wished to demonstrate.

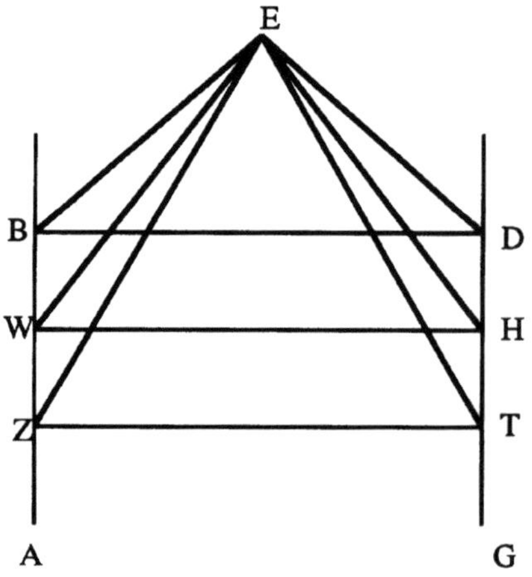

FIGURE 6

A	B	G	D	E	W	Z	H	T	I	K	L	M	N	S	O	F
ا	ب	ج	د	ه	و	ز	ح	ط	ى	ك	ل	م	ن	س	ع	ف

و

الخطوط المتوازية ترى من البُعد مختلفة العَرْض
فليكن اب جد متوازيين والعين ه
فأقول إنّ اب جد ترى خطوط العَرْض
التى بينهما مختلفةً والخطّ الأقرب من العين يرى أعظم
برهانه أنْ نخرِج الشعاع شعاعات هب هو هز هد هح هط
ونصل خطوط العَرْض بد وح زط
فزاوية بهد أعظم من زاوية وهح
فخطّ بد يرى أعظم من خطّ وح
فكذلك خطّ وح يرى أعظم من خطّ زط
فخطّا اب جد ترى
خطوط العَرْض بينهما مختلفةً
وذلك ما أردنا أنْ نبيّن ٭

---

(٢) ترى] + خطوط العرض الذى بينهما: م١ .
(٢) العرض] _ م١= العروض: فس م١ .
(٣) فليكن] مثاله فليكن: م٢ ، م٤ .
(٣) والعين ه] وموضع العين نقطة ه: م٣ .
(٤) ترى] يرى: م٢ ، م٣ ، م٥ .
(٤) خطوط العرض] الخطوط+ العرض: فس م١ .
(٥) التى] _ م٢ ، م٣ ، م٥ .
(٥) الخطّ] + العرض: هش م١= العرض: م٢ ، م٣ ، م٤ ، م٥ .
(٦) أن] أنا: م٥ .
(٦) الشعاع شعاعات] شعاع: م٢ ، م٣ ، م٤ ، م٥ .
(٦) هز] هو: م٣= _ م٤ . (٦) هط] + هز: فس م٤ .
(٧) العرض] + وهى: م٥ .
(٩) خطّ وح] وح: م٥ .
(١٠) فكذلك...اعظم] وكذلك يرى خط وح اعظم: م٢ ، م٣ ، م٤ ، م٥ .
(١١) فخطّا ... ترى] فخط ...ترى: م٢= فخطا ...يرى: م٣ .
(١٢) بينهما] بيينها: م٥ .
١٣) وذلك ما أردنا أن نبيّن] _ م٢= وذلك ما اردنا بيانه: م٥ .

7

Parallel lines [on a plane] lower than [that of] the eye
are seen from a vertical [distance] as unequal in width
For let the eye in the vertical [distance] be point A,
and [let] the two parallel lines be BG, DE;
and [let] the [separating] lines of width be WZ, HT, IK, [also] parallel.
Then I say that these [lines] are seen as unequal,
and the closest one is seen as greatest,
so that WZ, the closest one, is [seen as] greater than HT,
and HT is [seen as] greater than IK.
To demonstrate this we draw the ray as rays AW, AZ, AH, AT, AK, AI,
and we drop AL perpendicular to the plane BGDE;

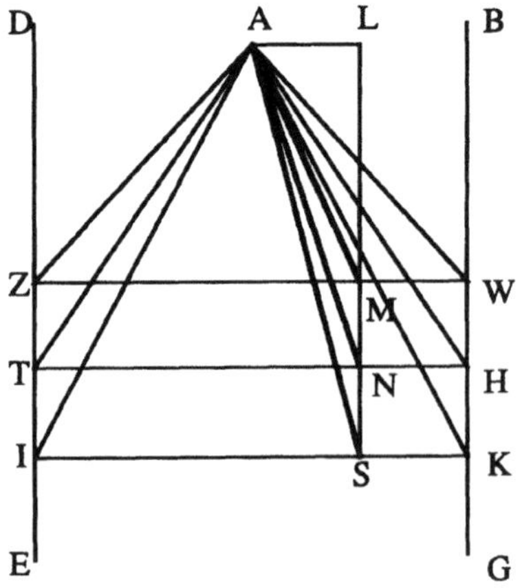

FIGURE 7

A	B	G	D	E	W	Z	H	T	I	K	L	M	N	S	O	F
ا	ب	ج	د	ه	و	ز	ح	ط	ى	ك	ل	م	ن	س	ع	ف

ز

الخطوط المتوازية المنخفضة عن العين
ترى من البُعد فى السُمْك مختلفة العَرْض
فليكن العين فى السُمْك علامة ا
والخطّان المتوازيان ب ج د ه
وخطوط العَرْض و ز ح ط ى ك وهى متوازيةٌ
فأقول إنّها ترى مختلفةً
ويرى الأقرب أعظم
فيرى و ز الأقرب أعظم من ح ط
و ح ط أعظم من ى ك
برهانه أنْ نخرِج الشعاع شعاعات ا و ا ز ا ح ا ط ا ك ا ى
ونسقِط ا ل عموداً على سطح ب ج د ه

---

(٣) ترى] يرى: م٣ ، م٥ .
(٣) البعد] + خطوط بينهما العرض المتوازية: م١ .
(٣) مختلفة العرض] مختلفة: م١ .
(٤) الخطّان] – م٢ ، م٣ ، م٤ .
(٥) د ه] و ه: م٥ .
(٥) وهى متوازية] – م٢ ، م٣ ، م٤ ، م٥ .
(٦، ١٠، ١٤، ١٦، ٢٣) ى ك] ك د: م٢ ، م٤ .
(٧) ترى] يرى: م٣ .
(٨) ويرى] يرى: م٥ .
(٨-٩) الأقرب أعظم فيرى] – م٢ ، م٤ .
(١٠) من] – م٤ .
(١١) الشعاع شعاعات] شعاع: م٢ ، م٤ ، م٥= الشعاع: م٣ .
(١١) ا و ا ز ا ح ا ط ا ك ا ى] او از اح اط اد اك: م٢ ، م٤= او از اى اك: م٣= او از والخطان المتوازيان ب ج وه وخطوط العرض وز ح ط د ك فاقول انها ترى مختلفة يرى الاقرب و هو وز اعظم من ح ط و ح ط اعظم من د ك اح اط اد اك: م٥ .
(١٢) ونسقط ا ل ... ب ج د ه] ونسقط اك ... ب ج: م٣= ب ج د و ه: م٥ .
(١٢) عموداً] تس م١ .

then we draw [line] LM perpendicular to WZ,
and we carry [line] LMNS through as a perpendicular to the two lines HT, IK;
and we join AM, AN, AS.
Now lines AM, AN, AS are perpendicular to lines WZ, HT, IK,
so the two angles ANT, AMZ are equal;
and the two lines NT, MZ are equal; and line AN is longer than line AM,
so angle MAZ is larger than angle NAT;
consequently line MZ is seen as greater than line NT,
and in the same way line MW is seen as greater than NH,
therefore all of the line WZ is seen as greater than all of the line HT,
and in the same way, line HT is seen as greater than line IK.
And that is what we wished to demonstrate.

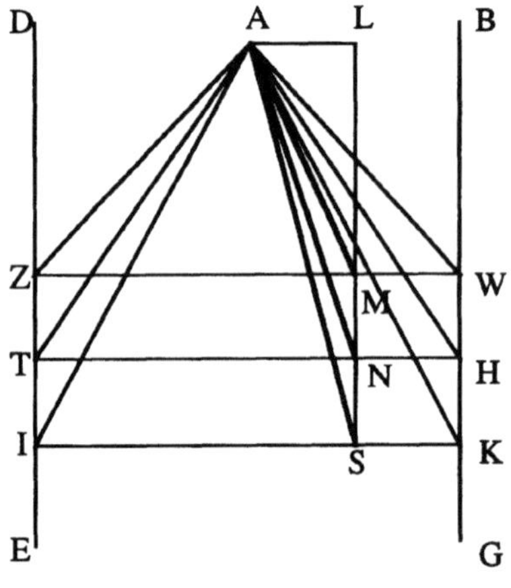

FIGURE 7

A	B	G	D	E	W	Z	H	T	I	K	L	M	N	S	O	F
ا	ب	ج	د	ه	و	ز	ح	ط	ي	ك	ل	م	ن	س	ع	ف

ونخرج ل م عموداً على وز
وننفذ ل م ن س عموداً على خطّى ح ط ى ك
ونصل ا م ا ن ا س
فخطوط ا م ا ن ا س أعمدةٌ على خطوط وز ح ط ى ك
فزاويتا ا ن ط ا م ز متساويتان
وخطّا ن ط م ز متساويان وخطّ ا ن أطول من خطّ ا م
فزاوية م ا ز أعظم من زاوية ن ا ط
فخطّ م ز يرى أعظم من خطّ ن ط
وكذلك خطّ م و يرى أعظم من ن ح
فجميع خطّ وز يرى أعظم من جميع خطّ ح ط
وكذلك خطّ ح ط يرى أعظم من ى ك
وذلك ما أردنا أنْ نبيّن ٥°

---

(١٣) ونخرج ل م عموداً] ونخرج عمود ل م عموداً: م١= ونخرج ل م عمود: م٢ .

(١٣) ل م] ا م: م٣ .

(١٤) عموداً] م١ ، م٢ .

(١٤، ١٦، ٢٣) ى ك] د ك: م٢ ، م٤ ، م٥ .

(١٦) فخطوط] بخطوط: م٣ .

(١٦) أعمدة] واقعة: م٣ .

(١٧) ا ن ط ا م ز] ا ن ا م: م٣ .

(١٨) وخطّا ... متساويان] هش م١= وخط ... متساويان: م٢ .

(١٨، ٢٠) ن ط] ب ط: م٥ .

(١٩) ن ا ط] ب ا ط: م٥ .

(٢٠) خطّ] _ م٣ .

(٢١) خطّ] _ م٥ .

(٢٢) يرى] _ م١ .

(٢٢-٢٣) أعظم...يرى أعظم] هش م١ .

(٢٢) جميع] _ م٣ .

(٢٣) خطّ ح ط] ح ط: م٤ ، م٥ .

(٢٣) ى ك] خط ى ك: م٣ .

(٢٤) أن نبيّن] بيانه: م٥ .

8

When equal magnitudes are placed on different locations [of the same line]
they are seen as unequal in size.
For let there be two equal magnitudes AB, GD upon the same line, namely AD,
their distance from the eye, which is at point E, being unequal;
and let the ray[s] EA, ED be [incident] on them,
and let ray EA be longer than ray ED.
Then I say that magnitude GD is seen as greater than magnitude AB.
So let the ray[s] EB, EG fall,
and we draw around triangle AED a circle, AED;
and we draw the two lines BZ, HG as rectilinear extensions to lines EB, EG,
and we draw from the two points B and G two perpendiculars,
namely, BT, GI.

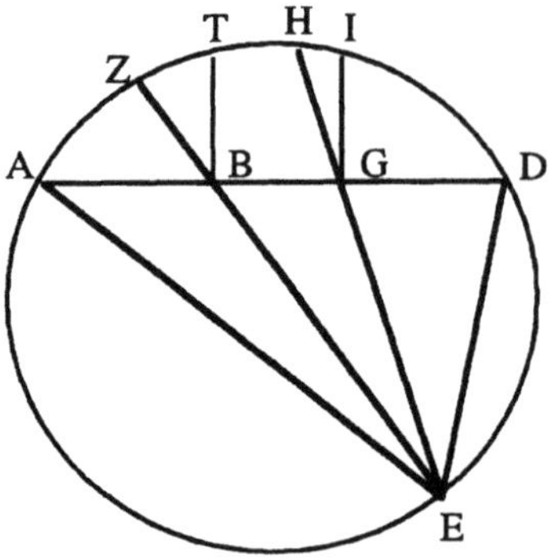

FIGURE 8

A	B	G	D	E	W	Z	H	T	I	K	L	M	N	S	O	F
ا	ب	ج	د	ه	و	ز	ح	ط	ى	ك	ل	م	ن	س	ع	ف

ح

المقادير المتساوية إذا وُضِعت على أماكن متفرقة
ترى مختلفةً فى العظَم
فليكن مقدارين متَساويين اب جد على خطّ واحد و هو اد
ويكون بُعدهما عن العين التى هى نقطة ه مختلفاً
وليكن عليهما شعاع ها هد
وليكن شعاع ها أطول من شعاع هد
فأقول إنّ مقدار جد يرى أعظم من مقدار اب
فليقع شعاع هب هج
ونخطّ على مثلّث اهد دائرةً عليها اهد
ونخرج خطّى ب ز ح ج متّصلين بخطّى هب هج على استقامة
ونخرج من نقطتى ب و ج عمودين
عليهما ب ط ج ى

---

(١) على أماكن] فى اماكن: م٥ .
(٣) ترى] يرى: م٣ .
(٤) مقدارين متساويين] + عليهما: م٢ ، م٤= مقداران متساويان عليهما: م٥ .
(٤) اد] خط اد: م٣ .
(٥) نقطة] ـ م٥ .
(١٠) ونخطّ ... عليها اهد] هش م١ .
(١١) ونخرج ... ح ج] ونخرج منها خطى ب ز ج ح: م٤= ونخرج ... ج ح: م٥ .
(١١) بخطّى هب هج] ـ م٥ .
(١٢) ب وَ ج] ب ج: م٢ ، م٣ ، م٤ ، م٥ .
(١٣) ب ط ج ى] ب ط ج ل: م٣ .

Now since line AB is equal to line GD,
and angle ABT equal to angle DGI,
then arc AT is equal to arc ID;
So arc ID is greater than arc ZA,
hence arc HD is much greater than arc ZA;
yet arc ZA subtends angle AEZ, and arc HD subtends angle HED,
so angle HED is greater than angle AEZ;
moreover, magnitude AB is seen through angle AEB
and magnitude GD is seen through angle GED;
therefore, magnitude GD is seen as greater than magnitude AB.
And that is what we wished to demonstrate.

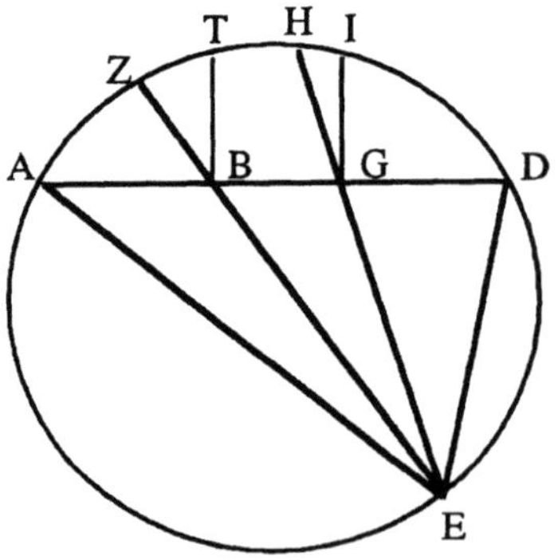

FIGURE 8

A	B	G	D	E	W	Z	H	T	I	K	L	M	N	S	O	F
ا	ب	ج	د	ه	و	ز	ح	ط	ى	ك	ل	م	ن	س	ع	ف

فلأنّ خطّ اب مساوٍ لخطّ جد
وزاوية ابط مساويةٌ لزاوية دجى
فقوس اط مساوية لقوس ىد
فقوس ىد إذاً أعظم من قوس زا
فقوس حد إذاً أعظم كثيراً من قوس زا
لكنّ قوس زا توتر زاوية اهز وقوس حد توتر زاوية حهد
فزاوية حهد أعظمَ من زاوية اهز
ولكنّ قدر اب يرى من زاوية اهب
وقدر جد يرى من زاوية جهد
فقدر جد يرى أعظم من قدر اب
وذلك ما أردنا أنْ نبيّن ٥

----

(١٥) دجى] د جل: م٣ .
(١٦) فقوس] تكون قوس: م٥ .
(١٦) ىد] لد: م٣ =د ك: م٥ .
(١٨) فقوس حد اذاً] فقوس جد: م٥ .
(١٩) لكن] ولكن: م٢ ، م٤ .
(١٩) لكن قوس زا توتر] فوتر: م٥ .
(٢٠) فزاوية حهد] ـ م٣ .
(٢٠) حهد] + وقوس حد توتر زاوية اهز: م١ .
(٢١) ولكنّ] لكن: م١.
(٢١) اهب] اهز: م٥ .
(٢١-٢٢) اهب وقدر جد يرى من زاوية] ـ م٣ .
(٢٣) فقدر] + خطّ: فس م١ .
(٢٣) من قدر] + خطّ: فس م١ .
(٢٣) يرى] ـ م١ = ـ م٣ .
(٢٣) اب] + فى الزاوية: م١ .
(٢٤) أردنا أن نبيّن] اردناه: م٢= اردنا بيانه: م٥ .

9

[For] equal parallel magnitudes with locations unequally distant [from the eye],
their difference of aspects is not seen
according to the [ratio of the] magnitudes of their distances.
For let there be two equal magnitudes AB, GD,
being unequally distanced from the point of the eye, which is point E.
Then I say that in aspect, [the ratio of] magnitude GD to AB
is not the same as [the ratio of] magnitude BE to DE.
So we draw ray[s] AE, EG,
and we inscribe around center E and radius EZ a segment of a circle HZT.
Now because triangle EZG is greater than sector EZH
and triangle EZD is smaller than sector EZT,
then [the ratio of] the magnitude of triangle EZG to sector EZH
is greater than [the ratio of] triangle EZD to sector EZT;
convertendo, [the ratio of] the size of triangle EZG to triangle EZD
is greater than [the ratio of] sector EZH to sector EZT;

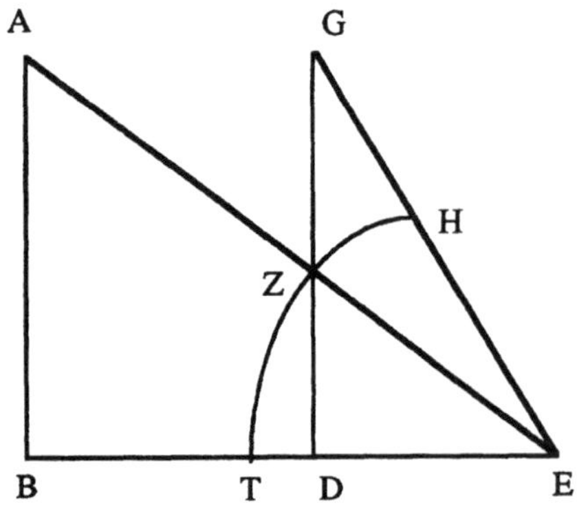

FIGURE 9

A	B	G	D	E	W	Z	H	T	I	K	L	M	N	S	O	F
ا	ب	ج	د	ه	و	ز	ح	ط	ى	ك	ل	م	ن	س	ع	ف

ط

الأقدار المتساوية المتوازية التى أماكنها مختلفة فى البُعد
لا يرى اختلاف مناظرها
على أقدار أبعادها
فليكن المقادير المتساوية اب جد
وتكون مختلفة البُعد عن نقطة البصر التى هى نقطة ه
فأقول إنّ قدر جد عند اب فى المنظر
ليس كقدر به عند ده
فنخرج شعاع اه ه‍ج
وندير على مركز ه وببعد ه‍ز قطعةً من دائرة وهى حزط
فلأنّ مثلث ه‍زج أعظم من قطاع ه‍زح
ومثلث ه‍زد أصغر من قطاع ه‍زط
فقدر مثلث ه‍زج إذاً عند قطاع ه‍زح
أعظم من قدر مثلث ه‍زد عند قطاع ه‍زط
وعلى عكس ذلك يكون قدر مثلث ه‍زج عند مثلث ه‍زد
أعظم من قدر قطاع ه‍زح عند قطاع ه‍زط

---

(٢) المتوازية] ـ م٤ .
(٣) لا يرى] لا ترى: م٥ .
(٤) أقدار] قدر: م٢ .
(٥-٦) فليكن المقادير المتساوية ...مختلفة البعد] فلتكن المقادير المتساوية ...البعد: م٤= فلتكن الاقدار المتساوية اب جد مختلفة البعد: م٥ .
(٧) فى المنظر] ـ م٢ ، م٣ ، م٤ ، م٥ .
(٨) به عند ده] به...ده: م٢= دبه...ده: م٣= بد ...ده: م٤ ، م٥ .
(٨) به] بد: م٤ = + به: فس م٤ .
(٩) فنخرج] لنخرج: م٣ .
(١٠) وندير...قطعةً] وعلى مركز ه وببعد ه‍ز ندير قطعةً: م٢= ونخط ...نقطة: م٥ .
(١٠) و هى حزط] + من دائرة: م١= حزط: م٣ .
(١١) ه‍زج] ه‍زد: م٢ ، م٤ ، م٥ .
(١٣-١٤) فقدر مثلث ه‍زج...عند قطاع ه‍زط] بقدر...ه‍زج عند قطاع ه‍زط: م٣= فقدر...اعظم من قدر مثلث ه‍زج اذاً عند قطاع ه‍زح: م٥ .
(١٥) على عكس ذلك] + نعين به التبديل ذلك: م١ .

addendo, [the ratio of] the size of triangle EGD to triangle EZD
is greater than the [the ratio of] sector EHT to sector EZT;
but [the ratio of] the size of triangle EGD to triangle EZD
is the same as line GD to line ZD; and GD is equal to AB,
and [the ratio of] magnitude AB to ZD is the same as that of BE to ED;
consequently, [the ratio of] magnitude BE to ED
is greater than [the ratio of] sector EHT to sector EZT;
but [the ratio of] sector EHT to sector EZT
is the same as [the ratio of] angle HET to angle ZET,
and [the ratio of] BE to ED is greater than that of angle HET to ZET;
thus [the ratio of] line BE to ED is greater than that of line GD to AB by angle.
So it is demonstrated that [for] equal, parallel, and unequally distant magnitudes,
their difference of aspects is not seen
according to [the ratio of] the magnitude of their distances.
And that is what we wished to demonstrate.

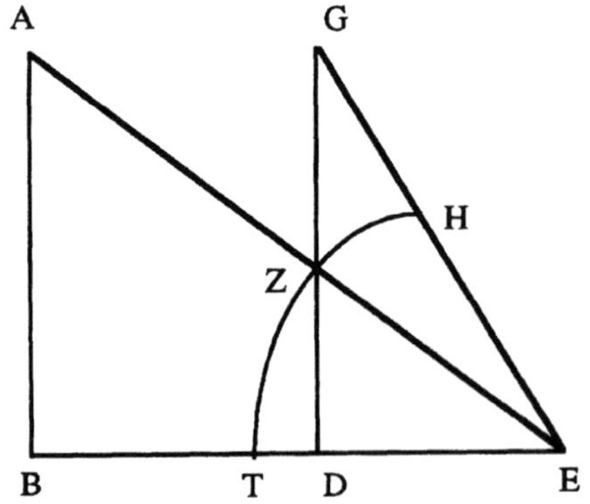

FIGURE 9

A	B	G	D	E	W	Z	H	T	I	K	L	M	N	S	O	F
ا	ب	ج	د	ه	و	ز	ح	ط	ى	ك	ل	م	ن	س	ع	ف

وعلى جهة التركيب يكون قدر مثلث ہ ج د عند مثلث ہ ز د
أعظم من قدر قطاع ہ ح ط عند قطاع ہ ز ط
ولكنّ قدر مثلث ہ ج د عند مثلث ہ ز د
كقدر خطّ ج د عند خطّ ز د و ج د يساوى ا ب
وقدر ا ب عند ز د كقدر ب ہ عند ہ د
فقدر ب ہ إذاً عند ہ د
أعظم من قدر قطاع ہ ح ط عند قطاع ہ ز ط
لكنّ قدر قطاع ہ ح ط عند قطاع ہ ز ط
مثل قدر زاوية ح ہ ط عند زاوية ز ہ ط
فَ ب ہ إذاً عند ہ د أعظم من قدر زاوية ح ہ ط عند زاوية ز ہ ط
وخطّ ب ہ عند ہ د أعظم من خطّ ج د عند خطّ ا ب فى الزاوية
فقد تبيّن أنّ الأقدار المتساوية المتوازية المختلفة فى البُعد
لا يرى اختلاف مناظرها
على قدر أبعادها
وذلك ما أردنا أنْ نبيّن ہ°ہ

---

(١٧) عند مثلث ہ ز د] عند مثلث ہ ز ج: م٢ ، م٥ = قدر مثلث ہ ز ج: م٤ .

(١٨) عند قطاع ہ ز ط] فس م٤ .

(١٨) قدر] فس م٥ .

(٢٠) ز د] ب د: م٥ .

(٢٢) فقدر ب ہ إذاً عند ہ د] فقدر د ب ہ . . . ہ د: م٣ = فقدر ب ہ عند ہ د اذاً: م٥ .

(٢٤) لكنّ . . . ہ ز ط] ــ م٢ ، م٣ = لكن . . . ح ہ ط . . . ہ ز ط: م٤ = ولكن . . . ہ ز ط: م٥ .

(٢٥) مثل قدر] كقدر: م٥ .

(٢٦) فَ ب ہ . . . ز ہ ط] ــ م٣ .

(٢٦-٢٧) قدر زاوية ح ہ ط عند زاوية ز ہ ط وخطّ ب ہ عند ہ د اعظم من] ــ م١ = قدر . . . فخط ب ہ عند د ہ: م٣ ، م٤ ، م٥ .

(٢٧) وخطّ ب ہ عند ہ د] فخط ب ہ اذاً عند ہ د: م٢ ، م٤ = فخط ب ہ د: م٥ .

(٢٧) عند خطّ ا ب] عند ا ب: م٣ .

(٢٧) فى الزاوية] ــ م٢ ، م٣ ، م٤ ، م٥ .

(٣٠) قدر] اقدار: م٤ ، م٥ .

(٣١) وذلك ما أردنا أن نبيّن] ــ م٢ = وذلك ما اردنا بيانه: م٥ .

10

Right-angled figures, when seen from a [far] distance, are seen as circular.

For let there be a right-angled figure, ABGD, at a far distance from the eye.
Then I say that it is seen as circular.
To demonstrate this, because sight from far away does not pause
at a single point, but rather moves,
then it is demonstrated that as it falls upon point E, it falls upon point Z,
and as it falls upon point H, it falls upon point T,
and it [also] falls upon the remaining points;
and because of this, that which is seen is figure EZHTKLMN,
while the [points] in between are not seen;
this is because what is hidden from sight is that of least magnitude in the figure,
that is the angles; and so the figure's angles are not seen;

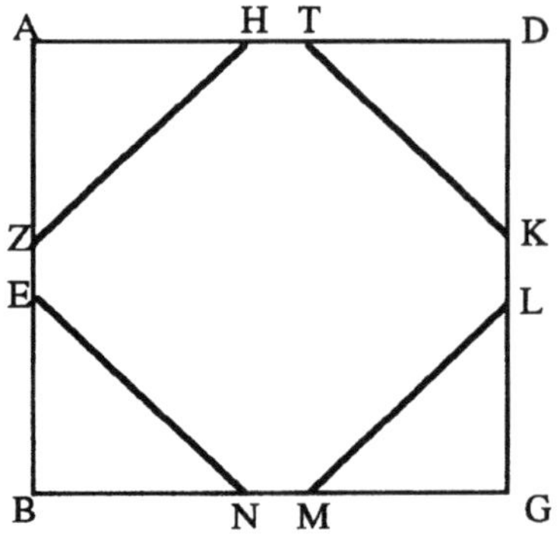

FIGURE 10

A	B	G	D	E	W	Z	H	T	I	K	L	M	N	S	O	F
ا	ب	ج	د	ه	و	ز	ح	ط	ى	ك	ل	م	ن	س	ع	ف

ي

الأشكال القائمة الزوايا إذا أُبصرت من بُعد ترى مستديرةً
فليكن الشكل القائم الزوايا شكَل اب جـ دً بعيداً عن البصر
فأقول إنّه يرى مستديراً
برهانه ولأنّ البصر من البُعد لا يقف
على نقطة واحدة ولكنّه ينتقل
فتبيّن أنّهً بينما يقع على نقطة ه وبينما يقع على نقطه ز
وبينما يقع على نقطة ح وبينما يقع على نقطة ط
وبينما يقع على بقيّة من النقط
فلذلك يكون الذى يرى شكل ه ز ح ط ك ل م ن
ولا يرى ما بين ذلك
لأنّ الذى يخفى عن البصر هو أقلّ قدراً من الشكل
وهو زوايا فإذن زوايا الشكل لا ترى

---

(٣) بعيداً] بعيد: م٢ ، م٤ .

(٣) شكل] _ م٥ .

(٣) القائم] القائمة: م٢ .

(٥) برهانه] _ م١ ، م٣ .

(٥) ولأنّ] لان: م٢ ، م٥ .

(٥) من البعد] من البعيد: م٢= من بعد: م٥ .

(٧) أنّه] _ م٤ .

(٧-٩) نقطة ه ...النقط] نقطة ه وبينما يقع على نقطة ط وبينما يقع على بقية النقط: م٢ ، م٤= نقطة ه ...بقية النقط: م٣= ه وبين ما يقع على ط و... بقية النقط: م٥ .

(٩) على] + نقطه: م١ .

(١٠) ه ز ح ك ل م ن] ه ز ح ط ك ل ن: م١ .

(١٢) لأنّ] + اصغر مما يكون فى الجسم من ابعاضه هو اول ما يغيب عن البصر اذا بعد فاصغر ما فى الشكل اب جـ د من ابعاضه زواياه فكذلك تغيب زوايا اب جـ د وكذلك حكم زوايا ح ط ك ل م ن؟ وعلى هذا زوايا فيما بعد ذلك فالشكل القائم الزوايا الذى عليه اب جـ د من البعد مستديراً: هش م١ .

(١٢-١٥) لأنّ الذى يخفى عن البصر ... من البعد مستديرة] _ م٢ ، م٣ ، م٤ ، م٥ .

and that figure which has no angles, is circular, therefore all figures are seen circular from a distance, and the right-angled figure which is ABGD is seen as circular from a distance.

And that is what we wished to demonstrate.

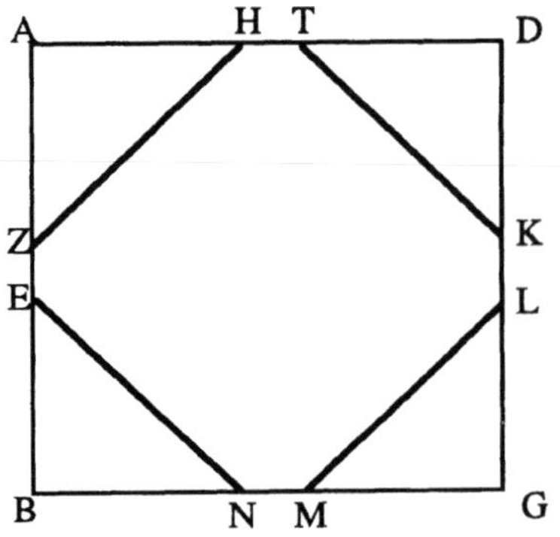

FIGURE 10

A	B	G	D	E	W	Z	H	T	I	K	L	M	N	S	O	F
ا	ب	ج	د	ه	و	ز	ح	ط	ى	ك	ل	م	ن	س	ع	ف

وما ليس له زوايا من الأشكال هو مستدير فإذن الأشكال كلّها ترى من البُعد مستديرة والشكل القائم الزوايا الذى عليه ا ب ج د يرى من البُعد مستديراً وذلك ما أردنا أنْ نبيّن ٥٥

---

(١٨) وذلك ما اردنا أن نبيّن] ـ م٢ = وذلك ما اردنا بيانه: م٥ .

11

[Of] planes below the eye, the most distant one is seen as the highest.
For let the eye's position be point A, which is than plane BEDG,
and let the ray[s] AE, AB, AD, AG fall.
Then I say that GD is seen as higher than DE, and DE [is seen as] higher than BE.
So let us cut line BE in half at point Z, and we draw ZH at a right angle.
Now because the visual ray falls first upon ZH before falling upon ZG,
then ray AG falls upon line ZH at point H, and ray AD [falls] at point T,
and ray AE [falls] at point I;
then all the rays fall upon line HZ;
but because point H is higher than point T, and point T is higher than point I,
and the ray that has point H on it [also] has point G on it,

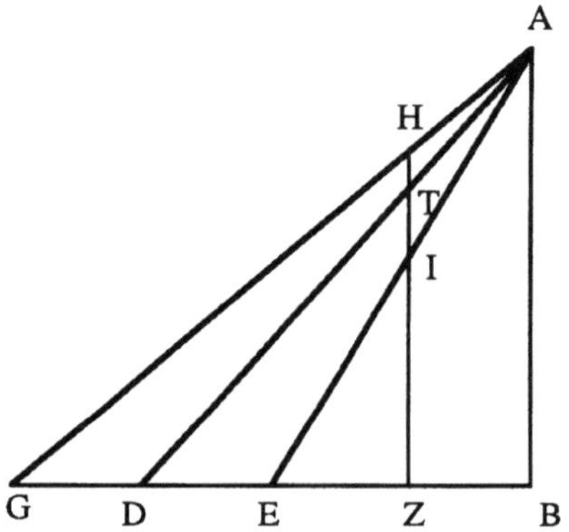

FIGURE 11

A	B	G	D	E	W	Z	H	T	I	K	L	M	N	S	O	F
ا	ب	ج	د	ه	و	ز	ح	ط	ى	ك	ل	م	ن	س	ع	ف

يا

السطوح التى تكون تحت البصر أبعدها يرى أرفع

فليكن موضع البصر نقطة ا ويكون أرفع من سطح ب ه د ج

وليقع شعاع ا ه ا ب ا د ا ج

فأقول إنّ جـ د يرى أرفع من د ه و د ه أرفع من ب ه

فلنقطع خطّ ب ه بنصفين على نقطة ز ونخرج ز ح على زاوية قائمة

فلأنّ شعاع البصر يقع أولاً على ز ح قبل وقوعه على ز ج

فيقع شعاع ا ج على خطّ ز ح على نقطة ح وشعاع ا د على نقطة ط

وشعاع ا ه على نقطة ى

فجميع الشعاعات وقعت على خطّ ح ز

فلأنّ نقطة ح أرفع من نقطة ط ونقطة ط أرفع من نقطة ى

والشعاع الذى عليه نقطة ح فعليه نقطة ج

---

(٢) السطوح التى تكون] السطوح التى: م١ = كل السطوح التى تكون: م٥ .

(٣) ويكون أرفع من سطح] ويكون ... سطوح: م١ = وليكون ... سطوح: م٣ .

(٣) ب ه د ج] ب ج: م٢ ، م٤ ، م٥ = ب ه ج: م٣ .

(٥) يرى] ايرىى: م١ .

(٥) فاقول ... ب ه] فاقول ان جه يرى ارفع من ب ه: م٣ .

(٦) فلنقطع] فيقع: م١ = فيقطع: م٥ .

(٦) على نقطة ز] نقطة ز: م١ = على نقطة م: م٥ .

(٦) زاوية] زوايا: م٣ .

(٧) شعاع البصر] البصر: م٣ .

(٧) ز ج] ز د: م٢ ، م٤ ، م٥ .

(٨) فيقع] فليقع: م٢ ، م٤ ، م٥ .

(٨) ح وشعاع ا د على نقطة] هش م١ .

(٩، ١١، ١٣) نقطة ى] نقطة ل: م٣ .

(١٠) فجميع ... ح ز] وعلى خطّ ح ز فعلى نقطة الاولى: م٢ ، م٤ ، م٥ = على خطّ ز ح: م٣ .

(١١) نقطة ح] نقطة ح ز: م١ .

(١١) ونقطة ط ارفع] رفع: م٢ .

(١٢) الشعاع الذى عليه نقطة ح فعليه نقطة ج] ــ م١ .

(١٢) والشعاع الذى عليه] ولكن الشعاع الذى عليه: م١ = ولكن الشعاع الذى على: م٣ .

and the ray that has point T on it [also] has point D on it, and the ray that has point I on it [also] has point E on it, and since GD is seen because of HT, and DE is seen because of TI, then GD is seen as higher than DE, and in the same way, DE is also seen as higher than EB. And that is what we wished to demonstrate.

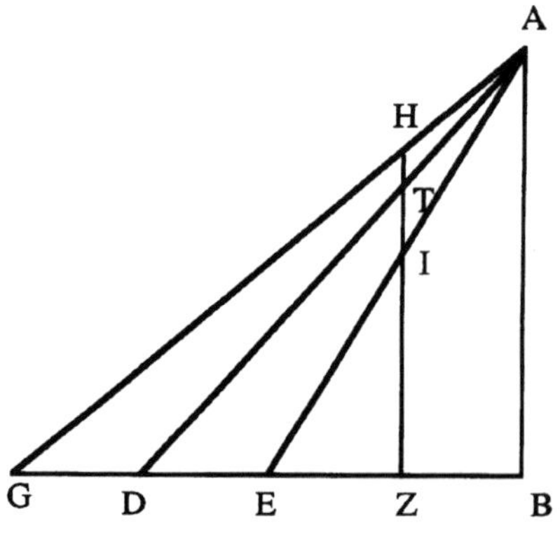

FIGURE 11

A	B	G	D	E	W	Z	H	T	I	K	L	M	N	S	O	F
ا	ب	ج	د	ه	و	ز	ح	ط	ى	ك	ل	م	ن	س	ع	ف

والشعاع الذى عليه نقطة ط فعليه نقطة د
والشعاع الذى عليه نقطة ى فعليه نقطة ه
فمن أجْل ح ط يرى جد ومن أجْل ط ى يرى د ه
فَ جد إذن يرى ارفع من د ه
وكذلك أيضاً يرى د ه أرفع من ه ب
وذلك ما أردنا أنْ نبيّن ٥°

----

(١٥) فمن أجل] فمن إجل ان: م٥ .
(١٥) ومن أجل ط ى] ومن اجل ط ل: م٣ .
(١٥) يرى ... يرى] يرا ... يرا: م٣ .
(١٦) إذن يرى ارفع من د ه] ارفع من ه د: م٢ ، م٤ ، م٥ .
(١٧) يرى] _ م٢ ، م٤ ، م٥ .
(١٧) ه ب] ه د: م٢= د ه: م٤ ، م٥ .
(١٨) أردنا أن نبيّن] اردناه: م٢= اردنا بيانه: م٥ .

12

[Of] planes higher than the eye, the most distant one is seen as lowest.
For let the eye be point A, being lower than the position of plane BG,
and let the ray[s] AB, AD, AE, AG fall,
Then I say that GE is seen as lower than DE, and ED [is seen as] lower than DB.
Now because of what we set forth in the preceding proposition,
ray AG is lower than ray AE, and ray AE is lower than ray AD,
and ray AD is lower than ray AB;
however, GE is seen by the two rays EA, AG,
and DE is seen by the two rays EA, AD,
and DB is seen by the two rays DA, AB;
therefore GE is seen as lower than DE, and ED is [seen as] lower than DB.
And that is what we wished to demonstrate.

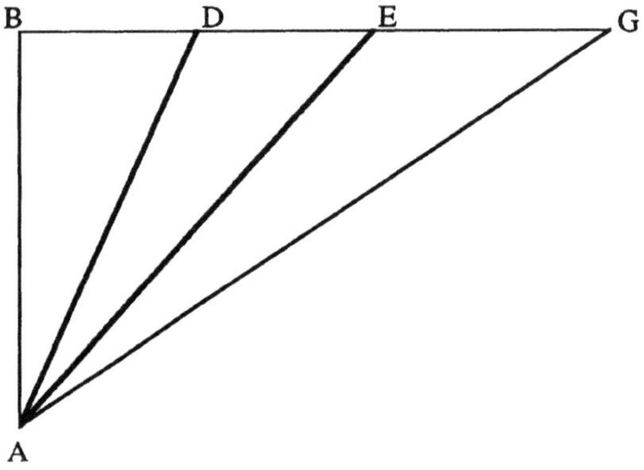

FIGURE 12

A	B	G	D	E	W	Z	H	T	I	K	L	M	N	S	O	F
ا	ب	ج	د	ه	و	ز	ح	ط	ى	ك	ل	م	ن	س	ع	ف

يب

السطوح التى تكون أعلا من البصر فإنّ أبعدها يرى أكثر انخفاضاً

فليكن البصر نقطة ا ويكون منخفضاً عن موضع سطح ب ج

وليقع شعاع ا ب ا د ا ه ا ج

فأقول إنّ ج ه يرى أكثر انخفاضاً من د ه وَ ه د أكثر انخفاضاً من د ب

فمن أجل ما قدّمنا فى الشكل المقدّم

يكون شعاع ا ج أخفض من شعاع ا ه وشعاع ا ه أخفض من شعاع ا د

وشعاع ا د أخفض من شعاع ا ب

ولكنّ ج ه يرى بشعاعى ه ا ا ج

وَ د ه يرى بشعاعى ه ا ا د

ود ب يرى بشعاعى د ا ا ب

فَ ج ه يرى أخفض من د ه وَ ه د أخفض من د ب

وذلك ما أردنا أنْ نبيّن ٥٥

----

(٢) أعلا] اعلى: م٢ ، م٥ .

(٢) فإنّ . . . يرى] هش م٢ .

(٣) عن موضع] على موضع: م٥ .

(٥) من د ه] من ه د: م٢ .

(٥) وَ ه د أكثر انخفاضاً من د ب] ـ م٢ ، م٣ ، م٤ ، م٥ .

(٨) وشعاع ا د] و ا د: م١ .

(٨) وشعاع ا د أخفض من شعاع ا ب] و ا ب اخفض من ا د: م١ .

(٩) بشعاعى] فس م٤ .

(٩-١٠) ه ا ا ج وَ د ه] جا اه وه د: م٢ ، م٣ ، م٤ ، م٥ .

(١٠) وَ د ه يرى بشعاعى ه ا ا د] ـ م١ .

(١١) شعاعى] شعاعين: م١ .

(١١) د ا ا ب] دا دب: م٥ .

(١٢) يرى] ـ م١ .

(١٢) د ه وَ ه د] ه د و ه د: م٢ ، م٣ ، م٤ ، م٥ .

(١٢) أخفض] ـ م٢ ، م٤ .

(١٣) وذلك ما اردنا أن نبيّن] ـ م٢ = وذلك ما اردنا بيانه: م٥ .

13

[Of] those magnitudes distant from the eye and facing it,
those to the right are seen to the left and those to the left are seen to the right.
For let there be two visible magnitudes, namely BD, AG,
and [let] the eye's postion be E,
and let us draw from it ray[s] ET, EI, ED, EZ, EH, EG.
Then I say that EZ, EH, EG are thought to be to the left,
and ET, EI, ED are thought to be to the right.
Now because EZ is to the right of EH, and EH is also to the right of EG,
then EG is thought to be to the left of EH,
and EH is seen as [being] to the left of EZ;
and in the same way ED, EI, ET are thought to be moved to the right.
And that is what we wished to demonstrate.

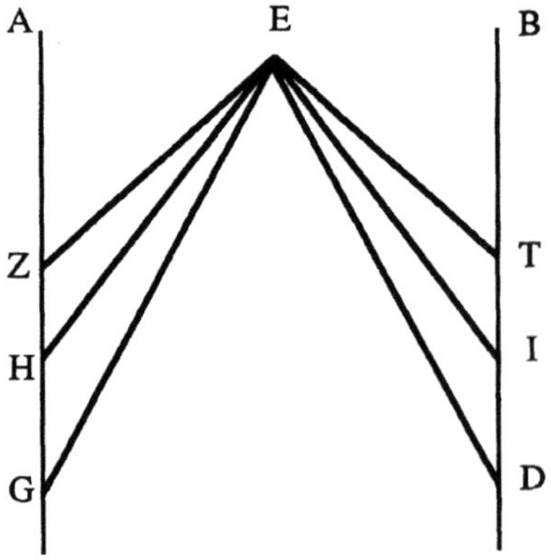

FIGURE 13

A	B	G	D	E	W	Z	H	T	I	K	L	M	N	S	O	F
ا	ب	ج	د	ه	و	ز	ح	ط	ى	ك	ل	م	ن	س	ع	ف

يج

الأقدار المتباعدة عن البصر المقابلة له
المتيامن منها يرى متياسراً والمتياسر منها يرى متيامناً
فليكن قدران يريان عليهما ب د ا ج
وموضع البصر نقطة ه
ولنخرج منه شعاع ه ط ه ى ه د ه ز ه ح ه ج
فأقول إنّ ه ز ه ح ه ج يُظَنّ أنّها متياسرة
وَ ه ط ه ى ه د يُظَنّ أنّها متيامنة
فلأنّ ه ز متيامن عن ه ح وه ح أيضاً متيامن عن ه ج
فلذلك يُظَنّ أنّ ه ج متياسر عن ه ح
وَ ه ح يرى متياسراً عن ه ز
ومثل ذلك ه د وَ ه ى وَ ه ط يُظَنّ أنّها منتقلة إلى اليمين
وذلك ما أردنا أنْ نبيّن ه

---

(٣) المتياسر منها] المتياسر: م٢، م٤، م٥ .
(٤) ب د ا ج] اب جد: م٢، م٤، م٥ = اج ب د: م٣ .
(٦) فأقول ... ه ج] — م٣ .
(٦) ه ز ه ح ه ج] ه ز ه ح: م٥ .
(٦-٧) متياسرة ... أنّها] — م١ .
(٧) وَ ه ط] ه ط: م٣ .
(٩) فلذلك ... ه ح] — م٢، م٤، م٥ .
(٩) يرى] — م٢ .
(١٠) ومثل ذلك] ومثل: م٢ .
(١٠) ه ى] د ى: م١ .
(١١) أن نبيّن] بيانه: م٥ .

14

When equal magnitudes are below the eye on the same side, then the one that is [most] distant is seen as higher than the ones beneath it. For let there be three equal magnitudes AB, GD, EZ, and [let] the eye be point H, which is higher than the magnitudes, and we draw ray[s] HA, HG, HE. Then I say that AB is seen as higher than GD, and GD seen as higher than EZ. Now because AH is higher than GH and HG higher than HE, and where there are the ray[s] HA, HG, HE, then there are points A, G, E, and it is clear that these [points] are also extremities of magnitudes AB, GD, EZ, therefore magnitude AB is seen as higher than magnitude GD, and magnitude GD is seen higher than EZ. And that is what we wished to demonstrate.

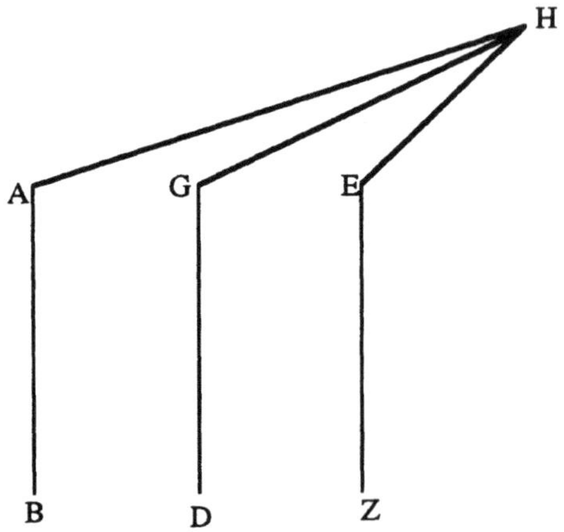

FIGURE 14

A	B	G	D	E	W	Z	H	T	I	K	L	M	N	S	O	F
ا	ب	ج	د	ه	و	ز	ح	ط	ى	ك	ل	م	ن	س	ع	ف

يد

الأقدار المتساوية إذا كانت تحت البصر على سمتٍ واحدٍ فالذى بَعُد منها يرى أعلى مما دونه

فليكن ثلثة أقدار متساوية اب جد ه ز والبصر نقطة ح وهو عال عن الأقدار فنخرج شعاع ح ا ح ج ح ه فاقول إنّ اب يرى أعلى من جد وَ جد يرى أعلى من ه ز فلأنّ اح أعلى من جح وَ حج أعلى من ح ه وحيث شعاع ح ا ح ج ح ه فثمّ نقط ا ج ه و بيّن أيضاً ثمّ ينتهى عِظَم اب جد ه ز فقدر اب يرى أعلى من قدر جد وقدر جد يرى أعلى من ه ز وذلك ما أردنا أنْ نبيّن ٥°

---

(٣) أعلى] اعلا: م١، م٣، م٤، م٥ .

(٣) فالذى] التى: م٥ .

(٤) فليكن] مثاله فليكن: م٢ = فلتكن: م٤، م٥ .

(٤) ثلثه] ثلاثه: م٤، م٥ .

(٤) متساوية] + عليها: م٢، م٤، م٥ .

(٥) عال عن] على: م١ = عايل عن: م٢ = عالى عن: م٣، م٤ .

(٧) يرى] فس م١، ـ م٢، م٣، م٤، م٥ .

(٧) جد] ج: م٥ .

(٧، ٨، ١١، ١٢) اعلى] اعلا: م٣، م٤ .

(٨) حج] جح: م٥ .

(٩) فثمّ نقط] فثمّ نقطة: م٢، م٣ = وثمّ نقط: م٤ .

(٩) حيث] حيث ان: م٤ .

(١٠) بيّن] تبين: م٢، م٣ .

(١٠) أيضاً]+ جد اعلى من ه ز: م١ = + انه: م٣، م٤، م٥ .

(١٠) ثم ينتهى] لم ينتهى: م٣ = لم ينتهه: م٥ .

(١٢) وقدر...ه ز] ـ م١، م٣ = وقدر...من قدر ه ز: م٥ .

(١٣) وذلك ما أردنا أن نبيّن] ـ م٢ = وذلك ما اردنا بيانه: م٥ .

15

[Of] equal magnitudes higher than the eye,
that which is [most] distant is seen as the lowest.
For let there be three equal magnitudes AB, GD, EZ, higher than the eye,
and [let] the eye be point H.
Then I say that AB is seen as lower than GD, and GD [seen as] lower than EZ.
So we draw ray[s] HB, HD, HZ.
Now because ray HB is lower than ray HD, and HD is lower than HZ,
and where there are extremities of the ray[s], HB, HD, HE,
then there are points B, D, Z,
and these [points] are [also] extremities of magnitudes AB, GD, EZ,
therefore, AB is seen as lower than GD, and GD [is seen as] lower than EZ.
And that is what we wished to demonstrate.

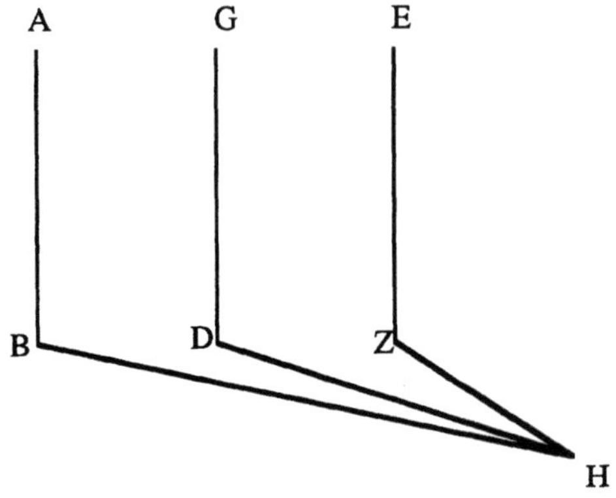

FIGURE 15

A	B	G	D	E	W	Z	H	T	I	K	L	M	N	S	O	F
ا	ب	ج	د	ه	و	ز	ح	ط	ى	ك	ل	م	ن	س	ع	ف

يه

الأقدار المتساوية إذا كانت أعلى من البصر
ما بَعُد منها يرى أخفض
فليكن ثلثة أقدار متساوية اب جد ەز أعلى من البصر
والبصر نقطة ح
فأقول إنّ اب يرى أخفض من جد وَ جد أخفض من ەز
ونخرج شعاع حب حد حز
فلأنّ شعاع حب أخفض من شعاع حد وَ حد أخفض من حز
وحيث ينتهى شعاع حب حد حز
فثمّ تكون نقط ب د ز
وثمّ تنتهى أقدار اب جد ەز
فَ اب اذن يرى أخفض من جد فَ جد من ەز
وذلك ما أردنا أنْ نبيّن ٥٥

(٢، ٤) أعلى] اعلا: م١، م٣، م٤ .
(٣) ما] فما: م٢ .
(٤) من] ــ م١ .
(٤) ثلثه] ثلاثه: م٥ .
(٤) متساويه] عليها: م٢، م٤= متساويه + عليها: م٥ .
(٥) من جد وَ جد أخفض] فس م١، ــ م٤ .
(٧) ونخرج] فنخرج: م٢، م٤، م٥ .
(٨) من شعاع حد وَ حد أخفض] هش م١ .
(١٠) تنتهى] تكون: م١، م٣= تكون تنتهى: م٢، م٤ .
(١١) إذن] ــ م٢، م٣، م٤، م٥ .
(١١) فَ جد من ەز] فمن ەز: م١= ــ م٢، م٣، م٤، م٥ .
(١١) وذلك ما أردنا أن نبيّن] ــ م٢= وذلك ما اردنا بيانه: م٥ .

16

When there are two magnitudes below the eye and one is greater than the other, the part seen of the greater [magnitude] while seeing the smaller one is less than that of the greater while the smaller is seen, when the eye is lowered.
For let the two magnitudes be lines AB, GD, AB [being] greater than GD, and [let] the eye be point E;
and the ray which we draw from point E passes through point G, namely EZ.
Now because what is seen of magnitude AB while [seeing] magnitude GD by means of ray EZ is magnitude BZ,
then magnitude AB is seen to be greater than magnitude GD by magnitude AZ;
after that, we lower the eye to the position of point H,
and we draw the ray through point G, and that is line HT;
so [what is seen of] magnitude TB while magnitude GD is seen is equal;
then magnitude AB is seen greater than magnitude GD by magnitude AT;

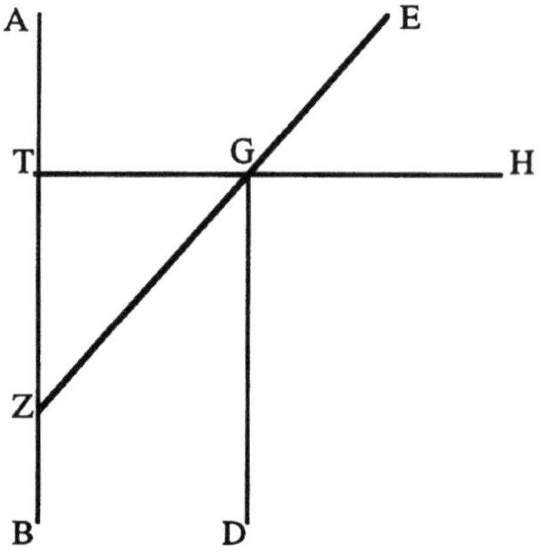

FIGURE 16

A	B	G	D	E	W	Z	H	T	I	K	L	M	N	S	O	F
ا	ب	ج	د	ه	و	ز	ح	ط	ى	ك	ل	م	ن	س	ع	ف

يو

إذا كان مقداران تحت البصر وأحدهما أعظم من الآخر
فإنّ الذى يرى من الأعظم مع رؤية الأصغر أصغر قدراً
ممّا يرى من الأعظم مع رؤية الأصغر إذا انخفض البصر
فليكن المقداران خطّى اب جد وَ اب أعظم من جد
والبصر نقطة ه
والشعاع الذى نخرِج من نقطة ه يمُرّ على نقطة ج وهو ه ز
فلأنّ الذى يرى من مقدار اب مع مقدار جد
وبشعاع ه ز هو مقدار ب ز
فقدر اب يرى أعظم من قدر جد بمقدار ا ز
ثمّ نخفض البصر إلى موضع نقطة ح
ونخرِج الشعاع على نقطة ج فيكون خطّ ح ط
فيرى قدر ط ب مع قدر جد متساويين
فيرى قدر اب أعظم من قدر جد بمقدار ا ط

(٣-٤) أصغر قدراً ...الأصغر] هش م١ .

(٤) ممّا] من ما: م٣ .

(٤) البصر] عن البصر: م٢ .

(٥) المقداران خطّى] القدران خطى: م٣= المقداران: م٥ .

(٦) والبصر نقطة ه] وموضع البصر نقطه ه: م٣ .

(٧) نقطة ج] نقطتى ج: م١ .

(٧) والشعاع ...ه] ـ م٣ .

(٨) مقدار] المقدار: م١ .

(٩) وبشعاع] بشعاع: م٢ ، م٣ ، م٤ ، م٥ .

(٩) هو] متساويان هو: م١= متساويا هو قال: م٢
متساويان وهو: م٣= متساوياً هو: م٤ ، م٥ .

(٩) مقدار ب ز] قدر زب: م١ ، م٣= مقدار زب: م٤ ، م٥ .

(١٠، ١٥، ١٦) از] اد: م٣ .

(١١) نقطة] ـ م٥ .

(١٢) فيكون خطّ ح ط] وهو ه زفيكون ح ط: م٥ .

(١٤) ا ط] طا: م٢ ، م٤ ، م٥ .

and from point E, it had been seen to be greater by magnitude AZ, and AZ is greater than AT.

Therefore, when the eye is lowered, the part seen of the greater magnitude, while <seeing> the smaller one, is more than what was seen before.

And that is what we wished to demonstrate.

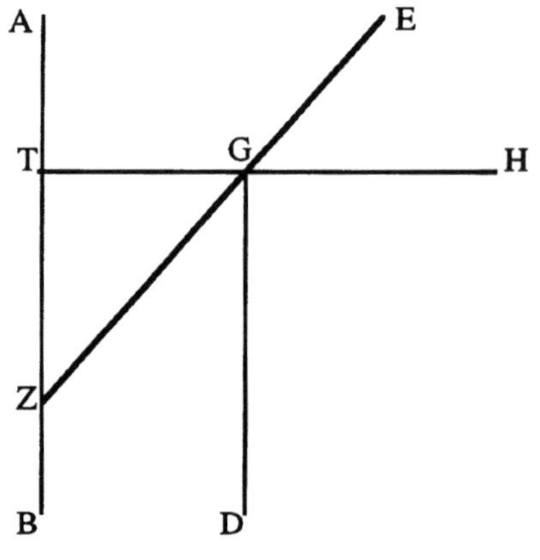

FIGURE 16

A	B	G	D	E	W	Z	H	T	I	K	L	M	N	S	O	F
ا	ب	ج	د	ه	و	ز	ح	ط	ى	ك	ل	م	ن	س	ع	ف

وكان يرى من نقطة ه أعظم منه بمقدار آز
و آز أعظم من اط
فإذا انخفض البصر يكون الذى يرى من المقدار الأعظم
مع ‹رؤية› المقدار الأصغر أعظم ممّا كان يرى قبل ذلك
وذلك ما أردنا أنْ نبيّن ه°ه

---

(١٧) فاذا] فاذ : م١ .
(١٧) المقدار] مقدار: م١ ، م٢ ، م٣ ، م٤ .
(١٨) المقدار] مقدار: م٢ ، م٣ ، م٤ .
(١٨) ممّا] فس: م١ .
(١٨) كان] فس: م٢ .
(١٩) وذلك ما اردنا أن نبيّن] ـ م٢ = وذلك ما اردنا بيانه: م٥ .

17

When there are two magnitudes above the eye and one is greater than the other, the part seen of the greater [magnitude] while [seeing] the smaller one is more than that of the greater, while the smaller is seen, when the eye is raised.

For let the two magnitudes be the two lines AB, GD,
AB being greater than GD, and [let] the eye be point E,
and the ray we draw from point E passes through point G, namely EZ.
Now because what is seen of magnitude AB while [seeing] magnitude GD
by means of ray EZ is magnitude BZ,
then magnitude AB is seen as greater than magnitude GD by magnitude AZ;
after that, we raise the eye to point H, and we draw the ray HT through point G,
so that what is seen of magnitude AB while [seeing] magnitude GD is TB,
and that which was seen by means of ray EZ of magnitude AB,
while [seeing] magnitude GD, was ZB,

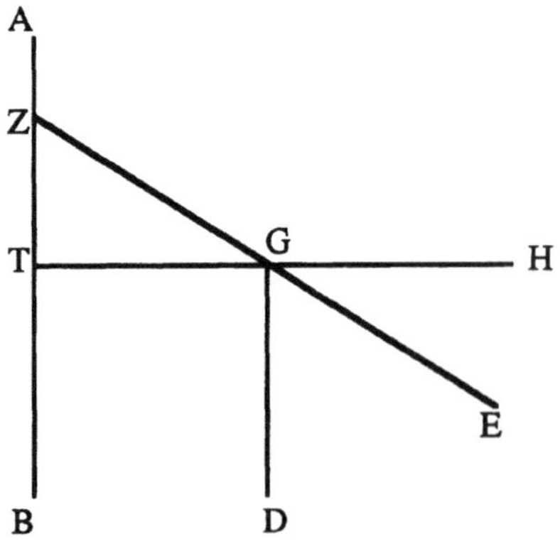

FIGURE 17

A	B	G	D	E	W	Z	H	T	I	K	L	M	N	S	O	F
ا	ب	ج	د	ه	و	ز	ح	ط	ى	ك	ل	م	ن	س	ع	ف

يز

إذا كان مقداران فوق البصر وأحدهما أعظم من الآخر
فإنّ الذى يرى من الأعظم مع رؤية الأصغر أعظم قدراً
ممّا يرى من الأعظم مع رؤية الأصغر إذا ارتفع البصر
فليكن المقداران خطّى اب جد
واب أعظم من جد والبصر نقطة ه
والشعاع الذى نخرج من نقطة ه يمُرّ على نقطة ج وهو ه ز
فالذى يرى من قدر اب مع قدر جد
بشعاع ه ز هو قدر زب
فَ اب يرى أعظم من جد بمقدار از
ثمّ نرفَع البصر إلى نقطة ح ونخرِج شعاع حط على نقطة ج
فيكون الذى يرى من قدر اب مع قدر جد طب
وكان الذى يرى بشعاع ه ز من مقدار اب
مع مقدار جد هو زب

---

(٣) يرى] هش م٢ .
(٣) من] فس م٤ .
(٣) الأعظم مع رؤية الأصغر] الاصغر مع رؤية الاعظم: م١ ، م٣ ، م٥ .
الاعظم من مع رؤية الاصغر: م٤ .
(٣) رؤية الأعظم] + اذا انخفض البصر: م٥ .
(٣) قدراً] + اذا انخفض البصر: هش م١ .
(٥) المقداران] القدران: م٣ .
(٧) يمرّ] فس م١ ، ـ م٢ ، م٤ ، م٥ .
(٧) على نقطة ج] ـ م٢ .
(٨) جد] جز: م٥ .
(٩) بشعاع ...زب] شعاع ...جب: م٣= بشعاع ...قدر زب: م٥ .
(١٠) از] اد: م٣ .
(١١) ثمّ] فس م١ . (١١) نرفع] نخفض: م٢ ، م٣ ، م٥ .
(١٢) الذى] ـ م٢ ، م٤ ، م٥ .
(١٣) وكان] فكان: م١ .
(١٤) هو زب] وهو دب: م١= وهو زب: م٣ .

and ZB is greater than TB, Therefore when the eye is raised, the part seen of the greater magnitude while <seeing> the smaller one is less than what was seen before. And that is what we wished to demonstrate.

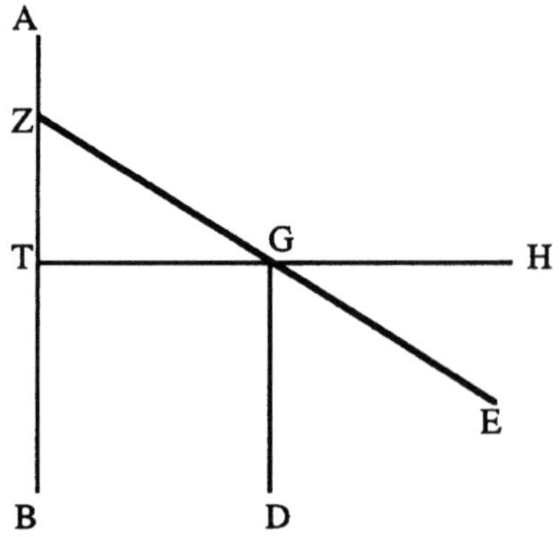

FIGURE 17

A	B	G	D	E	W	Z	H	T	I	K	L	M	N	S	O	F
ا	ب	ج	د	ه	و	ز	ح	ط	ى	ك	ل	م	ن	س	ع	ف

فَ زب أعظم من طب

فإذا ارتفع البصر يكون الذى يرى من المقدار الأعظم مع <رؤية> المقدار الأصغر أصغر ممّا كان يرى قبل ذلك وذلك ما أردنا أنْ نبيّن ∘∘

---

(١٥) فَ زب أعظم من طب] ــ م١ = فَ زب اصغر من اط: م٢ ، م٥ = فَ زا اعظم من اط: م٣ ، م٤ .

(١٦) من] ــ م١ .

(١٦) المقدار] مقدار: م١ .

(١٦-١٧) فإذا ... ذلك] ــ م٢ ، م٣ ، م٤ ، م٥ .

(١٧) ذلك] ــ م١ .

(١٨) وذلك ما أردنا أن نبيّن] وذلك ما اردناه: م٢ = وذلك: م٥ .

18

When unequal bodies [with uneven upright alignment] are on a straight line, [such that] the smaller one is closer to the eye, then they are seen by the same difference [in magnitude] from close and afar. For let the two unequal bodies be AB, GD, AB being greater than GD, and [let] the eye be point Z, and we draw ray ZGE through point G. Then I say that [whether] the eye gets close or distant on that straight line, the two bodies are seen by the same difference [in magnitude], and AB is viewed greater than GD by AE. So we move away the point of the eye to [point] H on the straight line, and we draw from point H ray HZZGGE. Then the two quantities are seen with the original difference in magnitude. And that is what we wished to demonstrate.

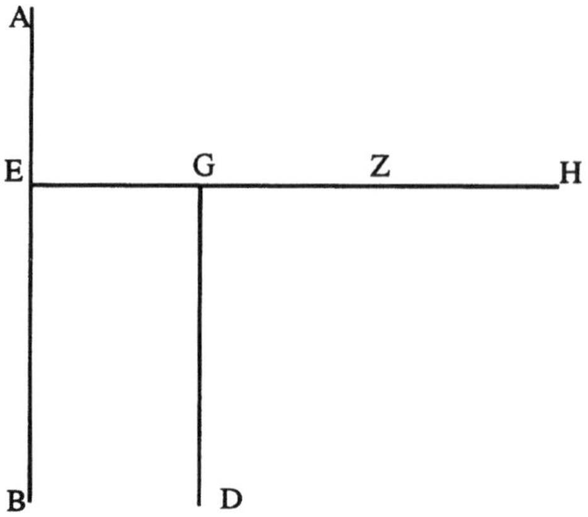

FIGURE 18

A	B	G	D	E	W	Z	H	T	I	K	L	M	N	S	O	F
ا	ب	ج	د	ه	و	ز	ح	ط	ى	ك	ل	م	ن	س	ع	ف

يح

إذا كانت الأجسام المتفاضلة على خطٍّ مستقيمٍ
وكان الأصغر أقرب إلى البصر
فإنّهما تريان على القُرب والبُعد على تفاضلٍ واحد
فليكن الجسمانِ المتفاضلان اب جد وَ اب أعظمُ من جد
والبصر نقطة ز ونخرج على نقطة ج شعاع زجه
فأقول إنّ البصر قَرُبَ أو بَعُد على هذا الخطّ المستقيم
يرى الجسمين على تفاضلٍ واحد
وَ اب يرى اعظم من جد ب اه
ونبعد نقطة البصر إلى ح على الخطّ المستقيم
ونخرج من نقطة ح شعاع ح زجه
ويرى العظمين على قدر التفاضل الأوّل
وذلك ما أردنا أنْ نبيّن ه ه

---

(٢) على] فس م٢ .

(٤) تريان] يريان: م٥ .

(٤) على القرب والبعد] على البعد والقرب: م١ ، م٤ ، م٥ .

(٥) المتفاضلان] المتفضلان: م٢ .

(٦) نقطه ج] ـ م٢ .

(٧) قرب أو بعد] قرب ام بعد: م٢ ، م٤ ، م٥ .

(٧) على] عن: م٢ .

(٧) المستقيم] + الذى: م١ .

(٩) وَ اب] فاب: م٢ ، م٤ .

(٩) ب اه] ـ م١ ، م٣ .

(١٠) على الخطّ المستقيم] ـ م٥ .

(١١) نقطة ح] نقطة: م٤ .

(١١) ح زجه] ـ م٢= حزد ج ج ٥: م٣= حزجه: م٤ ، م٥ .

(١٢) على قدر تفاضل] ـ م٢.

(١٢) ويرى ...ألأوّل] فيرى العظم الاوّل والعظمين المتساويين و... ؟: م٤= ويرى العظمين الاولين: م٥ .

(١٣) وذلك ما أردنا أن نبيّن] ـ م٢= وذلك ما اردنا بيانه: م٥ .

19

[What to do] when we wish to know a body's height by means of the sun:

Let the body be AB, and let the ray that issues from the sun be GD, and that falls upon point A and on the ground upon point D, the shadow of the body BA on the ground being BD; and we place a known body EZ parallel to body AB, and ray GD falls upon point Z. Now there had fallen within triangle ABD another triangle, namely EZD, and so the two are similar, and [the ratio of] magnitude DE to EZ is the same as [the ratio of] DB to BA; but [the ratio of] magnitude DE to EZ is known, so [the ratio of] magnitude DB to BA is known; and magnitude BD is known, therefore AB is [also] known. And that is what we wished to demonstrate.

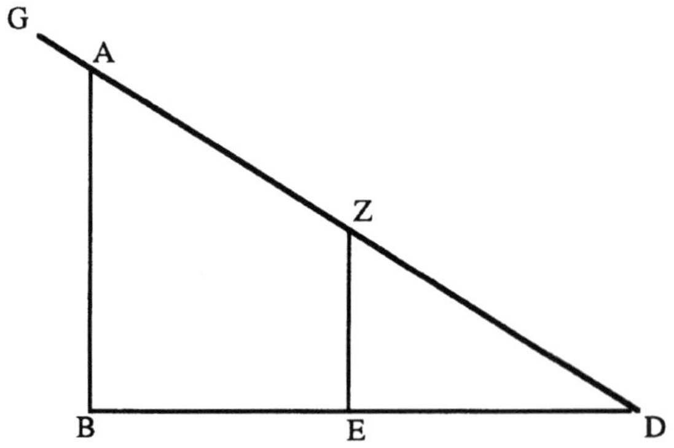

FIGURE 19

A	B	G	D	E	W	Z	H	T	I	K	L	M	N	S	O	F
ا	ب	ج	د	ه	و	ز	ح	ط	ى	ك	ل	م	ن	س	ع	ف

يط

إذا أردنا أنْ نعلَم كمّيّة ارتفاع جسمٍ بالشمس
فليكن الجسم اب ويكون شعاع الذى يخرُج من الشمس جـ د
ويقع على نقطة ا وعلى الأرض على نقطة د
ويكون ظلّ جسم با على الأرض ب د
ونضَع جسماً معلوماً وهو ه ز موازياً لجسم اب
ويقع شعاع جـ د على نقطة ز
فقد وقع فى مثلث اب د مثلث آخر وهو ز ه د
وهما متشابهان
فقدر د ه عند ه ز كقدر د ب عند ب ا
ولكنّ قدر د ه عند ه ز معلوم
فقدر د ب عند ب ا معلوم
وقدر ب د معلوم
فَ اب إذاً معلوم
وذلك ما أردنا أنْ نبيّن ٥٥

----

(٢) بالشمس] _ م٢، م٣، م٤ .
(٣) فليكن] مثاله فليكن: م٢، م٤ .
(٥) ظلّ ... ب د] جسم الارض الذى هو ب ا ب د: م١، م٣= ظل جسم ب ا ب د: م٢، م٤ .
(٦) موازياً لجسم اب] _ م٢، م٣، م٤، م٥ .
(٨) آخر] اجز: م٢ .
(٨) ز ه د] ز د ه: م٥ .
(٩) وهما متشابهان] _ م٢، م٣، م٤، م٥ .
(١١) د ه] ه د: م٥ .
(١٢-١٣) فقدر ... ب د معلوم] وقدر ب د معلوم فقدر ... معلوم: هش م١، م٣= وقدر د ب عند ب ا معلوم: م٢= فقدر ب د عند ب ا معلوم وقدر د ب عند ب ا معلوم: م٤= فقدرب د عند قدر ب ا معلوم= م٥ .
(١٤) فَ اب إذاً معلوم] _ م٥ .
(١٥) أردنا أن نبيّن] اردناه: م٢= اردنا بيانه: م٥ .

20

[What to do] when we wish to know a body's height without the sun:
Let the body be AB, and [let] the eye's point be G,
and we place mirror ED, and we draw ED rectilinearly to point B,
and we draw from G a visual ray such as GH;
and let the ray be deflected from the position of point H
until it falls on A, the farthest [part] of body AB;
and from point E we draw a line rectilinearly along ED, namely ET,
and from point G, we draw a perpendicular to line ET, namely GT.
Now because ray GH falls upon point H and is deflected to point A,
its deflection is at equal angles,
so angle GHT is equal to angle AHB;
but angle ABT is equal to angle GTB,
so the remaining angle HGT is equal to angle HAB;

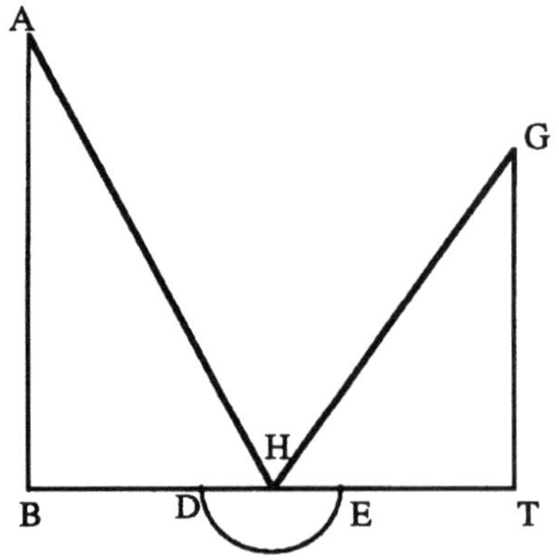

FIGURE 20

A	B	G	D	E	W	Z	H	T	I	K	L	M	N	S	O	F
ا	ب	ج	د	ه	و	ز	ح	ط	ى	ك	ل	م	ن	س	ع	ف

ك

إذا أردنا أنْ نعلَم كمّيّة ارتفاع جسمٍ من غير شمس
فليكن الجسم اب ونقطة البصر ج
ونضَع مرآة ەد ونخرِج ەد على استقامة إلى نقطة ب
ونخرِج من نقطة ج شعاع البصر مثل جح
ولينعطف الشعاع من موضع نقطة ح
حتّى يقع على أقصي جسم اب الذى هو نقطة ا
ونخرِج من نقطة ە خطّاً على استقامة خطّ د ە وهو ەط
ونخرِج من نقطة ج عموداً على خطّ ەط وهو جط
فلأنّ شعاع جح وقع على نقطة ح وانعطف إلى نقطة ا
يكون انعطافه على زوايا متساوية
فزاوية جحط مساوية لزاوية احب
ولكنّ زاوية ابط مساوية لزاوية جطب
فبقيت زاوية حجط مساوية لزاوية حاب

---

(٢) إذا] واذا: م١ ، م٣ ، م٤= وان: م٢ .

(٢) أن نعلم] _ م٣ ، م٤ ، م٥ .

(٢) أردنا] + معرفة: م٣ ، م٤ .

(٢) غير] _ م٣ .

(٣) مرآة ەد] على بسيط العرض: فس م١= من اه دز: م٢ .
مرآة د ە: م٣ ، م٥= مرآة دز: م٤ .

(٤) ەد] د ە: م١= خط ەد: م٣ ، م٤= خطّ د ە: م٥ .

(٦) ولينعطف] ولنعطف: م٣ ، م٤ ، م٥ .

(٦) موضع] _ م١ .

(٧) أقصى] اقصا: م٢ ، م٤ ، م٥ .

(٨) خطّاً ... ەط] خطاً وهو خط ەط على استقامة خط د ە: م٢ ، م٤ ، م٥ .

(٩) على] الى: م٢ ، م٤ ، م٥ .

(١٠) إلى] على: م١ ، م٥ .

(١١) انعطافه] انعطافها: م١ ، م٢ ، م٣ ، م٤ .

(١٢) لزاوية] _ م٢ .

therefore triangle AHB has angles equal to triangle GHT;
and triangles with equal angles have proportionate sides,
so [the ratio of] magnitude GT to TH is the same as [that of] AB to BH;
but [the ratio of] magnitude GT to TH is known,
so [the ratio of] magnitude HB to BA is known,
but HB is known, so AB is [also] known.
And that is what we wished to demonstrate.

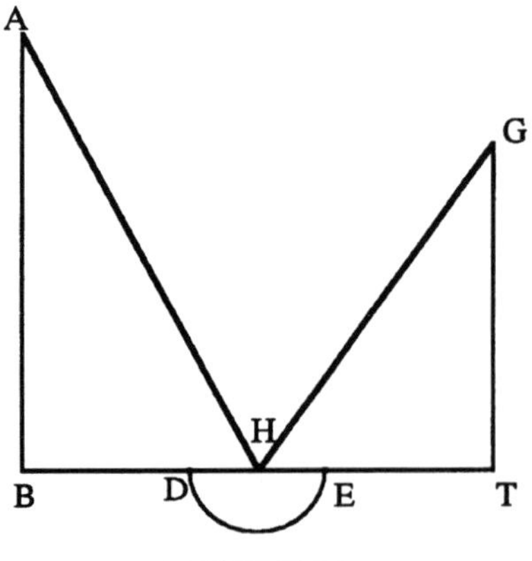

FIGURE 20

A	B	G	D	E	W	Z	H	T	I	K	L	M	N	S	O	F
ا	ب	ج	د	ه	و	ز	ح	ط	ى	ك	ل	م	ن	س	ع	ف

فمثلث ا ح ب مساوية زواياه لزاوية مثلث جـ ح ط
والمثلثات المتساوية الزوايا أضلاعها متناسبة
فقدر جـ ط عند ط ح كقدر ا ب عند ب ح
ولكنّ قدر جـ ط عند ط ح معلوم
فقدر ح ب عند ب ا معلوم
ولكنّ ح ب معلوم فَ ا ب إذن معلوم
وذلك ما أردنا أنْ نبيّن ٥

(١٤-١٥) لزاوية ... متساوية] ـ م١ = لزوايا ... متساوية: م٥ .
(١٦) أضلاعها] فاضلاعها: م١ ، م٣ .
(١٨-٢٠) ولكنّ ...إذن معلوم] ولكن قدر جد عند طح معلوم
فقدر جـ ط عند ط ح معلوم فقدر جـب عند با معلوم
ولكن جـب معلوم فَ ب ا معلوم: م٥ .
(١٩) معلوم] ايضاً معلوم: م٢ .
(٢٠) إذن] اذاً: م٢ ، م٤ ، ـ م٣ .
(٢١) وذلك ما أردنا أن نبيّن] ـ م٢ = وذلك ما اردنا بيانه: م٥ .

21

[What to do] when we wish to learn an object's depth:
Let the depth be AD, the position of the eye being point E;
we wish to know the amount of [depth] AD.
So let us draw the visual ray, ED,
and let this fall on the [ground] surface at point B and in depth at point D;
and we draw line BZ at right angles to AD,
and we draw from E a perpendicular to line BZ, [line] EZ.
Now because angle EZB is equal to angle BAD,
and angle ABD is equal to EBZ,
then the remaining angle that is enclosed by BEZ
is equal to the [remaining] angle that is enclosed by ADB;
so triangle ADB has equal angles to [those of] triangle BEZ,
and their sides are proportionate,

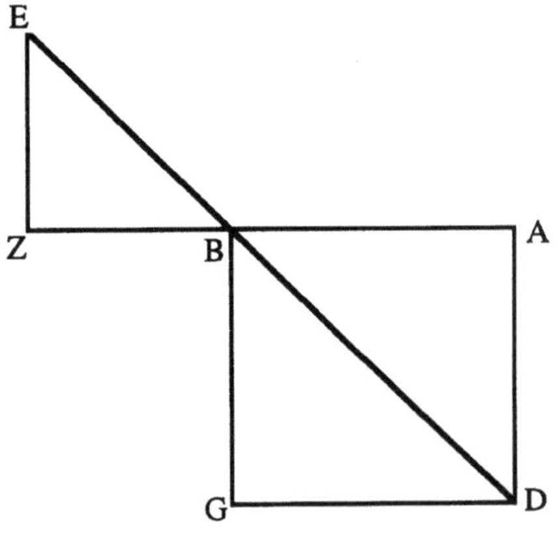

FIGURE 21

A	B	G	D	E	W	Z	H	T	I	K	L	M	N	S	O	F
ا	ب	ج	د	ه	و	ز	ح	ط	ى	ك	ل	م	ن	س	ع	ف

كا

إذا أردنا أنْ نعرف كمّيّة عمق شيء
فليكن العمق اد ويكون موضع البصر نقطة ه
ونريد أنْ نعلَم كمّية اد
فلنخرِج شعاع البصر عليه ه د
وليقع على البسيط على نقطة ب وعلى العمق على نقطة د
ونخرِج خطّ ب ز على زوايا قائمة من اد
ونخرِج من نقطة ه عموداً إلى خطّ ب ز عليه ه ز
فلأنّ زاوية ه ز ب مساوية لزاوية ب ا د
وزاوية ا ب د مساوية لزاوية ه ب ز
فالزاوية الباقية التى يحيط بها ب ه ز
مساوية للزاوية التى يحيط بها ا د ب
فمثلث ا د ب متساوى الزوايا لمثلث ب ه ز
وأضلاعهما متناسبةٌ

---

(٢) أن نعرف] معرفة: م١ ، م٣ ، م٤ ، م٥ .
(٣) اد ويكون موضع] دا و موضع: م٥ .
(٤) ونريد] فنريد : م٢ ، م٤ .
(٥) شعاع البصر] الشعاع : م٥ .
(٦) البسيط] بسيط الكرة: م١ .
(٧) ونخرج ... اد] _ م١ .
(٨) عموداً ... ه ز] عموداً الى خط ب ا عليه: م١ = عموداً الى ب ز عليه ه ز: م٥.
(٨) نقطة ه عموداً] نقطة ه عمود : م٢ = ه عموداً: م٥ .
(١٠) ا ب د] ا ب : م١ .
(١٠) لزاوية ه ب ز] لزاوية ه ب د : م٣ = ه ب ز: م١ ، م٢ ، م٣ .
(١١) فالزاوية] والزاوية: م٢ ، م٤ ، م٥ ، _ م٣ .
(١١، ١٣) ب ه ز] ب ه د : م٣ .
(١٢) مساوية للزاوية التى] فهى مساوى للتى: م١ = مساوية لزاويته الباقيه : م٣ = مساوية للزاوية الباقيه: م٥ . (١٢) بها] فس م١ .
(١٣) متساوية الزوايا لمثلث] مساوية زواياه لزوايا مثلث : م٥ .
(١٤) أضلاعهما] اضلاعها: م١ ، م٣ = اضلاعه: م٢ .

so the ratio of EZ to ZB is the same as the ratio of DA to AB; but the ratio of EZ to ZB is known, so the ratio of DA to AB is known, and BA is known, so the magnitude AD is [also] known.

And that is what we wished to demonstrate.

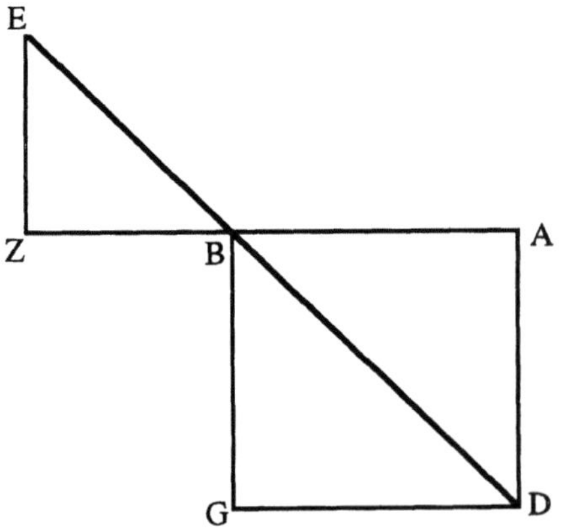

FIGURE 21

A	B	G	D	E	W	Z	H	T	I	K	L	M	N	S	O	F
ا	ب	ج	د	ه	و	ز	ح	ط	ى	ك	ل	م	ن	س	ع	ف

فنسبة هز عند زب كنسبة دا عند اب
ولكنّ نسبة هز عند زب معلومة فنسبة دا عند اب معلومة
ولكنّ با معلوم فقدر اد إذن معلوم
وذلك ما أردنا أنْ نبيّن ٥٥

---

(١٥) فنسبة] نسبة: م٢ ، م٣ ، م٤ .

(١٥) زب] د ب: م٥ .

(١٥، ١٦) نسبة ... عند] نسبة ... الى: م٥ .

(١٥-١٦) كنسبة دا عند اب ولكن نسبة هز عند زب] هش م١ .

(١٧) ولكن با معلوم فقدر اد إذن معلوم] هش م١= واذا ...
اذا معلوم: م٣ .

(١٧) إذن] اذاً: م٢ ، م٣ ، م٤ ، م٥ .

(١٨) وذلك ما أردنا أن نبيّن] ـ م٢= وذلك ما اردنا بيانه: م٥ .

22

[What to do] when we wish to know an object's length on the [ground] surface:
Let the long object be AB, and [let] the position of the eye be point G;
we wish to know the amount of length AB.
So let there be ray[s] GA, GB,
and we mark a point D on the ray, close to the position of the eye,
and we draw from point D a line parallel to line AB, [line] DE.
Now because line DE is drawn inside triangle ABG parallel to the base BA,
[the ratio of] magnitude GD to DE becomes the same as that of GA to BA;
but [the ratio of] magnitude GD to DE is known,
so [the ratio of] magnitude GA to AB is known,

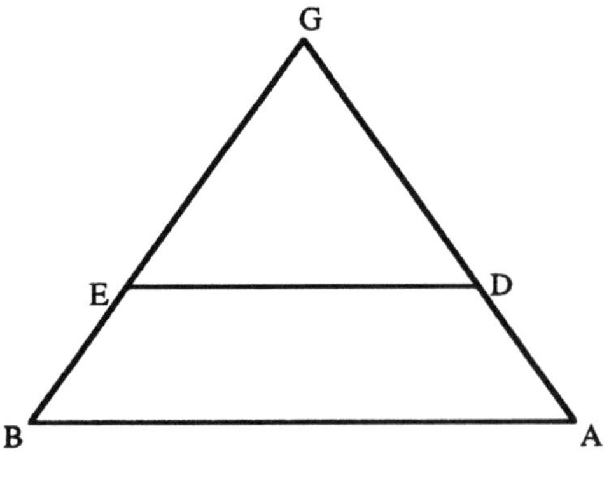

FIGURE 22

A	B	G	D	E	W	Z	H	T	I	K	L	M	N	S	O	F
ا	ب	ج	د	ه	و	ز	ح	ط	ى	ك	ل	م	ن	س	ع	ف

كب

إذا أردنا أنْ نعلَم كمّيّة طول شيء فى بسيط
فليكن الشي الطويل ا ب وموضع البصر نقطة جـ
ونريد أنْ نعلَم كمّية طول ا ب
فليكن الشعاع جـ ا جـ ب
ونتعلّم قريباً من موضع البصر على الشعاع نقطة د
ونخرج من نقطة د خطّاً موازياً لخط ب ا عليه د ه
فلانّ مثلث ا ب جـ أُخرج فيه خطّ د ه موازياً لقاعدة ب ا
يصير قدر جـ د عند د ه كقدر جـ ا عند ا ب
لكنّ قدر جـ د عند د ه معلوم
فقدر جـ ا عند ا ب يكون معلوماً

---

(٢) إذا] واذا: م٢ ، م٤ ، م٥ .
(٢) طول] فس م٥ .
(٣) فليكن] مثاله فليكن: م٢ ، م٤ .
(٤) ونريد] وبريد: م٥ .
(٥) الشعاع] + الذى نخرج: م٣ .
(٦) نتعلّم] يتعلم: م٤ .
(٦) البصر] + نقطه جـ ونريد ان نعلم كمية طول ا ب فليكن الشعاع جا جب ويتعلم قريبا من البصر: م٤ .
(٦) د] فس م١ .
(٧) ونخرج من نقطه د] ـ م٣: ونخرج من نقطه ط: م٥ .
(٧) ب ا] ا ب: م٢ ، م٣ ، م٤ ، م٥ .
(٨، ٧) د ه] ه د: م٥ .
(٨) موازياً لقاعدة ب ا] موازيا لـ ب ا: م٥ .
(٩) يصير قدر] تصير نسبة: م٥ .
(٩-١٠) كقدر جا عند ا ب ... د ه] ـ م٣ .
(٩، ١٠، ١١) قدر ... عند ...] نسبة ... الى ...: م٥ .
(١٠) لكن] ولكن: م٢ ، م٤ .
(١٠) معلوم] معلومة: م٥ .
(١١) فقدر جا عند ا ب] فقدر ا ب عند اجـ: م٢ ، م٤ .
(١١) يكون معلوماً] معلوم: م٢ ، م٣ ، م٤= معلومة: م٥ .
(١١) معلوماً] + لكن جا عند اب يكون معلوما: م١ .

and GA is known, so AB is known. 12

And that is what we wished to demonstrate.

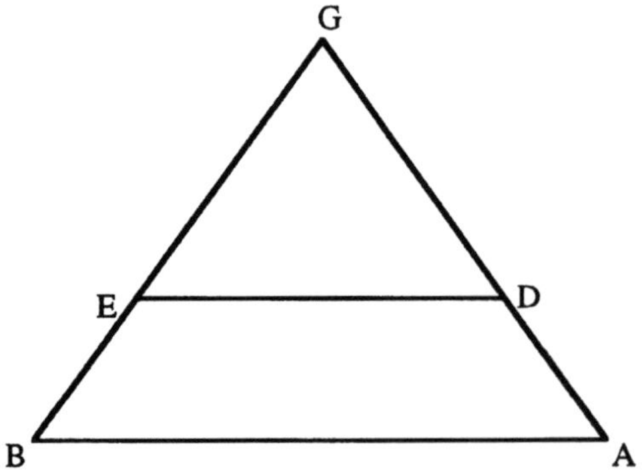

FIGURE 22

A	B	G	D	E	W	Z	H	T	I	K	L	M	N	S	O	F
ا	ب	ج	د	ه	و	ز	ح	ط	ى	ك	ل	م	ن	س	ع	ف

لكنّ جا معلوم فَ اب معلوم ١٢
وذلك ما أردنا أنْ نبيّن ٥٥

---

(١٢) لكنّ جا] ولكن جا: م٢، م٤ = فاذا كان جا: م٣ .

(١٢) اب] اب اذا: م٣ .

(١٣) وذلك ما أردنا أن نبيّن] _ م٢ = وذلك ما اردنا بيانه: م٥ .

23

When a segment of a circle lies below the eye [and] in the same plane, then the circle's segment is seen in the status of a straight line. For let the segment of a circle be GB, and [let] the eye's position be point A, and we draw ray[s] AB, AD, AE, AZ, AH, AT, AG. Then I say that arc BG is seen as a straight line. For let the center of the arc's circle be point K, and we join lines KB, KD, KE, KZ, KH, KT, KG. Now since KB is seen from angle KAB, and KD [is seen] from angle KAD, then BK is seen as greater than DK; and in the same way DK is seen as greater than EK, and KE is seen as greater than KZ; and similarly, from the other side KG is seen as greater than KT, and KT greater than KH, and KH greater than KZ. And because of that, line BG becomes a base for perpendicular AZ, and the same thing also occurs in the arc's interior [side]. And that is what we wished to demonstrate.

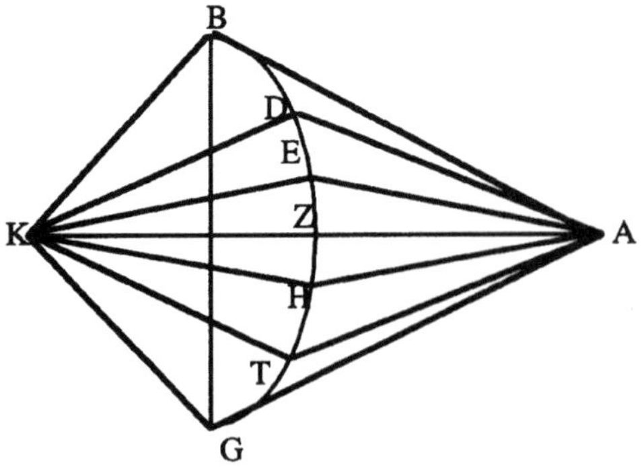

FIGURE 23

A	B	G	D	E	W	Z	H	T	I	K	L	M	N	S	O	F
ا	ب	ج	د	ه	و	ز	ح	ط	ى	ك	ل	م	ن	س	ع	ف

كج

إذا وقع تحت البصر قطعةٌ من دائرة فى سطحٍ واحدٍ
فإنّ قطعة الدائرة ترى بمنزلة خطّ مستقيمٍ
فليكن قطعة الدائرة ج ب وموضعاً البصر نقطة آ
ونخرج شعاع اب اد اه از اح اط اج
فأقولَ إنّ قوس ب ج يرى خطّاً مستقيماً
فليكن مركز دائرة القوس عليه نقطة ك
ونصل خطوط ك ب ك د ك ه ك ز ك ح ك ط ك ج
فلأنّ ك ب يرى من زاوية ك ا ب و ك د من زاوية ك ا د
فَ ب ك يرى إذن أعظم من د ك
وكذلك د ك يرى أعظم من ه ك
ك ه يرى أعظم من ك ز
وكذلك يرى من الناحية الأُخرى ك ج أعظم من ك ط
وَ ك ط أعظم من ك ح وك ح أعظم من ك ز
فمن أجْل ذلك يكون خطّ ب ج قاعدةً لعمود آ ز
ومثل ذلك يعرض أيضاً فى باطن القوس
وذلك ما أردنا أنْ نبيّن ००

---

(٣) ترى بمنزلة خطّ مستقيم] ترى بمنزلة خط خط مستقيم: م٥ .
(٤) فليكن] مثاله فليكن: م٢ ، م٤ = فلتكن: م٥ .
(٤) موضع] موقع: م٢ .
(٥) اح] ـ م٥ .
(٨) ك ه ك ز] ك ز ك ه: م١ . (٨) ونصل] ولنصل: م٣ ، م٤ .
(٩-١٠) ك ب ...د ك] ك ج يرى من زاوية ك ا ب و ك ب يرى من زاوية ك ا ج فَ ب د ويرى اعظم من د ك: م٣ .
(١٠) إذن] ـ م٥ .
(١٢) وك ه ... من ك ز] وه ك ... ك ز من: م١ = و ك ه من ك ز: م٣ = و ه ك ... ك ز: م٥ .
(١٥) ذلك] + يعنى قوس ك ج: م٢ ، م٣ ، م٤ ، م٥ .
(١٥) قاعدة لعمود آز] قاعدة العمق وان: م٣ = قاعدة العمود ازك: م٤ ، م٥ .
(١٦) ومثل‌ذلك يعرض أيضاً] بمثل ذلك ايضاً يعرض: م٥ .
(١٧) وذلك ما أردنا أن نبيّن] ـ م٢ = وذلك ما اردنا بيانه: م٥ .

24

When a sphere is viewed from any position,
then that which is seen of it is less than a hemisphere,
and that which is seen of it is enclosed by a circle.
For let there be a sphere, its center point A, and the eye's position point B,
and we join BA, and we draw the surface through BA;
so we cut the great circle which is circle GDB,
and we join BG, BD, AD, AG.
Now because AGB is a semicircle, angle AGB is a right angle,
and so is BDA; and we join GB, BD, and they become tangents to the circle,
and we draw line HT through point A parallel to line GD,
and angle K is a right angle.
Now if we revolve triangle BGD around the right angle, meaning angle K,
line AB being fixed <in> its location,
then the triangle revolves and then returns to the position it started from;

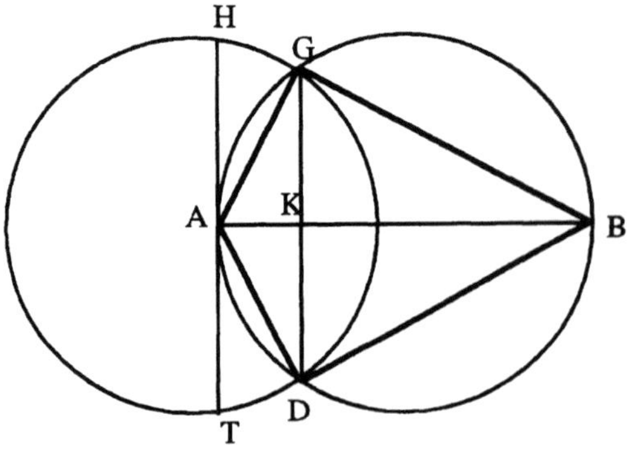

FIGURE 24

A	B	G	D	E	W	Z	H	T	I	K	L	M	N	S	O	F
ا	ب	ج	د	ه	و	ز	ح	ط	ى	ك	ل	م	ن	س	ع	ف

كد

إذا كانت كرةٌ من ایّ موضع من المواضع نُظر إليها
فإنّ الذى يرى منها يكون أقلّ من نصف اَلكرة
والذى يرى منها يحيط به دائرة
فليكن كرة مركزها نقطة آ ويكون موضع البصر نقطة ب
ونصل ب آ ونخرِج البسيط الذى على ب آ
فنقطع الدائرة العظمى التى عليها جـ د ب
ونصل ب جـ ب د اد اجـ
فلأنّ اجـ ب نصف دائرة تكون زاوية اجـ ب قائمة
وكذلك ب د آ ونصل جـ ب ب د يماسّان الدائرة
ونخرِج من نقطة آ خطّ ح ط موازياً لخطّ جـ د
وزاويَة كَ قائمة
فإنْ أدرنا مثلث ب جـ د على الزاوية القائمة أعنى زاوية كَ
وكان خطّ اب ثابتاً <فى> مكانه
فإنّ المثلّث يدور ثمّ يعود إلى الموضع الذى ابتدا منه

---

(٢) أيّ موضع من المواضع] اى المواضع: م٢، م٤، م٥.
(٢) نظر] + بصر: هش م٢.
(٣) يكون] يكو: م٣، _ م٥.
(٤) دائرة] هش م١.
(٥) فليكن] فلتكن: م٥.
(٥) نقطة] _ م١.
(٧) فنقطع] فيقطع: م٥.
(٧) التى] الذى: م٥.
(٨) ب جـ ب د اد اجـ] ب جـ ب د اد: م١.
(١١) من نقطة] على نقطة: م٢، م٤، م٥.
(١١) خطّ ح ط] خطاً خطاً ح ط: م١= خطاً: م٢، م٤، م٥= خط ح ا ط: م٣.
(١٢) وزاوية] فزاوية: م١.
(١٣) فإن أدرنا] وان ادر: م١= وان ادرنا: م٢، م٣= و؟ اردنا: م٤.
(١٣) ب جـ د] ب جـ ى: م٢، م٤، م٥.
(١٣) أعنى] _ م٤= وهى: م٥.

and line BG becomes tangent to the sphere in the same amount,
and DG cuts a circle from the sphere;
then the sphere is seen in the status of a section of a circle,
and what is seen of it is less than a hemisphere,
because arc HGDT is a semicircle,
so GD is less than a semicircle, and it is seen by the ray[s] GB, BD.
And that is what we wished to demonstrate.

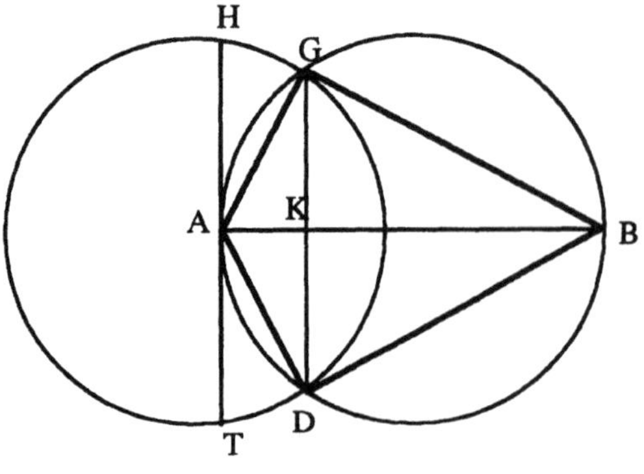

FIGURE 24

A	B	G	D	E	W	Z	H	T	I	K	L	M	N	S	O	F
ا	ب	ج	د	ه	و	ز	ح	ط	ى	ك	ل	م	ن	س	ع	ف

ويكون خطّ ب ج يماسّ الكرة بقدرٍ واحدٍ
ود ج يقطَع من الكرة دائرةً
فترى الكرة بمنزلة خطّ دائرةٍ
ويكون الذى يرى منها أقلّ من نصف كرة
لأنّ قوس ح ج د ط نصف دائرة
فَ ج د أقلّ من نصف دائرة ويرى بشعاع ج ب ب د
وذلك ما أردنا أنْ نبيّن ٥°٥

----

(١٦) ب ج] د ب: م٢ = د: م٤ .
(١٦) يماس] ماس: م١ .
(١٦-١٨) بقدر واحد ... فترى الكرة] _ م٣ .
(١٧) ود ج] د ى ج: م٤ = و ؟ج: م٢ = و ى ج: م٥ .
(١٩) لأنّ قوس] لا قوس: م٥ .
(٢٠) ح ج د ط] ح ج د ى: م٢ + نقص من: م٣ .
(٢٠) من] + من: م٣ .
(٢١) فَ ج د] و ج د: م٤ = ب ج: م٥ .
(٢١) يرى بشعاع] يرى من شعاع: م٢ ، م٤ = يرى من شعاعى: م٣ = ترى من شعاع: م٥ .
(٢٢) أن نبيّن] بيانه: م٥ .

25

When the eye approaches a sphere, what is seen of it is less, but it is thought through vision to be greater.

For let there be a sphere, its center point A, and the eye's position point B; so we draw line AB, and we draw around line AB a circle, GBD, and we draw from point A a line at right angles to line AB, line EZ, and we draw the plane through EZ, GD such that it is a segment of a circle, and we mark it GEZD, and we join GA, AD, DB, BG, GD.

Now because of what preceded, the two angles at G, D are right angles, and the two lines BG, BD, which are rays, are tangent to the circle, and magnitude GD, which is a segment of the sphere, is seen from point B; then we change the eye's position, making it become closer to the sphere as T, and we inscribe around line AT circle AKTL,

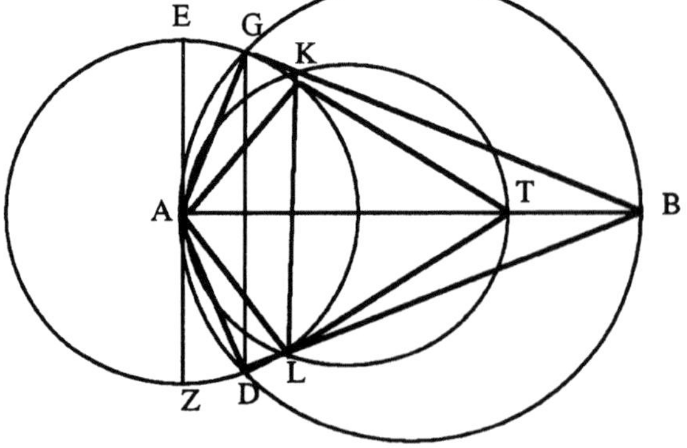

FIGURE 25

A	B	G	D	E	W	Z	H	T	I	K	L	M	N	S	O	F
ا	ب	ج	د	ه	و	ز	ح	ط	ى	ك	ل	م	ن	س	ع	ف

كه

إذا دنا البصر من الكرة يكون ما يرى منها أقلّ ويُظَنّ فى الرؤية أنّه أعظم

فليكن كرةٌ مركزها نقطة ا وموضع البصر نقطة ب فنخرجِ خطّ اب ونخطّ على خطّ اب دائرةً عليها جـ ب د ونخرِج من نقطة ا خطّاً على زاوية قائمة عند خطّ اب عليه ه ز ونخرِج السطح الذى هو على ه ز جـ د فيكون قطعةَ دائرةٍ ونتعلّم عليها جـ ه ز د ونصل جا اد د ب ب جـ جـ د فمن أجْل ما تقدّم يكون الزاويتان اللتان عند جـ د قائمتين ويكون خطّا ب جـ ب د اللذان هما الشعاعان مماسّين للدائرة ويكون عظَم جـ د الذى هو قطعة الكرة ترى من نقطة ب ونحوّل مَوضع البصر فنصيره أقرب إلى الكرة وعليه ط ونَدير على خطّ اط دائرة اك ط ل

---

(٣) فى الرؤية] من الرؤية: م١ .

(٤) فليكن كرة] فليكن الكرة: م١= فلتكن كرة: م٥ .

(٤) البصر] هش م٢ .

(٥) فنخرج خطّ اب] فنجى؟ اب: م٢= فنجى؟ بخط اب: م٥ .

(٦) من نقطة ا] من خطّ ا: م٢ .

(٦) زاوية] زوايا: م٢ ، م٣ ، م٤ ، م٥ .

(٧) هو على] على: م٢ ، م٤ ، م٥= هو: م٣ .

(٧) ه ز جـ د فيكون] ه ز د جـ ليكون: م٣ .

(٧) جـ د] اب: م٢ ، م٤ ، م٥.

(٨) عليها] ــ م٢ ، م٥ .

(٨) جا اد] دا اجـ: م٢ ، م٤ ، م٥ .

(٩) يكون] تكون: م٥ .

(٩) الزاويتان اللتان] الزوايا التى: م٢ ، م٤ ، م٥ .

(١٠) خطّا] خطّ: م٢ ، م٥ .

(١١) من] ــ م٢ ، م٥ . (١١) نقطه] ــ م٤ .

(١١) قطعه] نقطه: م٣ .

(١٢) ط] ط د: م١= نقطه ط: م٣ .

(١٣) اك ط ل] فس م٤= اط ك ل: م٥ .

and we join lines KT, TL, KA, AL;
similarly, magnitude KL, which is a segment of the sphere, is seen
from the position of point T,
while magnitude GD was seen from the position of point B,
and KL is less than GD.
Therefore, as the eye approaches, that which is seen is in actuality smaller,
but it is thought by vision to be greater,
because angle KTL is greater than angle GBD.
And that is what we wished to demonstrate.

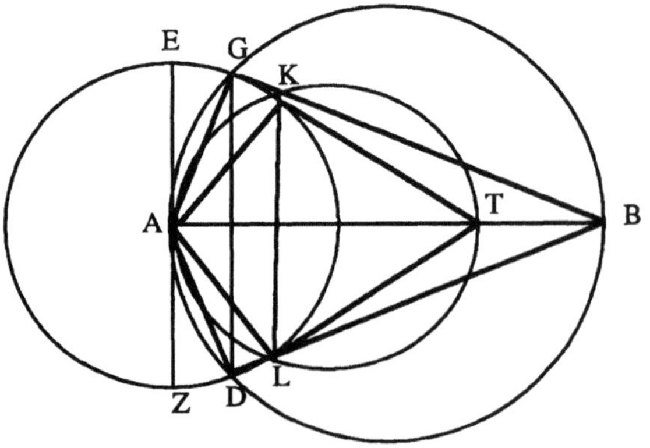

FIGURE 25

A	B	G	D	E	W	Z	H	T	I	K	L	M	N	S	O	F
ا	ب	ج	د	ه	و	ز	ح	ط	ى	ك	ل	م	ن	س	ع	ف

ونصل الخطوط كـط طل كـا ال
فعلى مثل ذلك يرى عِظَم كـل الذى هو قطعة الكرة
من موضع نقطة ط
وكان عظَم جـد يرى من موضع نقطة ب
وكـل أَصغر من جـد
فإذا دنا البصر يكون الذى يرى بالحقيقة أصغر
ويُظَنّ بالرؤية أنّه أعظم
لأنّ زاوية كـطـل أعظم من زاوية جـب د
وذلك ما اردنا أنْ نبيّن ٥°

---

(١٤) كـط طل كـا ال] طل لا كا: م٢= م٤= كـط طل لا كا: م٣= طل ا لكا: م٥ .

(١٥) فعلى] نصل: م٣ .

(١٥) كـل] هش م٢ .

(١٨) نقطة ب] + او كل يرى من موضع نقطة ط الذى هو اقرب من ب: تس م١ .

(١٩) فإذا دنا] فاذا ادرنا: م٢= م٤= فاذا اردنا ادرنا: م٥ .

(٢١) لأنّ] لانه: م٢ .

(٢١) جـب د] جد: م٤= جـ د ى: م٥ .

(٢٢) أردنا أن نبيّن] اردناه: م٢= اردنا بيانه: م٥ .

26

When a sphere is seen by both eyes, and the distance between the eyes is equal to the sphere's diameter, then that which is seen of the sphere is half of it entirely. For let there be a sphere, the center of which is point A, and we draw around point A inside the sphere a circle, GB, BG being its diameter, and we draw from points B [and] G, two lines rectilinearly falling on the points of the two eyes, namely D [and] E. Then I say that the whole of the hemisphere is seen. So we join DE, and we draw from point A a line parallel to both BD and GE, namely AZ, so that each of the planes AD GZ is a parallelogram; then if we keep line AZ fixed, and we revolve one of the planes BZ, GZ, meaning plane BZ, until it returns to its position, the figure produced starts from point B and comes to point G,

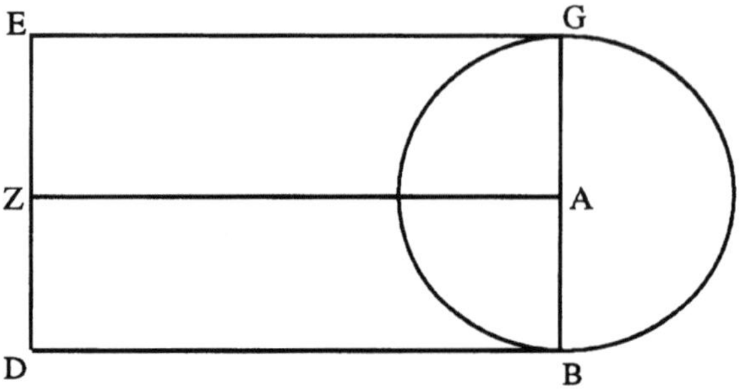

FIGURE 26

A	B	G	D	E	W	Z	H	T	I	K	L	M	N	S	O	F
ا	ب	ج	د	ه	و	ز	ح	ط	ى	ك	ل	م	ن	س	ع	ف

كو

إذا أُبصرت كرةٌ ببصرين
وكان ماَ بين البصرين مساوياً لقطر الكرة
فإنّ الذى يرى من الكرة يكون نصفها كلّه
فلتكن الكرة مركزها نقطة آ
ونخطّ فى الكرة على مركز آ دائرةً عليها جـ ب
ويكون قطرها ب جـ
ونخرِج من نقطتى ب جـ خطّين على استقامة
يقعانِ على نقطتى البصرين اللذين عليهما د ه
فأقول إنّ نصف الكرة يرى كلّه
فنصل د ه
ونخرِج من نقطة آ خطّاً موازياً لكلّ واحد من ب د جـ ه عليه آ ز
فكلّ واحد من سطحى آ د جـ ز متوازى الاضلاع
فإن اثبتنا خطّ آ ز وأدرنا واحداً من سطحى ب ز جز
أعني سطح ب ز حتّى يعود إلى موضعه
فإنّ الشكل الحادث يبدى من نقطة ب وصار الى نقطة جـ

---

(٣) مساوياً] متساو: م٢ ، م٤ .
(٤) كلّه] ـ م١ .
(٥) نقطة] ـ م١ ، م٤ ، م٥ .
(٧) ب جـ] فس م٢ .
(١٢) من نقطة] على نقطة: م٣ ، م٥ .
(١٢) لكلّ واحد من ب د جـ ه عليه آ ز] ل ب جـ د ه عليه ا ز : م١ = لخطى ب د جـ ه عليه ا د: م٣ .
(١٣) فكلّ واحد] وكل واحد: م٥ .
(١٤) فإن] فاذا: م٥ .
(١٤) وأدرنا] + ب ا: م٣ ، ـ م٥ . (١٤) واحداً] واحد: م٣ .
(١٤-١٥) سطحى ب ز جـ ز أعنى ب ز] سطحين اد جز: م١ = سطحى اب ز جد: م٣ .
(١٦) يبدى من] يرى به: م١ = يرى من: م٢ = يبدى به من: م٣ = يرى منه من: م٤ ، م٥ .
(١٦) نقطة جـ] + يحدث مثل الاول: هش م١ .

so the whole of the semicircle is seen, and the same amount is also seen of the sphere. Therefore, the entire hemisphere is seen from the position of the two eyes, [points] D [and] E. And that is what we wished to demonstrate.

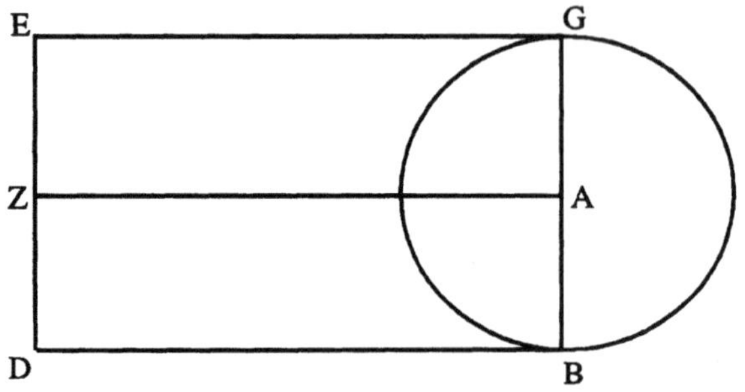

FIGURE 26

A	B	G	D	E	W	Z	H	T	I	K	L	M	N	S	O	F
ا	ب	ج	د	ه	و	ز	ح	ط	ى	ك	ل	م	ن	س	ع	ف

فيرى نصف الدايرة كلّه
وبقدر ذلك يرى أيضاً من الكرة ١٨
فجميع نصف الكرة إذن يرى
من موضع البصرين اللذين عليهما د ه
وذلك ما أردنا أنْ نبيّن ٥٥ ٢١

---

(١٧) فيرى] فترى: م٣ .
(١٨) يرى] _ م٢ ، م٤ ، م٥ .
(١٩) إذن] اذا: م٢ ، م٣ ، م٤ ، م٥ .
(١٩) فجميع نصف الكرة اذن يرى] فجميع اذا ما يرى: م٥ .
(٢٠) د ه] + النصف: م٥ .
(٢١) وذلك ما أردنا أن نبيّن] _ م٢= وذلك ما اردنا بيانه: م٥ .

27

When the distance between the eyes is greater than a sphere's diameter, then that which is seen of the sphere is more than half of it. For let there be a sphere, the center of which is point A, and we inscribe around point A a circle, EDH, the positions of the two eyes being B and G, and the distance between them is greater than the diameter of the circle, and we join BG. Then I say that more than half of the sphere is seen. So let the two rays BE, GD fall, and we draw them rectilinearly; then it is clear that their two ends meet in the same point because the sphere's diameter is smaller than BG; so let the two [ends] meet at point Z. Now because two lines falling on the circumference of the circle

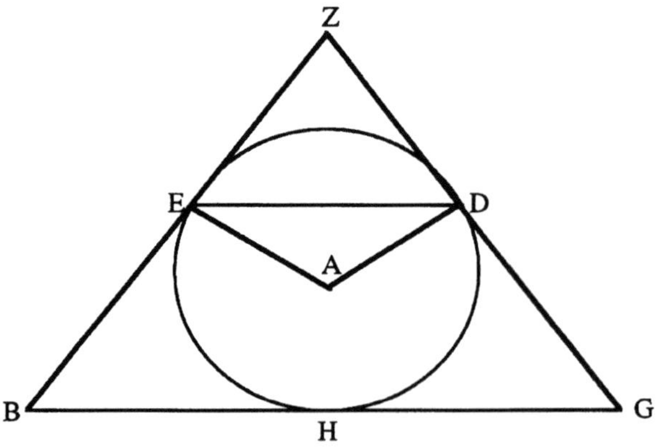

FIGURE 27

A	B	G	D	E	W	Z	H	T	I	K	L	M	N	S	O	F
ا	ب	ج	د	ه	و	ز	ح	ط	ى	ك	ل	م	ن	س	ع	ف

كز

إذا كان بُعد ما بين البصرين أعظم من قطر كرةٍ
فإنّ الذى يرى من الكرة أكثر من نصفها
فليكن كرةٌ مركزها نقطة آ
ونديز على مركز آ دائرةً عليها ه د ح
ويكون موضعا البصرين ب وَ ج
وبُعد ما بينهما أعظم من قطر الدائرة
ونصل ب ج
فأقول إنّه يرى من الكرة أكثر من نصفها
فليقع شعاعا ب ه ج د ونخرجهما على استقامة
فبيّن أنّ طرفيهما يلتقيان عَلى نقطة واحدةٍ
من أجْل أنّ قطر الكرة أصغر من ب ج
فلتقيان على نقطة ز
ولأنّه وقع على محيط الدائرة خطّان

---

(٢) إذا كان] واذا كان: م٤ .
(٢) بعد] ــ م٢ ، م٣ .
(٢) كرة] الكرة: م٢ ، م٤ ، م٥ .
(٣) أكثر] اعظم: م٢ ، م٥ .
(٤) فليكن] فلتكن: م٤ ، م٥ .
(٤) نقطة] ــ م١ ، م٣ ، م٤ ، م٥ .
(٦) موضعا] موضع: م١ ، م٣ ، م٥ .
(١٠) شعاعا ب ه ج د] شعاعا ب ك و ج ه: م٢ = شعاع ب ه جد: م٣ = شعاع جد ب ه: م٥ .
(١٣) من أجل ...نقطة ز] من اجل ز: م٢ ، م٥ = من اجل أن ك ه اصغر من قطر الدائرة: م٤ .
(١٢) فلتقيان] فيلتقيان: م٣ .
(١٤) ولأنّه] فلانه: م١ ، م٣ .
(١٤) خطّان] ــ م٣ .

are drawn from a point outside the circle, [as] lines ZD, ZE,
segment DE is less than a semicircle,
so segment EHD is greater than a semicircle,
but from the position of point Z segment ED is seen,
and from the positions of the two eyes, points B [and] G,
segment EHD is seen.
Therefore it is clear that [what] is seen from the two positions B, G
is greater than a semicircle, and that is the amount seen of the sphere.
And that is what we wished to demonstrate.

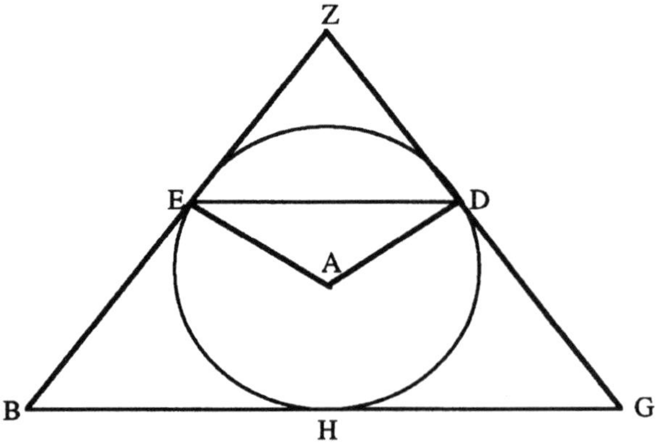

FIGURE 27

A	B	G	D	E	W	Z	H	T	I	K	L	M	N	S	O	F
ا	ب	ج	د	ه	و	ز	ح	ط	ى	ك	ل	م	ن	س	ع	ف

يُخرَجان من نقطة خارجة من الدائرة عليهما زد زه
يكون قطعة د ه أصغر من نصف دائرة
فقطعة ه ح د أعظم من نصف دائرةٍ
ولكنّ من موضع نقطة ز يرى قطعةً ه د
ومن موضعى البصرين اللذين عليهما ب ج
يرى قطعة ه ح د
فبيّن أنّها ترى من موضعى ب ج
أعظم من نصف دائرة وقدر ذلك يرى من الكرة
وذلك ما أردنا أنْ نبيّن ه°ه

----

(١٥) يخرجان من نقطة خارجة من الدائرة] هش م١ ، ــ م٣ .

(١٥) عليهما] عليها : م٥ .

(١٥) زد ز ه] د ه د ك : م٢ ، م٥= زه زد : م٣ .

(١٦) يكون] تكون : م٥ .

(١٦) د ه] ك ه : م٢ ، م٤ ، م٥ .

(١٧) فقطعة ... دائرة] ــ م٢ .

(١٨) ز] فس م١= د : م٢ ، م٤ ، م٥ .

(١٨) من موضع نقطة ز يرى] موضع نقطة ز يرى : م١= من موضع نقطة د : م٤ .

(١٨) ه د] ه ك : م٢ ، م٤ ، م٥ .

(١٨) يرى] ترى : م٥ .

(١٩) من] ــ م٢ ، م٤ .

(١٩) ه ح د] ه ج ك : م٢ ، م٤ ، م٥= ه ح : م٣ .

(٢١) فبيّن] فنبين : م٣ .

(٢٢) وذلك ما أردنا أن نبيّن] ــ م٢= وذلك ما اردنا بيانه : م٥ .

28

When the distance between the eyes is smaller than a sphere's diameter, then that which is seen of the sphere is less than half. For let there be a sphere, the center of which is point A, and we inscribe around point A a circle, BG; and let the distance between the eyes be DE, this being smaller than the sphere's diameter; and we draw from these two positions lines DB, EG [as] tangents to the circle, and these are rays. Then I say that which is seen of the sphere is less than half. So let us draw BD, GD rectilinearly, then it is clear that they meet at a single point because DE is smaller than the sphere's diameter, and the meeting point of the two [extensions] is at point Z. Now because the two lines ZG, ZB fall on the circle from a single point, segment GB is smaller than a semicircle;

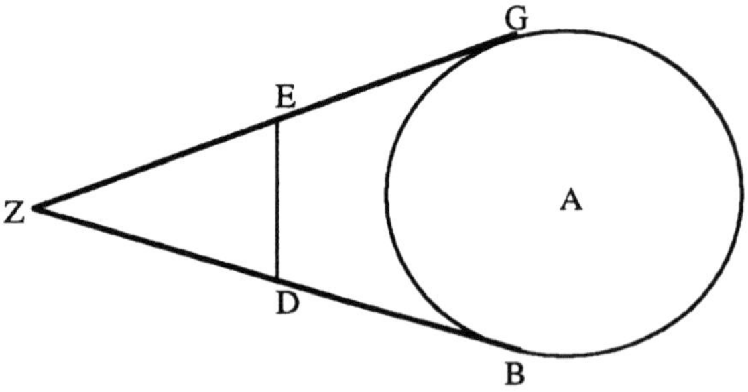

FIGURE 28

A	B	G	D	E	W	Z	H	T	I	K	L	M	N	S	O	F
ا	ب	ج	د	ه	و	ز	ح	ط	ى	ك	ل	م	ن	س	ع	ف

كح

إذا كان بُعد ما بين البصرين أقلّ من قطر الكرة
فإنّ الذى يرى من الكرة أقلّ من نصفها
فليكن كرةٌ مركزها آ
ونديرِ على مركز آ دائرةً عليها ب ج
وليكن بُعد ما بين البصرين د ه
وهو أصغر من قطر الكرة
ونخرجِ من موضعيهما خطّى د ب ه ج
مماسين للدائرة وهما الشعاعان
فأقول إنّ الذى يرى من الكرة أقلّ من نصفها
فلنخرجِ ب د ج ه على استقامة
فبيّن أنّهما يلتقيان على نقطة واحدة
من أجْل أنّ د ه أصغر من قطر الكرة
ويكون التقاؤهما على نقطة ز
فلأنّه وقع على الدائرة خطّا ز ج ز ب من نقطةٍ واحدةٍ
يكون قطعة ب ج أصغر من نصف دائرة

---

(٢) إذا كان] واذا كان: م٤ .
(٢) بعد ما بين] البعد بين: م٣ .
(٣) فان] + فان: م٢ .
(٤) فليكن] فلتكن: م٤ ، م٥ .
(٤) مركزها آ] مركزها نقطه ا: م٣ .
(٥) مركز آ] مركزها: م٣ .
(٦) د ه] ه د: م٥ .
(١٠) إنّ الذى] انه: م٣ .
(١١) ب د ج ه] ب د ج: م١ .
(١٤) يكون] ليكن: م٥ .
(١٥) وقع] قد وقع: م٥ .
(١٦) يكون] تكون: م٥ .

but where there is the circle BG, there are rays DB, EG,
and this is how much of the sphere the two rays enclose, 18
and it has been demonstrated that it is less than a hemisphere.
And that is what we wished to demonstrate.

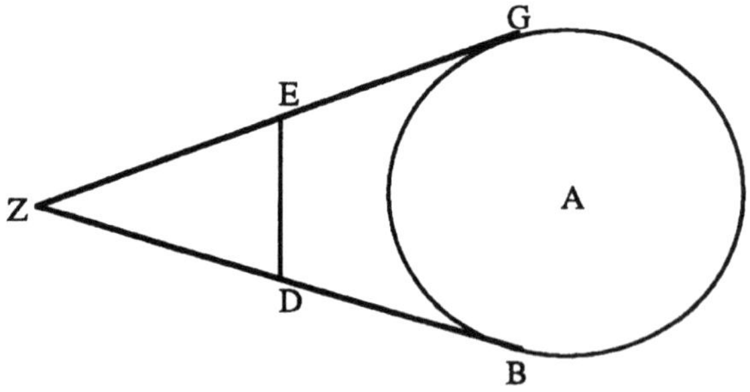

FIGURE 28

A	B	G	D	E	W	Z	H	T	I	K	L	M	N	S	O	F
ا	ب	ج	د	ه	و	ز	ح	ط	ى	ك	ل	م	ن	س	ع	ف

ولكنّ حيث دائرة ب ج فثمّ شعاعان د ب ه ج فبذلك القدر من الكرة يحيط الشعاعان وقد تبيّن أنّه أقلّ من نصف الكرة وذلك ما أردنا أنْ نبيّن ٥°

----

(١٧) شعاعان] شعاع: م٢، م٥ .

(١٨) الشعاعان] الشعاعات: م١ .

(٢٠) وذلك ما اردنا أن نبيّن] _ م٢= وذلك ما اردنا بيانه: م٥ .

29

When there is a cylinder viewed from any position,
then that which is seen of it is less than half.
For let there be a cylinder, the center of its base being A,
and we inscribe around point A a circle, GB,
and the position of the eye is point D,
and the point [representing] the eye and the plane of the cylinder's base
which is [marked] BG are in the same plane;
and we join point D to point A by line DA,
and we extend the two rays DB, DG from point D,
these being tangents to the circle;
and from the points B and G we draw the two sides of the cylinder
[as] lines BE, GZ at right angles;
and we draw the two planes through DB, BE and through GZ, DG.
And it is clear that neither of these two cuts the cylinder
because these two [planes] are tangents to the surface of the cyclinder

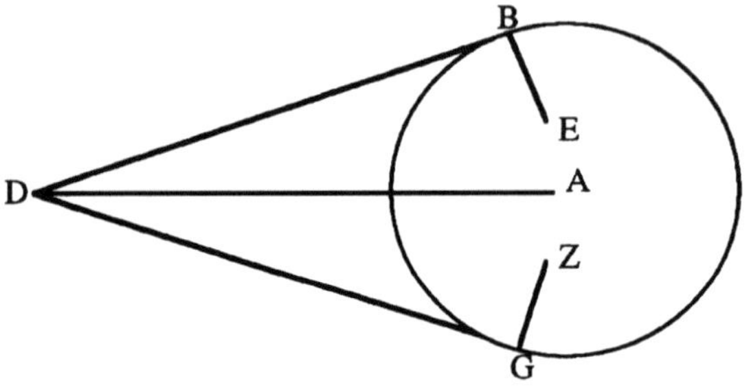

FIGURE 29

A	B	G	D	E	W	Z	H	T	I	K	L	M	N	S	O	F
ا	ب	ج	د	ه	و	ز	ح	ط	ى	ك	ل	م	ن	س	ع	ف

كط

إذا كانت أسطوانةٌ ونُظر إليها من أىّ المواضع كان فإنّ الذى يرى منها أقَلّ من نصفها فلتكن أسطوانةٌ ويكون مركز قاعدتها آ وندير على نقطة آ دائرةً عليها جب وموضع البصر نقطة د ويكون نقطة البصر وسطح قاعدة الأسطوانة التى عليها ب ج فى سطحٍ واحد ونصل نقطة د بنقطة آ بخطّ دآ ونخرِج من نقطة د شعاعين د ب د ج ويكونان مماسّين للدائرة ونخرِج من نقطتى ب وَ ج ضلعى الأسطوانة اللذين عليهما ب ه ج ز على زوايا قائمة ونخرِج السطحين الذين على د ب ب ه وعلى ج ز د ج فبيّن أنّه ليس واحد منهما يقطَع الأسطوانة لأنّهما مماسّان لبسيط الأسطوانة

----

(٢) المواضع كان] الموضع كان: م٢ ، م٤= المواضع: م٣= موضع كان: م٥ .

(٤) قاعدتها آ] قاعدتها نقطه ا: م٣ .

(٥) على] كل: م٣ .

(٥) عليها] _ م٢ ، م٤ ، م٥ .

(٧) قاعدة] قاعدته: م٢ ، م٤ .

(٧) يكون نقطة البصر] تكون نقطة البصر: م٤= تكون نقطة د البصر: م٥ .

(٨) التى عليها] الذى عليه: م١ ، م٣ ، م٥= التى عليه: م٢ .

(١٠) شعاعين] شعاع: م٢ ، م٤ ، م٥= شعاعى: م٣ .

(١٢) ب وَ ج] + فى بسيط الاسطوانة على خطين يكون مماسان قاعدة الاسطوانة عليها ب ه ج ز: هش م١ .

(١٢-١٣) ضلعى الأسطوانة الذين عليهما] الذين على الاسطوانه خطى: م٣ .

(١٣-١٤) على زوايا قائمة . . . ج ز د ج] _ م٥ .

(١٤) السطحين] البسيطين: م٢ ، م٣ ، م٤ .

(١٦) مماسّان] مماسين: م٥ .

since the two lines DB, DG are tangents and the two lines BE, GZ are also tangents. And segment BG is seen by the two rays DB, DG, and it is less than a semicircle. Therefore this is the amount by which the cylinder is seen. And that is what we wished to demonstrate.

FIGURE 29

A	B	G	D	E	W	Z	H	T	I	K	L	M	N	S	O	F
ا	ب	ج	د	ه	و	ز	ح	ط	ى	ك	ل	م	ن	س	ع	ف

من أجْل أنّ خطّى د ب د جـ مماسّان
وخطّى ب ه جـ ز أيضاً مماسّان
وقطعة ب جـ ترى بشعاعى د ب د جـ
وهى أقلّ من نصف دائرة
فبقدر ذلك يكون الذى يرى من الأسطوانة
وذلك ما أردنا أنْ نبيّن ٥٥

---

(١٦) لأنّهما ...الأسطوانة] لانهما مماس بسيط الاسطوانة: هش م٢ = لانهما مماسين لبسيط الاسطوانه: م٣ .

(١٧) مماسّان] مماسين: م٥ .

(١٨) أيضاً مماسان] مماسان: م٢ ، م٤ = مماسين: م٥ .

(١٩) بشعاعى] شعاعين: م١ = من شعاعى: م٢ ، م٣ ، م٤ ، م٥ .

(١٩) د ب] ب د: م٢ ، م٤ ، م٥ .

(٢٠) هى] هو: م١ .

(٢٢) وذلك ما أردنا أن نبيّن] ــ م٢ = وذلك ما اردنا بيانه: م٥ .

30

And we draw a circle, its center being point A,
and point Z being fixed as the position of the eye;
and we join point A to point Z by line AZ,
and from point A we extend line GD on both sides of AZ at right angles,
so line GD is the diameter of the circle;
and we inscribe a circle around line AZ, circle ABZE,
and we join AB, BZ, ZE, EA;
then the two lines BZ, ZE are tangents to the circle
because the two angles B, E are right angles.
Now because the two rays ZB, ZE had fallen
upon the circumference of the circle from point Z,
then the side of the circle, BE, is seen;

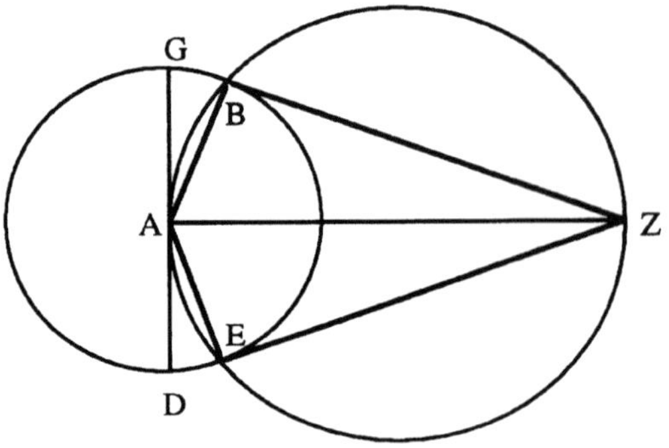

FIGURE 30

A	B	G	D	E	W	Z	H	T	I	K	L	M	N	S	O	F
ا	ب	ج	د	ه	و	ز	ح	ط	ى	ك	ل	م	ن	س	ع	ف

ل

ونخطّ دائرةً ويكون مركزها نقطة آ ويكون نقطة ز ثابتةً وهو موضع البصر ونصل من نقطة آ بنقطة ز بخطّ آز ونخرج من نقطة آ عن جنبتى آز خطّ جد على زوايا قائمة فخطّ جد قطر الدائرة وندير على خطّ آز دائرةً عليها اب زه ونصل اب ب ز زه ه‌ا فخطّا ب ز زه هما مماسّان للدائرة من أجْل أنّ زاويتى ب ه قائمتان فلانّه قد وقع شعاعا زب زه على محيط الدائرة من نقطة ز يكون يرى ناحية الدائرة التى عليها ب ه

----

٢) يكون] تكون: م٥ .
(٣) ثابتة] قائمة: م٢ ، م٣ ، م٤ ، م٥ .
(٣) وهو موضع البصر] ـ م٢ ، م٣ ، م٤= وهى موضع البصر: م٥ .
(٧) اب زه] اب هز: م٥ .
(٨) زه] د ه: م٣ .
(٩) فخطّا ب ز زه] خطا اب د ه: م١= بخطا ب ز زه: م٣ .
(٩) هما مماسّان للدائرة] مماسان: م٢ ، م٤ ، م٥ = مماسان الدائرة= م٣ .
(١١) قد] ـ م٢ ، م٣ ، م٤ .
(١١) شعاعا] شعاعين: م١= شعاعى: م٣ .
(١٢) نقطة ز] نقطة ب: م٢ ، م٥ .
(١٣) يرى] ب ز فى: م٢ ، م٤ ، م٥ .
(١٣) ناحية] محيط: م٥ .

but GBED is a semicircle, so BE is then less than a semicircle. And we have set up this proposition for pillars and cylinders, so if from the two points B, E the sides of pillars and cylinders are drawn at right angles, then they would be tangents to them on the side that the ray[s] fall on them; and the side BGDE is concealed from the eye, and that which is seen is side BE, which is less than a semicircle and this is the amount by which the pillars and cylinders are seen. And that is what we wished to demonstrate.

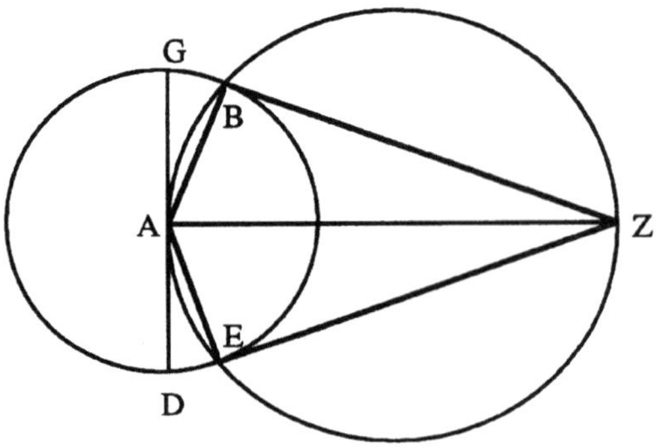

FIGURE 30

A	B	G	D	E	W	Z	H	T	I	K	L	M	N	S	O	F
ا	ب	ج	د	ه	و	ز	ح	ط	ى	ك	ل	م	ن	س	ع	ف

ولكنّ جبهد نصف دائرة
فَ به إذاً أقلّ من نصف الدائرة
ووضعنا هذا الشكل من أجْلِ العُمُد والأساطين
فإنّه إنْ أُخرِجَ من نقطتى ب و ه
أضلاع العُمُدَ والأساطين على زوايا قائمة
تكون مماسّةً لها فى الناحية التى يقع عليها الشعاع
ويستتر عن البصر ناحية بجده
ويكون الذى يرى ناحية به
التى هى أقلّ من نصف الدائرة
وبقدر ذلك يكون ما يرى من العُمُد والأساطين
وذلك ما اردنا أنْ نبيّن ٥٥

----

(١٤) هد] ده: م١ .

(١٥) إذا] اذر: م١ .

(١٧-١٨) فإنّه...الأساطين] ـ م٢ ، م٤ .

(١٨) العمد] والعمد: م٢ .

(١٨) على] عن: م٥ .

(١٨) أضلاع] ـ م٢ ، م٣ .

(١٩) يكون] تكون: م٥ .

(٢٠) يستتر] يستر: م١ ، م٣ ، م٥ .

(٢١) ناحية] من ناحية: م٤ ، م٥ .

(٢١) به] فس م٢ .

(٢٤) وذلك ما أردنا أن نبيّن] ـ م٢= وذلك ما اردنا بيانه: م٥ .

31

When the eye approaches a cylinder, that which is seen of it and enclosed by the ray is in actuality less, but it is thought to be more than what is seen of it when the eye is remote from it. For let there be a cylinder, its base being circle BG, and [let] the circle's center be point A, and [let] the eye's position be point E; then we join point E to the center by line EA, and let the two rays EB, EG fall; and we erect from the two points B and G the two lines BZ, GH at right angles to the surface of the cylinder. Now because of what preceded, it will be clear that ZBGH is less than half of the cylinder; so we view it from the position of point E, then we move the eye close to the cylinder and set it at the position of point T; now whatever the eye encloses from the position of point T

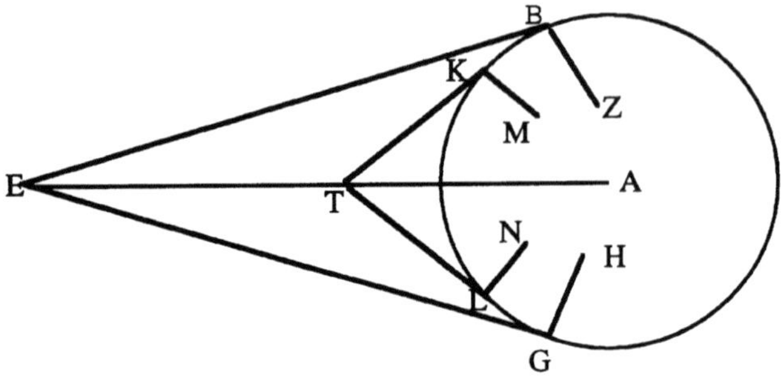

FIGURE 31

A	B	G	D	E	W	Z	H	T	I	K	L	M	N	S	O	F
ا	ب	ج	د	ه	و	ز	ح	ط	ى	ك	ل	م	ن	س	ع	ف

لا

إذا دنا البصر من أسطوانة
يكون ما يرى منها ويحيط به الشعاع أقلّ بالحقيقة
ويُظَنّ أنّه أعظم
ممّا كان يرى منها إذا كان البصر بعيداً منها
فلتكن أسطوانةٌ وقاعدتها دائرة ب ج
ومركز الدائرة نقطة آ وموضع البصر نقطة ه
ونصل نقطة ه بالمركز بخط ه آ
وليقع شعاعا ه ب ب ز
ونقيم من نقطتى ب و ج
خطّى ب ز ج ح مع بسيط الأسطوانة على زوايا قائمة
فمن أجْل ما قدّمنا ستبيّن أنّ
ز ب ج ح أقلّ من نصف الأسطوانة
وننظُر إليه من موضع نقطة ه
ثمّ ننقُل البصر الى قُرب الأسطوانة
فنجعَله فى موضع نقطة ط
فيكون الذى يحيط به البصر من موضع نقطة ط

----

(٢) إذا] فاذا: م١ = واذا: م٣ .

(٢) أسطوانة] الاسطوانة: م٢ ، م٣ .

(٥) بعيداً] بعيد: م٢ .

(٦) فلتكن] فليكن: م٣ .

(٧) البصر] ــ م٥ . (٨) نقطة] ــ م٥ .

(٨-٢٩) وليقع شعاعا ه ب ب ز ...أن نبين] ــ م٤ .

(٩) شعاعا ه ب ب ز] شعاع ه ب ه ج: م٢ .

(١٠) نقطتى] + ب ج خطى: هش م١ = خطى: م٣ .

(١١) ب ز ج ح] ب ز ه ح: م١ .

(١٣) ز ب ج ح] ب ج ز ح: م٢ ، م٥ . (١٤) من] ــ م١ .

(١٦) فنجعله ... ط] فنجعل فى نقطة ط: م٥ .

(١٧) ط] فس م٢ .

is thought to be seen as greater than ZB, GH;
so let two rays TK, TL fall,
and we erect from the positions of the two points K and L
two sides at right angles to the cylinder's surface, [as lines] KM, LN;
and it is clear that the two rays TK [and] TL enclose MKLN,
which is a part of the cylinder enclosed by the two rays TL, TK,
and the two rays EB, EG enclose the segment ZBGH;
but segment ZBGH is greater than KMLN,
but it is seen by eyesight as smaller,
because angle T is greater than angle E.
And that is what we wished to demonstrate.

FIGURE 31

A	B	G	D	E	W	Z	H	T	I	K	L	M	N	S	O	F
ا	ب	ج	د	ه	و	ز	ح	ط	ى	ك	ل	م	ن	س	ع	ف

يُظَنّ أنّه يرى أعظم من زب جح
فليقع شعاعا طك طل
ونقيم من موضعى نقطتى ك وَ ل
ضلعين مع بسيط الأسطوانة على زوايا قائمة عليهما كم لن
فبيّن أنّ شعاعى طك طل يحيط به مك لن
الذى هو جزء من الأسطوانة يحيط به شعاعا طل طك
وشعاعا ەب ەج يحيطان بقطعة زبجح
لكنّ قطعة زب جح أعظم من قطعة كملن
وترى فى مرآة العين أصغر
من أجْل أنّ زاوية طَ أعظم من زاوية ەَ
وذلك ما أردنا أنْ نبيّن ە°

----

(١٨) يظنّ] ويظن: م٥ .
(١٨) يرى] — م٣ .
(١٨) زب جح] زبح: م٢ .
(١٩) فليقع شعاعا] فليقع شعاعات: م٢ = فلتقع شعاعات: م٥ .
(١٩) طك طل] طكل: م١ = طد طل: م٢ ، م٥ .
(٢٠، ٢١، ٢٢، ٢٥) ك] د: م٢ ، م٥ .
(٢٠) كَ وَ لَ] كل: م٣ = دل: م٥ .
(٢١) عليهما] عليها: م٢ ، م٣ ، م٤ .
(٢٢) ان] اذاً ان: م٢ ، م٣ = ايضاً ان: م٥ .
(٢٣) جزء من] جزومن: م١ ، م٢ ، م٣ = جزوى: م٥ .
(٢٣) يحيط به] + يحيطان باقل مما يحيط به شعاعا ەب ەج
: فس م١ = التى يحيط بها: م٢ .
(٢٣-٢٤) شعاعا طل طك وَا] — م٢ ، م٥ .
(٢٤) شعاعا ەب ەج] شعاعات ە ەد: م١ .
(٢٤- ٢٥) يحيطان بقطعة زبجح لكن قطعة زبجح]
فقطعه بزجح: م٢ = يحيطان بقطعة بزجح: م٥ .
(٢٤) بقطعة] نقطة: م٣ .
(٢٥) لكنّ] ولكن: م٣ .
(٢٦) يرى] ترى: م٥ . (٢٦) مرآة] مرى: م٢ .
(٢٨) اردنا أن نبيّن] اردناه: م٢ = اردنا بيانه: م٥ .

32

When a cone with a circular base is viewed,
then what is seen of it is less than half.
For let there be a cone, and [let] its base be circle BG and its apex point A,
and [let] the position of the eye be point D, and let the two rays DB, DG fall.
Now as the two rays DB, DG are drawn from point D as tangents to circle BG,
[then] section BG is less than a semicircle,
because of our preceding demonstration;
then let us join point A, which is the cone's apex, to the two points B and G
as two sides of the cone, lines AB, AG;
and that which is enclosed by the two lines AB, AG and segment BG
is less than half the cone,
because segment BG is less than a semicircle;
therefore, what is seen of the cone is less than half.
And that is what we wished to demonstrate.

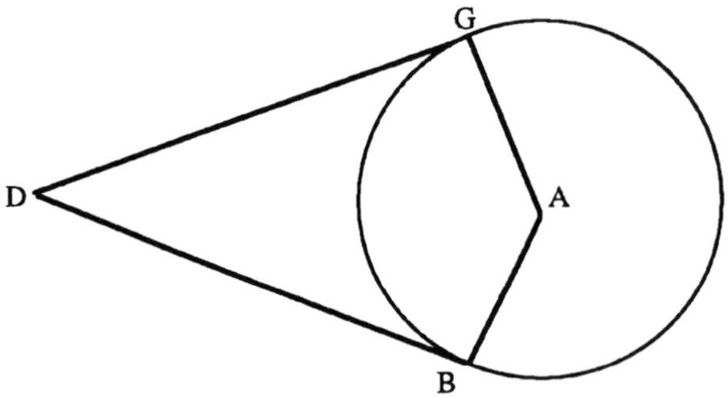

FIGURE 32

A	B	G	D	E	W	Z	H	T	I	K	L	M	N	S	O	F
ا	ب	ج	د	ه	و	ز	ح	ط	ى	ك	ل	م	ن	س	ع	ف

لب

المخروط الذى قاعدته دائرة إذا نُظر إليه
فإنّ الذى يرى منه أقلّ من نصفه
فليكن مخروط وقاعدته دائرة ب ج ورأسه نقطة آ
وموضع البصر نقطة د وليقع شعاعا د ب د ج
فلأنّه يُخرَج من نقطة د شعاعا د ب د ج مماسّين لدائرة ب ج
تكون قطعة ب ج أقلّ من نصف دائرة
من أجْل ما قدّمنا بيانه
فلنصل من نقطة آ التى هى رأس المخروط إلى نقطتى ب وَ ج
ضلعين للمخروط عليهما آ ب آ ج
والذى يحيط به خطّا آ ب آ ج مع قطعة ب ج
هو أقلّ من نصف المخروط
من أجْل أنّ قطعة ب ج هى أقلّ من نصف الدائرة
فالذى يرى من المخروط هو أقلّ من نصفه
وذلك ما أردنا أنْ نبيّن ٥

(٢-٥) المخروط ... نقطة د] _ م٤ .

(٢) دائرة] _ م٣ .

(٣) نقطة] _ م٢ ، م٥ .

(٣) مخروط] مخروطا : م٣ .

(٥) د] فس م١ . (٥) وليقع] وليكن : م١ .

(٥) شعاعا] شعاعات : م٢ ، م٤ .

(٦) فلأنّه ... لدائرة] فلان ... لدائرة : م١ = فلانه ... للدائرة : م٣ .

(٦) يخرج] خرج : م١ ، م٢ ، م٤ = قد خرج : م٥ .

(٨) من أجل ما] لما : م٢ .

(٨) بيانه] + هاهنا يسقط فان ماقبله فى الاسطوانه وما بعده فى المخروط : هش م٤ .

(٩) نقطتى ب وَ ج] نقطتين اللتين هما نقطتين ب ج : م١ = نقطة ب و ج : م٢ ، م٤ .

(١٠) ضلعين للمخروط] ضلعى المخروط : م٢ = ضلعى المخروط : م٥ .

(١١) والذى] فالذى : م٢ ، م٤ ، م٥ ، _ م٣ .

(١١) يحيط به خطا] يحيط به خط : م١ = يحيط بها خطا : م٣ .

(١٣) قطعة ب ج] + فقطعة ب ج : م٣ = مخروط قطعة ب ج : م٤ .

(١٥) وذلك ما أردنا أن نبيّن] _ م٢ = وذلك ما اردنا بيانه : م٥ .

33

When the eye approaches the cone, while being in the same plane with the base,
then the part seen of the cone that is enclosed by the eye is smaller,
but it is thought to be greater.
For let there be a cone, and [let] its base be circle AB and its apex point G,
the eye's position point D, the circle's center point K, and we join DK;
and let the two rays DA, DB fall,
and we join the cone's sides, which are AG, GB.
Now the part ABG of the cone is seen from the eye's position, point D,
by the two rays DA, DB, and that is less than half the cone;
and we move the eye to point E, which is closer to the cone,
and let the two rays EZ and EH fall; and we join the two sides ZG, HG;
then again, the part seen of the cone, which is ZGH,
is enclosed by the eye, E, [and] by the two rays EZ, EH,
and ZGH is less than ABG, but it is thought through vision to be greater,
because angle ZEH is greater than angle ADB.
And that is what we wished to demonstrate.

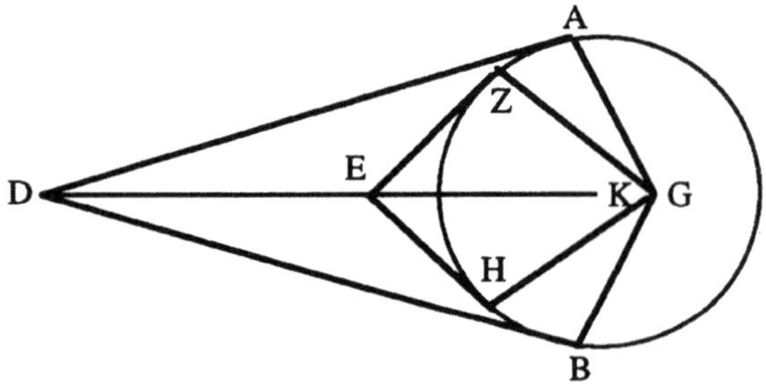

FIGURE 33

A	B	G	D	E	W	Z	H	Ṭ	I	K	L	M	N	S	ʿO	F
ا	ب	ج	د	ه	و	ز	ح	ط	ى	ك	ل	م	ن	س	ع	ف

لج

إذا دنا البصر من المخروط وكان مع سطح القاعدة فى سطحٍ واحدٍ
فإنّ الجزء الذى يرى من المخروط الذى يحيط به البصر أقلّ
ويُظَنّ أنّه أعظم
فليكن مخروطٌ قاعدته دائرة اب ورأسه نقطة ج
وموضع البصر نقطة د ويكون مركز الدائرة نقطة ك ونصل دك
وليقع شعاعا دا دب
ونصل أضلاع المخروط التى عليهما اج جب
فجزء ابج من المخروط يرى من موضع بصر نقطة د
بشعاعى دا دب وهو أقلّ من نصف المخروط
وننقُل البصر إلى نقطة ه ويكون أقرب الى المخروط
وليقع شعاعا هز هح ونصل ضلعى زج حج
فيكون أيضاً الجزء الذى يرى من المخروط الذى عليه زجح
يحيط به بصر ه بشعاعى هز هح
وزجح أقلّ من ابج ويُظَنّ فى الرؤيه أنّه أعظم
من أجْل أنّ زاوية زهح أعظم من زاوية ادب
وذلك ما أردنا أنْ نبيّن ٥٥

---

(٢) إذا] فاذا: م١، م٣ . (٣) الجزء] الجزو: م٢، م٣، م٤، ـ م٥ .
(٤) أعظم] اكبر: م٥ . (٥) مخروط قاعدته] مخروط وقاعدته: م٢، م٤، م٥= مخروطاً قاعدته: م٣ . (٥) دائرة] ـ م٥ . (٦) وموضع البصر نقطة د] ـ م٥ .
(٨) أضلاع] ضلعى: م١، م٣ .
(٨) التى عليهما] الكرين عليهما: م١= الذين عليهما: م٣= التى عليها: م٤ .
(٩) فجزء] فجزو: م١، م٢، م٣، م٤= فجز: م٥ .
(١٠-١١) بشعاعى ... إلى نقطة ه] بشعاعين ...الى موضع نقطة ه: م١، م٣ .
(١٢) زج حج] جز جح: م٢= زج جح: م٣، م٤، م٥ .
(١٣) الجزء] الجز: م١، م٥= الجزو: م٢، م٣، م٤ .
(١٣) يرى...زجح] يرا ...زججح: م٣ .
(١٤) بشعاعى] شعاعين: م١= شعاعى: م٣ .
(١٥) و زجح] و ججح: م١= و زججح: م٣ .
(١٧) وذلك ما أردنا أن نبيّن] ـ م٢= وذلك ما اردنا بيانه: م٥ .

34

When there is a cone with a circular base, and if, from the points of tangency of the lines issued from the eye to its base, two lines are drawn on the surface of the cone to its apex, after which on these lines and the lines issued from the eye planes are drawn; and [further] if the eye's position is on the meeting point of the two [planes], meaning if it is on their common section, on whatever point on that [line], then that which is seen of the cone is always the same if the [point of the] eye is produced on a plane parallel to the original plane.

For let there be a cone, and [let] its base be circle BG and its apex point A, and [let] the eye's position be point D; and from it we extend rays DG, DZ and draw from the two points of tangency Z, G to point A, the cone's apex, two sides of the cone [as] lines ZA, GA; and we draw the surface that

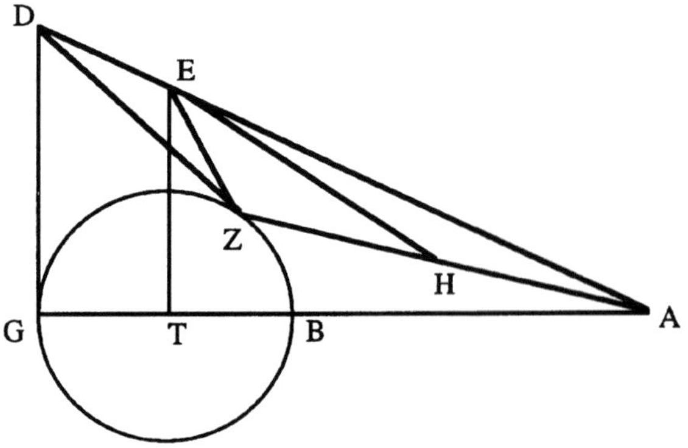

FIGURE 34

A	B	G	D	E	W	Z	H	T	I	K	L	M	N	S	O	F
ا	ب	ج	د	ه	و	ز	ح	ط	ى	ك	ل	م	ن	س	ع	ف

لد

إذا كان مخروطٌ قاعدته دائرة فإنّه إذا أُخرِجَ من مواضع تماسّ الخطوط التى تُخرَج من البصر إلى القاعدة خطّان على بسيط المخروط إلى رأسه ثم أُخرِجَ على تلك الخطوط وعلى الخطوط التى تُخرَج من البصر سطوح فإنّ البصر إذا كان على موضع التقائهما أعني إذا كان على تقاطعهما المشترك فى اىّ المواضع منه كان فإنّ الذى يرى من المخروط أبداً يكون متساوياً

إذا كان خروجِ البصر على سطحٍ موازٍ للسطح الأوّل فليكن مخروطٌ قاعدته دائرة ب ج ورأسه نقطة آ وموضع البصر نقطة د ونخرجِ منها شعاعيِ د ج د ز ونخرجِ من موضع تَماسّ نقطتيِ ز ج إلى نقطة آ التى هى رأس المخروط أضلاعاً ضلعى المخروط عليهما ز آ ج آ ونخرجِ البسيط الذى

---

(٣) مواضع] موضع: م٢ .

(٣) تماس]ساير: م١ ، م٣ ، م٥= تتماس: م٤ .

(٤) خطّان] خطين: م٢ .

(٤-٥) بسيط ...تلك] فس م١ .

(٤) المخروط] + وعلى الخطوط التى تخرج: م٤ .

(٦) سطوح] سطوحاً: م٣ .

(٩) أبداً] ـ م٥ .

(١٠) خروج] بخروج: م١ .

(١٠) مواز] موازى: م١ ، م٣ .

(١٣) شعاعى] شعاعين: م١ .

(١٣) د ج د ز] د ب ز ج: م٢= د ز د ج: م٤ ، م٥ .

(١٤) موضع] موضعى: م٣ ، م٤ ، م٥ .

(١٤) نقطة آ] النقط: م٣ .

(١٥) أضلاعاً] فس م١ ، ـ م٣ .

(١٥) ضلعى المخروط] ضلعين المخروط: م٣ ، م٥ .

passes through AZ, DZ and the one that passes through GD, GA. Then their common section is a line, so let it be AED. Then I say that if the eye is positioned on [any point on] line AED, what is seen of the cone is equal. So let the eye's position on line AED be point E, and we draw from it two rays to the cone, and it is demonstrated that these two fall on lines AZ, AG, because the eye is produced on a plane parallel to the original plane; and the two rays are drawn in straight lines, since if they were to fall outside AG, AZ, they would be bent toward the cone, and that is not possible; so let them be lines ET, EH. [Now] since the eye is on a plane parallel to the original plane, and the [cone of] ray[s] is drawn rectilinearly, and things seen under equal angles are seen as equal,

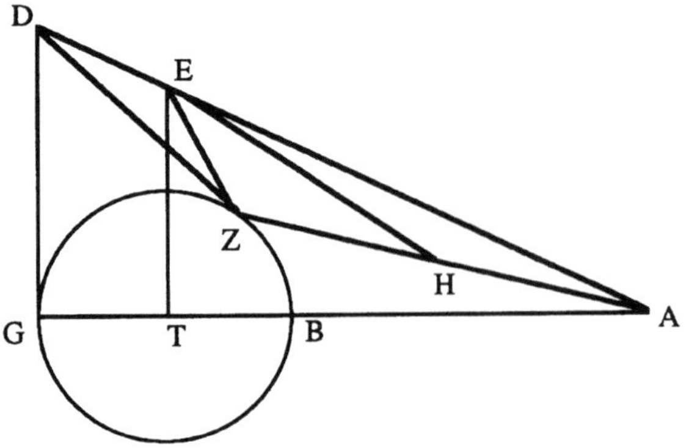

FIGURE 34

A	B	G	D	E	W	Z	H	T	I	K	L	M	N	S	O	F
ا	ب	ج	د	ه	و	ز	ح	ط	ى	ك	ل	م	ن	س	ع	ف

يمُرّ على ا ز د ز والذى يمُرّ على جد جا
فيكون قطعهما المشترك خطّاً فليكن خطّ اهد
فأقول إنّه إنْ وُضِع البصر على خطّ اهد
يكون ما يرى من المخروط متساوياً
فليكن موضع البصر نقطة ه على خطّ اهد
ونخرِج منها شعاعين إلى المخروط
فتبيّن أنّهما يقعان على خطّى از اج
من أجْل أنّ البصر يُخرَج على سطحٍ موازٍ للاوّل
والشعاعان يُخرَجان على خطوط مستقيمة
فإنّه إنْ كان يقع خارجاً من خطّى اج از
فإنّهما ينعطفان إلى المخروط وذلك ما لا يمكن
فليكن عليهما ەط ەح
فلأنّ البصر على سطحٍ موازٍ للسطح الأوّل
والشعاع يُخرَج على استقامة
والأشياء التى ترى من زوايا متساوية فهى ترى متساويةً

---

(١٦) ونخرج ...والذى يمرّ على] ـ م٣ .
(١٦) از دز...جد جا] ازد ...جد جا : م٤ ، م٥= ...زا ج ا : م٤.
(١٧) فيكون قطعهما المشترك خطّاً] فيكون تقاطعهما المشترك لهما خطّاً: م٥ .
(١٨) اەد فأقول إنّه] اه فاقول: م٥ .
(١٩) فليكن موضع] فليكن من موضع: م٤= وليكن من موضع: م٥ .
(٢٠) خطّ] ـ م٥ .
(٢٠) موضع البصر نقطة ه] من موضع نقطة ه: م٢ .
(٢١) شعاعين] شعاعان: م١ ، م٣ .
(٢٣-٢٤) مواز للأول وَ] موازى: م٢ ، م٤ ، م٥= موازى الاول و: م٣ .
(٢٤) الشعاعان] والشعاع : م٢= الشعاعين: م٤ ، م٥ .
(٢٥) إن كان] إن كانت: م٣ ، م٤= اذا كان: م٥ .
(٢٦) ينعطفان إلى] ينعطف الى: م٥ .
(٢٧) ما لا يمكن] مما لا يمكن: م٥ .
(٢٨) سطح مواز للسطح الأوّل] السطح مواز للاول: م٢= سطح موازى الاول: م٣= سطح للسطح الاول: سطح مواز للاول: م٤ ، م٥ .
(٣٠) فهى ترى] ترى: م٣= فهى: م٥ .

and any position of the eye on a plane parallel [to the original one] on line AED forms equal angles, then that which is seen of the cone is then the same [amount], because ZA [and] AG cut a smaller part of the cone according to the ratio that they cut the circle of the base. And that is what we wished to demonstrate.

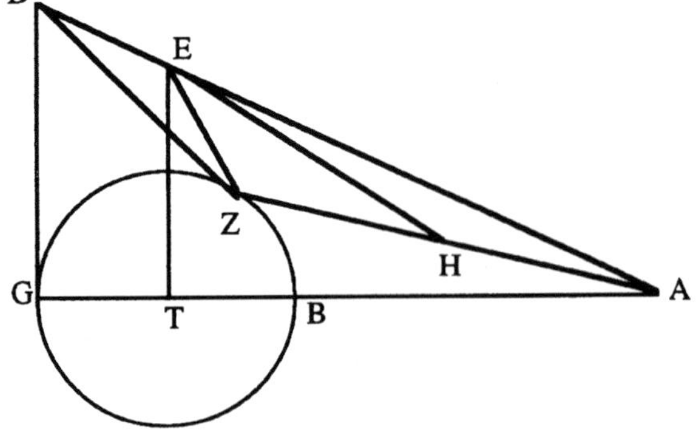

FIGURE 34

A	B	G	D	E	W	Z	H	T	I	K	L	M	N	S	O	F
ا	ب	ج	د	ه	و	ز	ح	ط	ى	ك	ل	م	ن	س	ع	ف

وكلّ بصر يوضع على سطحٍ موازٍ على خطّ ا ه د
فزواياه متساوية
فالذى يرى إذاً من المخروط متساوىٌ
من أجْل أنّ زا اج يقطَع الموضع الأصغر من المخروط
كمثل قطعها دائرة القاعدة
وذلك ما أردنا أنْ نبيّن ٥٥

---

(٣١) كلّ بصر...خطّ ا ه د] كل البصر موضع على خطّ ا ه د: م٣ .

(٣٣) متساوى] متساو: م١ ، م٢ ، م٣ .

(٣٤) قطعها] قطعهما: م٥ .

(٣٦) أن نبيّن] بيانه: م٥ .

35

When the eye is [on a line] equidistant from a cone,
if it is higher [on that line], what is seen of the cone is greater,
and if it is lower [on it], what is seen of it is less.
For let there be a cone, its base being circle BG and its apex point A,
and let its sides be BA, AG; and we join BG;
and we extend line GB rectilinearly to point H;
and we draw from point T a line TK, parallel to AB.
And wherever T is, I say if the eye is at the position of point T,
that which is seen of the cone is greater
than that which is seen [of it] when the eye is at the position of point K.
So let AK and AT be joined,
and we extend AT to point H and AK to point L.

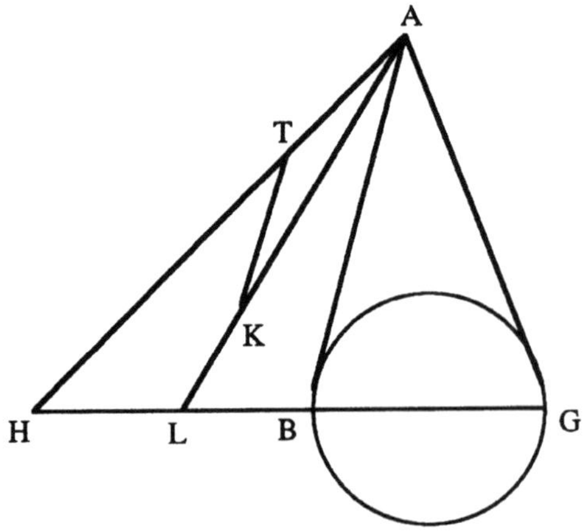

FIGURE 35

A	B	G	D	E	W	Z	H	T	I	K	L	M	N	S	O	F
ا	ب	ج	د	ه	و	ز	ح	ط	ى	ك	ل	م	ن	س	ع	ف

له

إذا كان البصر متباعداً عن المخروط باستواء
فإنّه إذا كان أرفع يكون ما يرى من المخروط أعظم
وإذا كان أخفض فإنّ ما يرى منه أصغر
فليكن مخروطٌ قاعدته دائرة ب ج ورأسه نقطة آ
وليكن أضلاعه عليها ب ا ا ج ونصل ب ج
ونخرِج خطّ جب على استقامة إلى نقطة ح
ونخرِج من نقطة ط خطّاً موازياً لخطّ ا ب عليه ط ك
فحيث ما كانت ط فأقول إنّه إذا كان البصر على موضع نقطة ط
فإنّ الذى يرى من المخروط أعظم
ممّا يرى إذا كان البصر على موضع نقطة ك
ولنصل ا ك ا ط
ونخرِج ا ط الى نقطة ح و ا ك إلى نقطة ل

----

(٢) إذا] واذا: م١، م٣.
(٢) باستواء] المستوى: م٢، م٤، م٥= باستوى: م٣.
(٣) كان] + باستوا: م٥.
(٣) يكون ما يرى] ما يكون يرى: م١.
(٤) فإنّ] كان: م٢، م٤، م٥= يكون: م٣.
(٧) جب] ب ج: م١.
(٨) موازياً لخطّ اب] مواز الخط اب: م١.
(٨) ط ك] ط: م٣.
(٩) فحيث ما كانت] حيث ما كان: م٢= فحيث ما كان: م٣= حيث ما كانت: م٤، م٥.
(٩) فاقول ... ط] ـ م١.
(١٠) المخروط] الخطوط: م٣.
(١١) إذا كان] ان لو كان: م١= الو كان: م٣.
(١١) موضع] + ترى: م١.
(١٢) ولنصل] فلنصل: م٢، م٤، م٥.
(١٣) اك اط] طا ال: م١.
(١٣) نقطة ح ...نقطة ل] ح ...ل: م٥.

And it is clear that if the eye is [placed] at point L [and then H], then what is seen of the cone is different [in each case], and what is seen of it from point H is greater and what is seen of it from point L is less. But what is seen from the position of point H is equal to what is seen from the position of point T, and what is seen from the position of point L is equal to what is seen from point K, as we demonstrated in what preceded. Thus, if the eye is placed on point T, what is seen of the cone is greater than what is seen of it if the eye is at the position of point K. And that is what we wished to demonstrate.

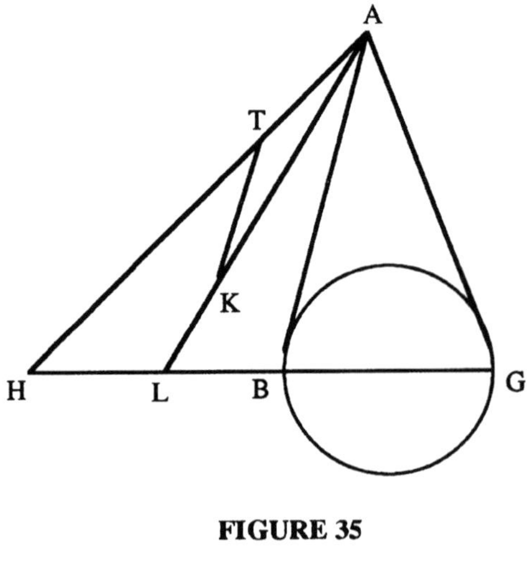

FIGURE 35

A	B	G	D	E	W	Z	H	T	I	K	L	M	N	S	O	F
ا	ب	ج	د	ه	و	ز	ح	ط	ى	ك	ل	م	ن	س	ع	ف

فبيّن أنّه إذا كان البصر على نقطة ل
فيكون ما يرى من المخروط مختلفاً
ويكون ما يرى منه من موضع نقطة ح أعظم
والذى يرى منه من موضع نقطة ل أقلّ
لكنّ الذى يرى من موضع نقطة ح
يكون مساوياً لما يرى من موضع نقطة ط
وما يرى من موضع نقطة ل
مساوياً للذى يرى من موضع نقطة ك
كما قد تبيّن فيما قدّمنا
فإذا وُضِع البصر على نقطة ط
كان ما يرى من المخروط أعظم
ممّا يرى منه إذا كان البصر فى موضع نقطة ك
وذلك ما اردنا أنْ نبيّن °°

---

(١٤) فبيّن أنّه إذا كان البصر على نقطة ل] ـ م١، م٣ .
(١٤) البصر] ـ م٥ .
(١٥) فيكون] ويكون: م٥ .
(١٦) موضع] + نقطة ح اعظم من الذى يرى من موضع نقطة: هش م١ .
(١٨) لكنّ] ولكنّ: م٣ .
(١٨) يكون] ـ م٣ .
(١٩-٢٠) نقطة ط ...نقطة ل] نقطة ل نقطة ك: م٢ ، م٤ ، م٥ .
(٢١) مساوياً] مساو: م٢ .
(٢١) ك] ـ م٥ .
(٢٢) قد] ـ م٥ .
(٢٣) على نقطة] اذا على نقطة: م٢ ، م٤ ، م٥ .
(٢٦) وذلك ما أردنا أن نبيّن] ـ م٢= وذلك ما اردنا بيانه: م٥ .

36

When a line is drawn upward from the center of a circle at right angles
to the surface of the circle, and the eye is placed on that line,
then the diameters drawn in this circle are seen as equal.
For let there be a circle, its center [being] point A,
and we draw from it line AB perpendicular to the circle's plane,
and [let] the eye's position on it [be] point B.
Then I say that the diameters are seen as equal.
So let two diameters be drawn as GD [and] EZ intersecting at point A,
and we join BG, BE, BD, BZ.
Now since line ZA equals line AG, and AB is common, and the angles are right,
the base ZB is equal to base BG;
and angles at the bases are equal,
[making] the angle ABG equal to angle ABZ

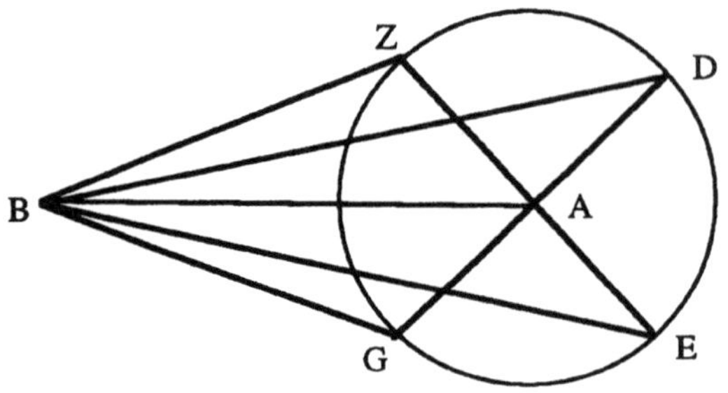

FIGURE 36

A	B	G	D	E	W	Z	H	T	I	K	L	M	N	S	O	F
ا	ب	ج	د	ه	و	ز	ح	ط	ى	ك	ل	م	ن	س	ع	ف

لو

إذا أُخرِج من مركز دائرة خطّ فى الهواء على زوايا قائمة
على بسيط الدائرة ووُضع البصر على ذلك الخطّ
فإنّ الأقطار المخرجة فى تلك الدائرة ترى متساويةً
فليكن دائرةٌ ومركزها نقطة آ
ونخرِج منه خطّ آب قائماً على سطح الدائرة
وموضع البصر عليه على نقطة ب
فأقول إنّ الأقطار ترى متساويةً
فلنخرج قطرين عليهما جد هز يتقاطعان على نقطة آ
ونصل ب ج ب ه ب د ب ز
فلأنّ خطّ زآ مساو لخطّ آج وخطّ آب مشترك والزوايا قائمة
تكون قاعدة زب مساوية لقاعدة ب ج
وتكون الزوايا التى تلى القاعدتين متساوية
زاوية آب ج مساوية لزاوية آب ز

----

(٢) أُخرِج] خرج: م٤ .

(٢) دائرة] الدائرة: م٢ .

(٢) خطّ فى الهواء] خطاً فى الهوا: م٢ ، م٣ ، م٥= خطاً فى الهواء: م٤ .

(٦) آب قائماً] ب ا قايم: م٢= اب قايم: م٣ ، م٤ ، م٥ .

(٧) عليه] ـ م٢ . (٨) فأقول إنّ] فان: م١ .

(٩) يتقاطعان على نقطة آ]+ وهى المركز: هش م١= يتقاطعان: م٢ ، م٥= يتقاطان: م٤ .

(١٠) ب ج ب ه ب د ب ز] ب ج ب ه ب ز ب د: م٢ ، م٤ ، م٥ .

(١١) فلانّ ...قائمه] فلان خطّ ا مشترك والزوايا قائمه: م٤ .

(١١) مساو لخطّ آج] مساو لخطّ اد: م١ ، م٢ ، م٥= مساوى لخطّ اد: م٣ .

(١١) آب] اد: م١ .

(١٢) ب ج] ب د: م١ ، م٢ ، م٣ ، م٤ ، م٥ .

(١٣) القاعدتين] القاعدة: م٢ ، م٤= القاعدة بين: م٥ .

(١٣) متساويه] ـ م١ ، م٣ ، م٤ ، م٥ .

(١٤) زاوية آب ز] زاوية اب: م١= زاوية اب د: م٢= زاوية اب ب ز: م٣ .

(١٤) آب ج] اب د: م٢ ، م٤= اب ب ج: م٣ ، م٥ .

and angle EBA equal to angle DBA; so angle GBD is equal to angle EBZ, and things seen under equal angles are equal. Therefore GD is <seen as> equal to EZ. And that is what we wished to demonstrate.

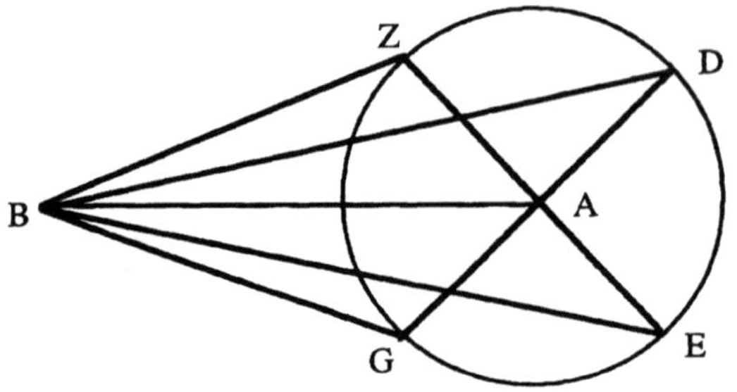

FIGURE 36

A	B	G	D	E	W	Z	H	T	I	K	L	M	N	S	O	F
ا	ب	ج	د	ه	و	ز	ح	ط	ى	ك	ل	م	ن	س	ع	ف

وزاوية ه ب ا مساوية لزاوية د ب ا
فزاوية جـ ب د مساوية لزاوية ه ب ز
والأشياء التى ترى من زوايا متساوية فهى متساوية
فَ جـ د إذاً <يرى> مساو لـ ه ز
وذلك ما اردنا أنْ نبيّن ٭

(١٤-١٥) لزاوية ا ب ز وزاوية ه ب ا مساوية] ـ م٤، م٥ .

(١٥-١٦) د ب ا ... جـ ب د ... ه ب ز] د ب ب ا ... جـ ب ب د ... ه ب ب ز: م٤ .

(١٩) وذلك ما أردنا أن نبيّن] ـ م٢ = وذلك ما اردنا بيانه: م٥ .

37

And I say when the line that is drawn from the center of the circle is not
at right angles to the plane of the circle,
but it is equal to half the circle's diameter,
then the diameters are seen as equal.
For let there be a circle, ABGD,
and we draw in it two diameters, AB, GD,
intersecting at right angles at the center of the circle;
and let ZE be equal to half the circle's diameter;
and we join ray[s] ZA, ZG, ZB, ZD.
Now because line BE is equal to line EZ and to line EA,
the three lines are equal, that is, lines EA, EZ, and EB;
and if we rotate the semicircle drawn on line AB, it falls on Z.
Then it is demonstrated from that that angle AZB is right,
and in the same way, GZD is also right;

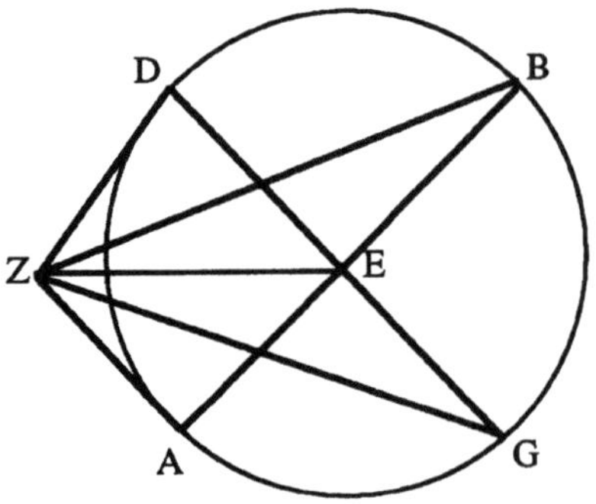

FIGURE 37

A	B	G	D	E	W	Z	H	T	I	K	L	M	N	S	O	F
ا	ب	ج	د	ه	و	ز	ح	ط	ى	ك	ل	م	ن	س	ع	ف

لز

وأقول إنّه إنْ لم يكن الخطّ الذى يُخرَج من مركز الدائرة على زوايا قائمة على سطح الدائرة ولكنّ يكون مساوياً لنصف قطر الدائرة فإنّ الأقطار ترى متساويةً فليكن دائرة عليها ا ب جـ د ونخرج فيها قطرين عليهما ا ب جـ د يتقاطَعان على مركز الدائرة على زوايا قائمة وليكن ز ه مساوياً لنصف قطر الدائرة ونصل شعاع ز ا ز جـ ز ب ز د فلأنّ خطّ ب ه مساو لخطّ ه ز و لخطّ ه ا فالثلثة الخطوط متساوية أعنى خطّ ه ا وخطّ ه ز وخطّ ه ب فإذا أدرنا نصف الدائرة المخطوطة على خطّ ا ب يقع على نقطة ز فتبيّن من ذلك أنّ زاوية ا ز ب قائمة وكذلك يكون زاوية جـ ز د أيضاً قائمة

---

(٢) إن] وان: م٢ ، م٤= اذا: م٥ .

(٢) يخرج] خرج: م٣ .

(٣) على سطح] عند سطح: م٢ ، م٤ .

(٥) ترى] يرى: م٢ ، م٣ .

(٦) فليكن دائرة عليها] فلتكن: م٤= فليكن دائرة: م٥ .

(٧) قطرين] قطرى: م١ ، م٣ .

(٨) على مركز الدائرة على زوايا قائمة] على زوايا قائمة: م٢ ، م٤= على زوايا قائمة على مركز الدائرة: م٣ .

(٩) وليكن] ولكنَ: م٣.

(٩) مساوياً] مساو: م١= مساوى: م٣ .

(١٠) زب] _ م٣ ، م٥ .

(١٢) فالثلثة] فالثلاثة: م٥ .

(١٢) خطّ ه ا] خطّ اه: م٢ ، م٣ ، م٤ ، م٥ .

(١٢) وخطّ ه ب] _ م٢ .

١٢) خطّ ه ز]+ وخطّ ه ز: م١ .

(١٣) المخطوطة على خطّ ا ب] المخروطة على ا ب: م٢ ، م٤= المخطوطة على ا ب: م٥ .

(١٥) ايضاً قائمة] قائمة ايضاً: م٢ ، م٥ =تكون زاوية جـ ز د قائمة ايضاً: م٤ .

and right angles are equal,

and things seen under equal angles are equal.

Therefore AB is seen as equal to GD. 18

And that is what we wished to demonstrate.

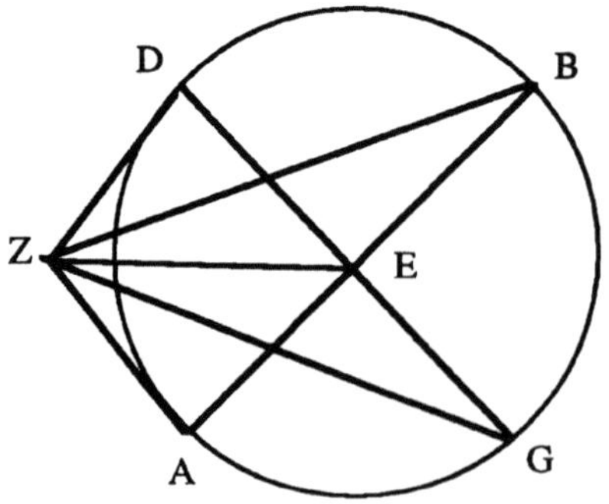

FIGURE 37

A	B	G	D	E	W	Z	H	T	I	K	L	M	N	S	O	F
ا	ب	ج	د	ه	و	ز	ح	ط	ى	ك	ل	م	ن	س	ع	ف

والزوايا القائمة متساوية
والأشياء التى ترى من زوايا متساوية فهى متساوية
فَ اب إذن يرى مساوياً لـ جد ١٨
وذلك ما اردنا أنْ نبيّن ٠َ٠

----

(١٧) ترى] يرى: م٣ .
(١٧) فهى متساوية] ترى متساوية: م٥ .
(١٨) إذن] اذاً: م٢ ، م٤= اذا: م٣ ، م٥ .
(١٨) مساوياً] مساو: م٢ ، م٥= متساوية: م٣= متساوياً: م٤ .
(١٨) إذن يرى مساوياً] يرى اذاً مساو: م٥ .
(١٩) وذلك ما أردنا أن نبيّن] ــ م٢ = + واقول إنّه إن لم يكن... أن نبين: م٤= وذلك ما اردنا بيانه: م٥ .

38

And when the line drawn from the eye's position to the circle's center is not at right angles to the plane of the circle, and neither is it equal to half the diameter nor inclined to any of the diameters such that the raised angles are equal, meaning the large one to the large one, and the small one to the small one, then the diameters in that situation are seen as unequal. For let there be a circle, AGBD, and we draw two diameters AB, GD in it [such that] their intersection is not at right angles; and let line AB be equal to line GD, and we draw a line from point E to point Z, which is the eye's position, line EZ, which is not perpendicular to the plane of the circle, nor is it equal to half the diameter, nor is angle GEZ equal to angle ZEB, nor is angle AEZ equal to angle DEZ;

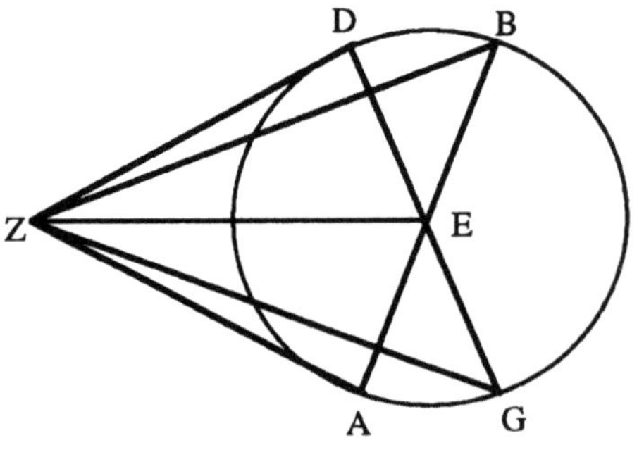

FIGURE 38

A	B	G	D	E	W	Z	H	T	I	K	L	M	N	S	O	F
ا	ب	ج	د	ه	و	ز	ح	ط	ى	ك	ل	م	ن	س	ع	ف

لح

وإنْ لم يكن الخطّ الذى يُخرَج من موضع البصر إلى مركز الدائرة
على زوايا قائمة من سطح الدائرة
ولا مساوياً لنصف قطرها
ولا مائل إلى أحد القطرين حيث تكون الزوايا العاليات متساوية
أعني الكبرى للكبرى والصغرى للصغرى
فإنّ الأقطار عند ذلك ترى مختلفةً
فليكن دائرةٌ عليها ا ج ب د
ونخرِج فيها قطرى ا ب ج د
ولا يكون تقاطعهما على زوايا قائمة
وليكن خطّ ا ب مساوياً لخطّ ج د
ونخرِج من نقطة ه خطّاً إلى نقطة ز وهو موضع البصر
وعليه ه ز يكون غير قائم على سطح الدائرة
ولا مساو لنصف قطرها
ولا تكون زاوية ج ه ز مساويةً لزاوية ز ه ب
ولا زاوية ا ه ز مساويةً لزاوية د ه ز

---

(٢) وإن لم يكن الخطّ] فان لم يكن الخط: م١ = وإن لم يكن البصر: م٥ .
(٤) مساوياً] مساو: م٢، م٤، م٥ =مساوى: م٣ .
(٤) أحد القطرين] القطر كبرى؟: م١ = القطر يرى: م٣ = القطر: م٥ .
(٤) القطر] + فانه؟: فس م١ .
(٥) حيث] حتى: م٢، م٤، م٥ . (٥) العاليات] + المايلات: م١ .
(٦) الكبرى ... للصغرى] الصغرى للصغرى والكبرى للكبرى: م٢ .
(٧) ترى] يرى: م٣ .
(٨) فليكن] فلتكن: م٤، م٥ . (٨) ا ج ب د] ا ج د ب: م٢، م٣، م٤، م٥ .
(١١-١٢) ا ب مساوياً لخطّ ج د وَ] ــ م٢، م٤، م٥ .
(١٢) نخرج] يخرج: م٤ . (١٢) خطّاً] ــ م٢، م٤، م٥ .
(١٣) يكون] ويكون: م٢، م٤، م٥ .
(١٥، ٢١) تكون] يكون: م١ .
(١٥-١٦) ولاتكون ج ه ز ... لزاوية د ه ز] ولايكون زاوية ج ه ز مساوية لزاوية د ه ز ولان زاوية ا ه د لزاوية: م١ = ولاتكون زاوية د ه ز: م٥ .

and we join ZG, ZD, ZA, ZB. Now either angle AEZ is greater than angle BEZ or angle BEZ is greater than angle AEZ; and either angle GEZ is greater than angle ZED or angle DEZ is greater than angle GEZ. And because of that, either angle AZB is greater than angle GZD or angle GZD is greater than angle AZB; therefore, the diameters are seen as unequal if the line drawn from the eye's position to the circle's center is in the state we described. And that is what we wished to demonstrate.

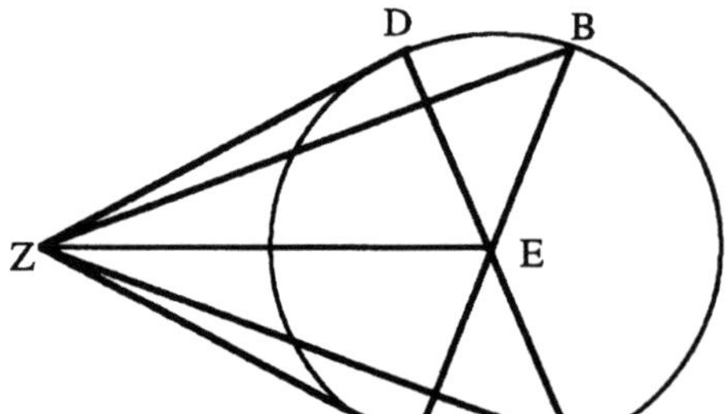

FIGURE 38

A	B	G	D	E	W	Z	H	T	I	K	L	M	N	S	O	F
ا	ب	ج	د	ه	و	ز	ح	ط	ى	ك	ل	م	ن	س	ع	ف

ونصل زجـ زد زا زب
فيكون إمّا زاوية اهز أعظم من زاوية ب هز
وإمّا زاوية ب هز أعظم من زاوية اهز
وزاوية جـ هز إمّا أنْ تكون أعظم من زاوية زهد
وإمّا أنْ تكون زاوية دهز أعظم من زاوية جـ هز
فمن أجْل ذلك يكون إمّا زاوية ازب أعظم من زاوية جـ زد
وإمّا زاوية جـ زد أعظم من زاوية ازب
فيكون إذاً الأقطار ترى مختلفةً
إذا كان الخطّ الذى يُخرَج من موضع البصر إلى مركز الدائرة
على الحال التى ذكرنا
وذلك ما أردنا أنْ نبيّن ٥°

---

(١٧) زجـ زد زا زب] زجـ زا زد زب: م٢، م٣، م٤، م٥ .
(١٨) إمّا] اه: م٣ .
(١٩-٢٠) اهز...زاوية] ــ م٣ .
(٢٠) زهد] زهب: م٢، م٥= ب زه: م٤ .
(٢١) زاوية دهز] دهز: م٢، م٤، م٥ .
(٢٢) يكون] تكون: م٥ .
(٢٢) ازب] ادب: م١، م٣ .
(٢٢) ازب اعظم من زاوية] ــ م٤ .
(٢٣) جـ زد ...ازب] زاب...جـ دز: م٢، م٥= زاب...جـ دب: م٤ .
(٢٤) إذاً] فس م١ .
(٢٥) إذا كان] اذا ما كان: م١، م٥= اما اذا كان: م٢، م٣، م٤ .
(٢٦) التى] الذى: م١ .
(٢٧) وذلك ما أردنا أن نبيّن] ــ م٢= وذلك ما اردنا بيانه: م٥ .

39

We draw a circle, its center being point A,
and [let] the eye's position be point B,
and let the perpendicular drawn from the eye's position
not fall on the point that is the circle's center but fall outside it,
and let it be line BG;
then we join point G to point A, by line GA,
and we also draw line BA from point A to point B.
Then I say that angle GAB is less than all the angles
formed by the lines that pass through point A and line BA.
So we draw DE, and from point G we draw a perpendicular to line DE
in that plane, GZ, and we join BZ,
then BZ is perpendicular to DE.
Now because angle GZA is a right angle,
angle AGZ is less than a right angle,
and the longer side subtends the greater angle,
so side AG is longer than side AZ,

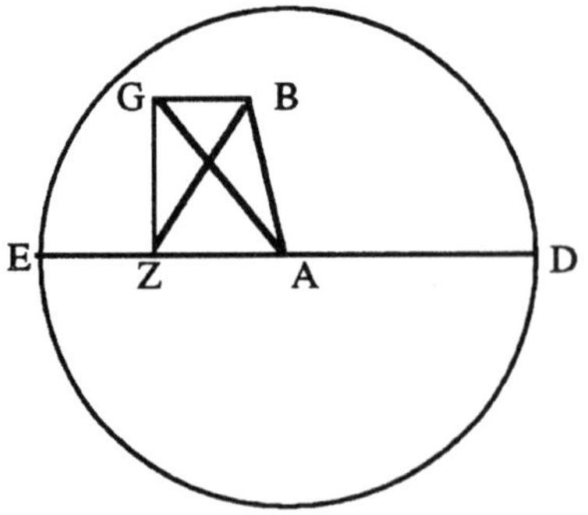

FIGURE 39

A	B	G	D	E	W	Z	H	T	I	K	L	M	N	S	O	F
ا	ب	ج	د	ه	و	ز	ح	ط	ى	ك	ل	م	ن	س	ع	ف

لط

نخطّ دائرةً يكون مركزها آ وموضع البصر نقطة ب ويكون العمود الذى يُخرَج من موضع البصر لا يقع على النقطة التى هى مركز الدائرة ولكنّ يقع خارجاً منها وليكن خطّ ب ج ونصل نقطة جـ بنقطة آ بخطّ جا ونخرِج أيضاً من نقطة آ إلى نقطة ب خطّ ب ا فأقولُ إنّ زاوية جاب أصغر من جميع الزوايا التى تكون من الخطوط التى تمرّ على نقطة آ مع خطّ ب ا فنخرِج د ه ونخرِج من نقطة جـ عموداً إلى خطّ د ه فى ذلك السطح عليه ج ز ونصل ب ز فَ ب ز إذاً عمود على د ه فلأنّ زاوية ج ز ا قائمة تكون زاوية ا ج ز أصغر من قائمة والزاوية العظمى يوترها الضلع الأطول فضلع ا ج اطول من ضلع ا ز

---

(٢) نخطّ دائرةً يكون مركزها آ] ونخط . . . ا: م١ ، م٣= خط . . . نقطة آ: م٥ .

(٤) منها] منه: م٢ ، م٤ .

(٥) نقطة التى] نقطة ا التى: م٤ ، م٥ .

(٥) هى مركز الدائرة ولكنّ يقع] فهى. . .ولكن يقع: م١= هى . . .ولكنّ: م٥ .

(٧-٨) ونصل . . . ب ا] ونصل خطّ جا ونصل ايضاً خطّ اب: م٥ .

(٩) جاب أصغر] جاب اذا اصغر: م٢ .

(١١) نقطة جـ] نقطة د: م٤ .

(١٣) عمود] عموداً: م٣ ، م٤ ، م٥ .

(١٤) ج ز ا] ج ب ا: م٢ ، م٤ ، م٥= ج ز: م٣ .

(١٦) الضلع] ضلع: م١ .

(١٦، ١٧) الأطول] الاعظم: م٢ ، م٤ ، م٥ .

(١٧) فضلع] ضلع: م١ .

(١٧) ا ج اطول من ضلع ا ز] اج . . .اب: م١= اجا . . .از: م٤ .

hence the ratio of BA to AZ is greater than its ratio to AG;
but angles ZGB and GZA are right angles,
and GB, BZ are unequal,
and the remaining angle ZAB is greater than angle GAB;
in the same way, it will be clear that angle GAB
is less than all the angles
that are [formed] by the lines that pass through point A
and that produce right angles with line AB;
and the square on BA is equal to the square on BG
and the square on BZ is greater than the square on BG
and [by subtraction] the remaining square on ZA is less than the square on GA.
Therefore angle BAG is less than angle BAZ.
And that is what we wished to demonstrate.

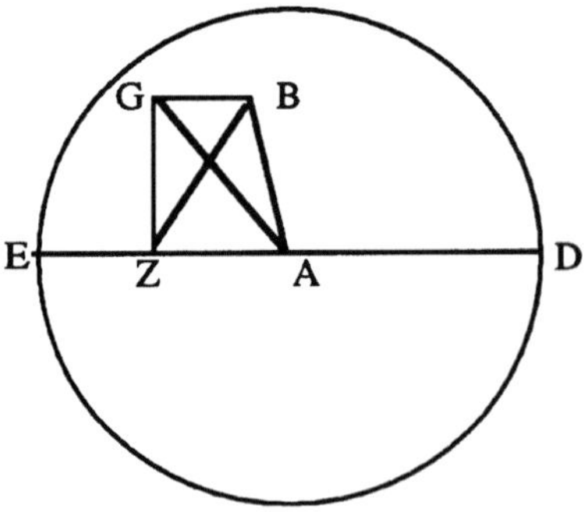

FIGURE 39

A	B	G	D	E	W	Z	H	T	I	K	L	M	N	S	O	F
ا	ب	ج	د	ه	و	ز	ح	ط	ى	ك	ل	م	ن	س	ع	ف

فنسبة ب ا إذن عند ا ز أعظم من نسبتها الى ا ج
ولكنّ زاوية ز ج ب وزاوية ج ز ا قائمتان
ويكون ج ب ب ز مختلفين
وزاوية ز ا ب الباقية اعظم من زاوية ج ا ب
وعلى مثل ذلك أيضاً ستبيّن إنّ زاوية ج ا ب
أصغر من جميع الزوايا
التى تكون من الخطوط التى تمُرّ على نقطة ا
التى تحدُث زوايا قايمةً مع خطّ ا ب
ومربّع ب ا مساو لمربّع ب ج
ومربّع ب ز أعظم من مربّع ب ج
فيبقى مربّع ز ا أصغر من مربّع ج ا
فزاوية ب ا ج أصغر من زاوية ب ا ز
وذلك ما أردنا أنْ نبيّن ٥°

---

(١٨) نسبة ب ا إذن] نسبة ب ا اذاً : م٢ ، م٣ ، م٤= فاذاً نسبة ب ا : م٥ .

(١٨) عند] إلى : م٥ .

(١٩) ولكنّ] ليكن : م١ . (١٩) ج ز ا] ج ز د : م٢ ، م٥= ج د : م٤ .

(١٩-٢٠) قائمتان ويكون] ويكون قائمتين : م١ ، م٢ ، م٣ ، م٤= يكون قائمتين : م٥ .

(٢١) ز ا ب] د ا ب : م٢= ز ا ا ب : م٣ . (٢١) ج ا ب] ج ا ز : م١= ج ا ا ز : م٣ .

(٢٢) وعلى ... ج ا ب] وعلى ... ج ا ا ب : م٣= على ... ج ا ب : م٥ .

(٢٢) ستبيّن] فستبين : م٣ .

(٢٥) قائمة] _ م٢ ، م٤ ، م٥ .

(٢٥) ا ب] ب ا : م٢ ، م٣ ، م٤ ، م٥ .

(٢٦) ب ج] ب ج ا : م= ج ا ب ز : م٢ ، م٤ ، م٥=ب ج ا ب ز ا : م٣ .

(٢٦) مساو] مساوى : م٣ .

(٢٧) ومربّع] ولمّا مربع : م١ .

(٢٧) ب ز] ب ز ا : م٣ .

(٢٨) فيبقى] فيبقا : م٣ ، م٤ .

(٢٨) مربّع ز ا] مربع ا : م٢= مربع : م٣= مربع ا ز : م٥ .

(٢٩) ب ا ج ... ب ا ز] ج ا ب ا د ... ب ا د : م٣ .

(٣٠) أن نبيّن] بيانه : م٥ .

40

We draw a circle, ABGD, and we draw in it two diameters,
AB, GD, intersecting at right angles;
and [let] the eye's position be point E;
and [let] the line we draw from point E be at right angles to line GD
but form with line AB an [arbitrary] angle, however it may fall,
and [let] line EZ be greater than half the circle's diameter;
and because line GD is at right angles to both AB [and] EZ,
all planes drawn through line GD are perpendicular to EZ, AB.
So we draw a perpendicular from point E to the plane [of the circle].
Then it is demonstrated that it falls on the common section of the planes
on AB; let it be EK;
and we draw the diameter of the circle, HT,
and let line LM be equal to the circle's diameter,
and we cut it in half at point N;
and we erect at point N a line, NS, at right angles to line LM,
and let NS be equal to line EZ.

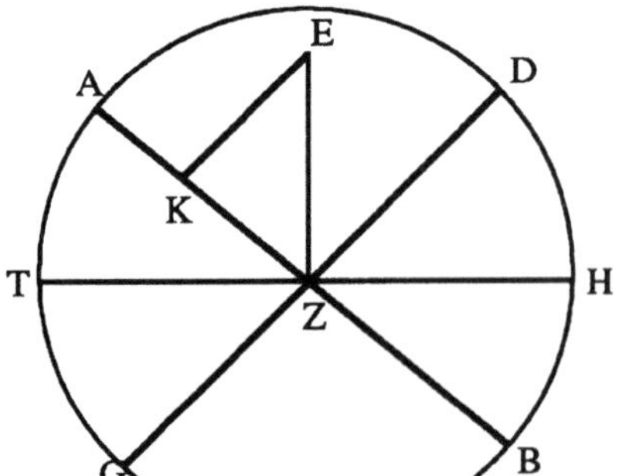

FIGURE 40.1

A	B	G	D	E	W	Z	H	T	I	K	L	M	N	S	O	F
ا	ب	ج	د	ه	و	ز	ح	ط	ى	ك	ل	م	ن	س	ع	ف

م

نخطّ دائرةً عليها ا ب جـ د ونخرِج فيها قطرين
عليهما ا ب جـ د يتقاطعان علِى زوايا قائمة
ويكون موضِع البصر نقطة ه
ويكون الخطّ الذى نخرِج من نقطة ه على زوايا قائمة عند خطّ جـ د
ويصير مع خطّ ا ب زاوِيَة كيف ما وقعت
ويكون خطّ ه ز أعظم من نصف قطر الدائرة
ولأنّ خطّ جـ د عند كلّ واحد من ا ب ه ز على زوايا قائمة
تكون جميع السطوح التي يُخرَج على خطّ جـ د قائمةً عند ه ز ا ب
فنخرِج من نقطة ه عموداً على السطح
فتبيّن أنّه يقع على تقاطِع السطوح المشتركة
على ا ب ويكون عليه ه ك
ونخرِج قطر الدائرة عليه ح ط
وليكن خطّ ل م مساوياً لقطر الدائرة
ونقطعه بنصفين على نقطة ن
ونقيم من نقطة ن خطاً على زوايا قائمة عند خطّ ل م عليه ن س
وليكن خطّ ن س مساوياً لخطّ ه ز

----

(٢) نخطّ] ونخط: م١، م٣ .
(٣) عليهما] الذى عليهما: م٣ .
(٦) ويصير مع خط ا ب] ونصير موضع نقطة: م٥ .
(٧) نصف] فس م٥ .
(٨) ولأنّ... واحد] فلان...واحد: م٢، م٣= ولان...واحدة: م٤، م٥ .
(٨-٩) على زوايا قائمة تكون جميع السطوح] _ م٤ .
(١٠) على] الى: م٢، م٣، م٤، م٥ .
(١٢) ه ك] ه د: م٣، م٤ .
(١٣) الدائره] _ م٤ .
(١٤) وليكن خطّ ل م] _ م٤ .
(١٤، ١٧) مساوياً] مساوى: م٣ .
(١٥، ١٦) من نقطة ن] على نقطة ن: م١= من نقطة ز: م٢، م٤، م٥ .
(١٦) عليه ن س] عليه ه ز ن س: م٢، م٤ .
(١٧) خطّ ن س] ن س: م٢، م٤ .
(١٧) لخطّ] به: م٢ .

Then I say that the segment enclosing line LM and passing through point S
is greater than a semicircle
because NS is greater than [either of] lines LN, NM.
So let it be LSM, and we join LS, SM,
and in this way, the angle enclosed by lines LS, SM
is equal to the angle [at] E, which is enclosed by DE, EG;
and let angle HZE be equal to angle LNO,
and we set NO equal to EZ, and we join LO, OM;
and we draw around triangle LOM a segment of a circle, LOM.
Then the angle at point O is equal to the angle contained by HE, ET.
And let angle BZE be equal to angle LNF,
and [let] line NF be equal to line EZ, and we join LF, FM,

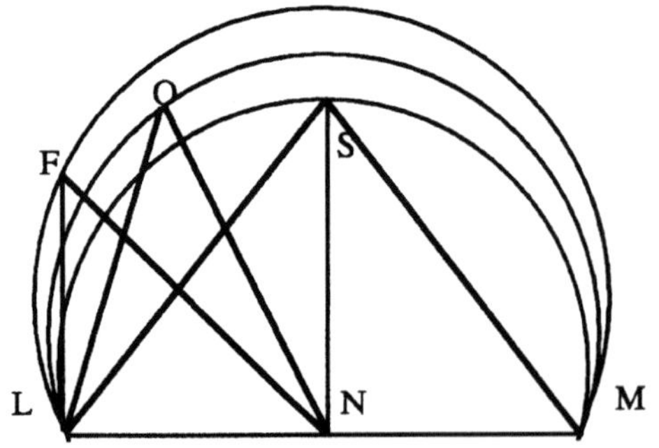

FIGURE 40.2

A	B	G	D	E	W	Z	H	T	I	K	L	M	N	S	O	F
ا	ب	ج	د	ه	و	ز	ح	ط	ى	ك	ل	م	ن	س	ع	ف

فأقول إنّ القطعة التى تحيط بخطّ ل م وتقع على نقطة س
هى أعظم من نصف الدائرة
من أجْل أنّ ن س أعظم من خطّى ل ن ن م
فليكن عليها ل س م ونصل ل س س م
وكذلك تكون الزاوية التى يحيط بها خطّا ل س س م
مساويةً لزاوية ه التى يحيط بها د ه ه ج
وليكن زاوية ح ز ه مساويةً لزاوية ل ن ع
ونجعَل ن ع مساوياً ل ه ز ونصل ل ع ع م
ونخطّ على مثلث ل ع م قطعة الدائرة عليها ل ع م
فتكون زاوية نقطة ع مساويةً للزاوية التى يحيط بها خطّا ح ه ه ط
وليكن زاوية ب ز ه مساويةً لزاوية ل ن ف
ويكون خطّ ن ف مساوياً لخطّ ه ز ونصل ل ف ف م

---

(٢٠) من خطّى ل ن ن م] من كل واحد من ا ب م ز: م٢= من كل من ا ب: م٥ .
(٢١) فليكن ... س م] تس م١ ، ـ م٣ .
(٢٢-٢٣) وكذلك ... د ه ه ج] فزاوية ل س التى من خطى ل س س م متساوياً لزاوية ه التى يحيط ه ج د: م٢= فزاوية ل س ... يحيط ب ه ج د: م٤= فزاوية ... ز ه ج د: م٥ .
(٢٣) بها] ـ م٣ .
(٢٣) ه] فس م١ .
(٢٣) د ه ه ج] فس م١ .
(٢٤) وليكن] ولتكن: م٤ .
(٢٤) ح ز ه] ح ز د ه: م٣= ج ز ه: م٥ .
(٢٤) لزاوية ل ن ع] ل ز ن ع: م٢ ، م٤= لزاوية ل ن ن ع: م٣ .
(٢٤-٢٥) ل ن ع ... ن ع] ل ز ب ع ... ب ع: م٥ .
(٢٥) ل ه ز] ل ن س: م٢= ل ب ز: م٣ ، م٤ ، م٥ .
(٢٦) الدائرة] ـ م٢= دايره: م٣ ، م٤ ، م٥ .
(٢٧) نقطة] ـ م٥ .
(٢٧-٢٨) للزاويه التى ... لزاوية ل ن ف] لزاوية ا ن ف: م٢= للزاويه التى ... لزاوية ا ن ن ف: م٣= لزاوية ز ن ف: م٤ ، م٥ .
(٢٩) مساوياً] مساو م١= متساوياً: م٢= مساوى: م٣ .

and we draw around triangle LFM a segment of a circle, LFM.
Then angle F becomes equal to angle AEB.
Now because angle S is greater than angle O,
and angle S is equal to angle GED, and angle O is equal to angle HET,
then GD is seen as greater than HT;
again, since angle O is equal to angle HET,
and angle F is equal to angle AEB,
[then] HT is seen as greater than AB.
Therefore GD is seen as greater than all the lines that pass through Z
and produce angles with line EZ, and AB is seen as smaller than all of them.
And that is what we wished to demonstrate.

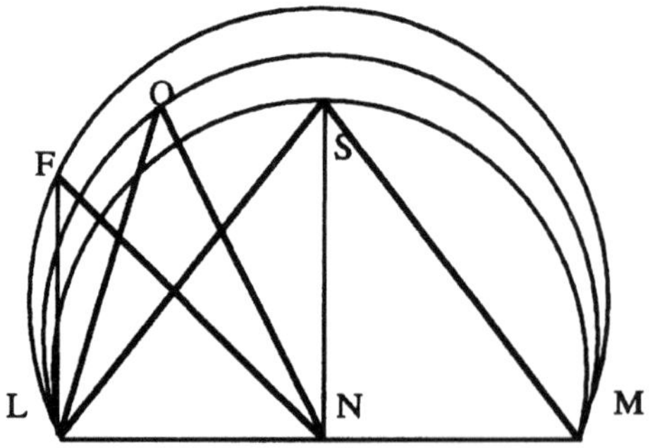

FIGURE 40.2

A	B	G	D	E	W	Z	H	T	I	K	L	M	N	S	O	F
ا	ب	ج	د	ه	و	ز	ح	ط	ى	ك	ل	م	ن	س	ع	ف

ونخطّ على مثلث ل ف م قطعة الدائرة عليها ل ف م
فتصير زاوية ف مساويةً لزاوية اه ب
فلأنّ زاوية س أعظم من زاوية ع
وزاوية س مساوية لزاوية جه د وزاوية ع مساوية لزاوية ح ه ط
فيكون جد يرى أعظم من ح ط
وأيضاً لأنّ زاوية ع مساوية لزاوية ح ه ط
وزاوية ف مساوية لزاوية اه ب
يكون يرى ح ط أعظم من ا ب
فيكون جد يرى أعظم من جميع الخطوط التى تمُرّ على نقطة ز
وتحدُث زوايا مع خطّ ه ز و ا ب يكون يرى أصغر من جميعها
وذلك ما اردنا أنْ نبيّن ه

---

(٣٠) الدائرة] دائرة: م٣ ، م٥ .
(٣٠) مثلث ل ف م] مثلث ل ع م: م٤ ، م٥ .
(٣١) اه ب] اه ز: م٢ ، م٥ .
(٣٢-٣٣) أعظم من زاوية ع وزاوية س] ـ م٤ .
(٣٣) وزاوية س] ولكن زاوية س: م٢ .
(٣٣) جه د] جه ب: م٢ ، م٤ ، م٥ .
(٣٣، ٣٥) ح ه ط] جه ط: م٢ ، م٤ ، م٥ .
(٣٤، ٣٧) ح ط] ج ط: م٥ .
(٣٤-٣٥) فيكون ...ح ه ط] ـ م٣ .
(٣٥-٣٦) زاوية ع متساوية لزاوية ح ه ط وَ] ـ م٤ .
(٣٧) يكون يرى ح ط اعظم] يكون ح ط يرى اعظم: م٣ .
(٣٨) يرى] ترى: م٤ .
(٣٨) فيكون] ويكون: م٥ .
(٣٩) وتحدث] فس: م١ .
(٤٠) أن نبيّن] بيانه: م٥ .

41

When the line drawn from the eye's position to the center is not
greater than half the diameter, but less,
then the opposite of what was seen in the case of the diameters occurs;
that is to say, that which was seen as the greater diameter
is seen as the smaller,
and the one which was seen as the smaller is seen as the greater in this case.
For let there be a circle, ABGD,
and we draw in it two diameters, AB, GD,
intersecting at right angles;
and we draw another arbitrary diameter, EZ,
the eye's position being point T,
and the line drawn from it to the center, HT,
is less than half the circle's diameter;
and let line KL be equal to the circle's diameter,
and we cut it in half at point M,

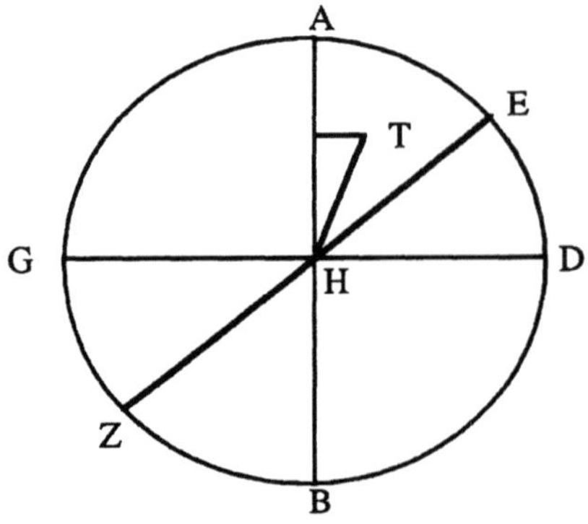

FIGURE 41.1

A	B	G	D	E	W	Z	H	T	I	K	L	M	N	S	O	F
ا	ب	ج	د	ه	و	ز	ح	ط	ى	ك	ل	م	ن	س	ع	ف

ما

إذا لم يكن الخطّ الذى يُخرَج من موضع البصر إلى المركز
أعظم من نصف القطر ولكنّ أقلّ
فيعرض فى الأقطار خلاف ما كان يرى
وذلكَ أنّه يكون الذى كان يرى أعظم من الأقطار
يرى أصغر
فالذى كان يرى أصغر يرى فى هذه الجهة أعظم
فلتكن دائرةٌ عليها ا ب جـ د
ونخرِج فيها قطرين عليهما ا ب جـ د
يتقاطَعان على زوايا قائمة
ونخرِج قطراً آخر كيف ما وقع عليه ه ز
ويكون موضع البصر نقطة ط
ويكون الخطّ الذى يُخرَج منها إلى المركز عليه ط ح
وهو أصغر من نصف قطر الدائرة
وليكن خطّ ك ل مساوياً لقطر الدائرة
ونقطعه بنصفين على نقطة م

---

(٢) إذا لم يكن] ثم لايكون: م١ ، م٥= لا يكون: م٣ ، م٤ .
(٢) إلى المركز] ـ م٥ .
(٢) نخرِج] يخرج: م٤ .
(٣) لكنّ] + يكون: م٢ .
(٦) يرى أصغر] يرى اعظم اصغر: م٤ .
(٧) فالذى كان يرى أصغر] والذى كان يرى اصغر: م٥ .
(٩) فيها] منها: م٢ .
(١١) قطراً] قطر: م١ ، م٢ ، م٣ ، م٤ .
(١١) ه ز] + ويكون موضع البصر نقطة ط و يكون ا لخطّ الذى خرج منهما الى المركز عليه ه ز: م٣ .
(١٣) منها] منهما: م٣ .
(١٤) الدائرة] دائرة: م١ .
(١٥، ٢٢) مساوياً] مساو: م٤ .

and from M we erect line MN at right angles, this being equal to line TH;
and we draw around line KL and point N a segment of a circle, KNL;
then this is less than a semicircle,
because MN is less than half the circle's diameter;
so angle N, which is contained by KNL,
is equal to angle T, which is contained by GTD.
Again, we set angle KMS equal to angle ZHT
and let MS be equal to HT;
and we draw around line KL and point S a segment of a circle, KSL:
then angle S, enclosed by KSL, is
equal to angle T, enclosed by ZTE.

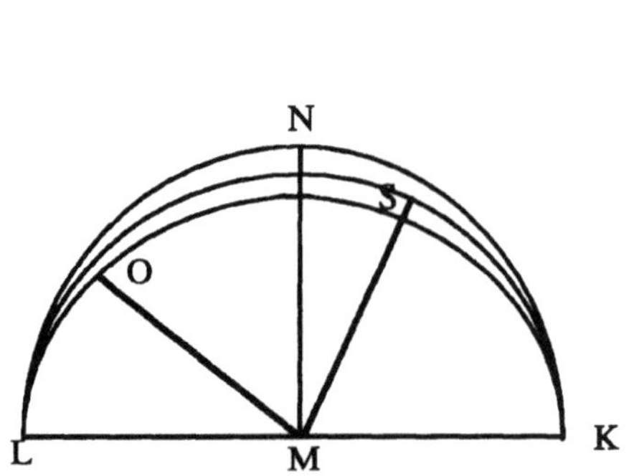

FIGURE 41.2

A	B	G	D	E	W	Z	H	T	I	K	L	M	N	S	O	F
ا	ب	ج	د	ه	و	ز	ح	ط	ى	ك	ل	م	ن	س	ع	ف

ونقيم من نقطة م خطّ م ن على زوايا قائمة ويكون مساوياً لخطّ ط ح

وندير على خطّ ك ل وَ نقطة ن قطعة دائرة عليها ك ن ل

فتكون أقلّ من نصف الدائرة

من أجْل أنّ م ن أصغر من نصف قطر الدائرة

فتكون زاوية ن التى يحيط بها ك ن ل

مساويةً لزاوية ط التى يحيط بها جـ ط د

ونجعَل أيضاً زاوية ك م س مساويةً لزاوية ز ح ط

وليكن خطّ م س مساوياً لخطّ ح ط

وندير على خطّ ك ل ونقطة س قطعة دائرة عليها ك س ل

فتكون زاوية س التى يحيط بها ك س ل

مساويةً لزاوية ط التى يحيط بها ز ط ه

---

(١٧) ويكون مساوياً] وتكون مساوياً: م٣ = يكون مساوياً: م٥ .

(١٨) ك ل وَ نقطة ن] ك ل و نقطة م: م٣ .

(١٨) ك ن ل] ك ل: م٢ ، م٣ ، م٤ ، م٥ .

(١٩) فتكون] يكون: م٢ = تكون: م٤ .

(١٩) الدائرة] دائرة: م١ ، م٣ ، م٥ .

(٢١) ن] _ م٣ .

(٢١) يحيط بها ك ن ل] ك ز ل مساويةً لها: م٥ .

(٢٢) جـ ط د] جـ ط ك: م٢ ، م٤ = خط ح ط ط ك: م٥ .

(٢٣) نجعل أيضاً] ايضاً نجعل: م٢ ، م٤ ، م٥ .

(٢٣) ك م س] ك م م س: م٣ .

(٢٣) ز ح ط] جـ ح ط: م٣ .

(٢٤) وليكن ...ح ط] _ م١ .

(٢٤) قطعة] _ م٤ .

(٢٤-٢٥) وليكن خطّ م س مساوياً لخطّ ح ط وندير على خط ك ل وَ نقطة س قطعة دائرة عليها ك س ل] _ م١ ، م٣ .

(٢٥) خطّ] _ م٣ .

(٢٥) ونقطة س قطعة دائرة] ونقطة س دائرة: م٤ .

(٢٧) لزاوية ط] + لزاوية لزاوية ع: م٢ = لزاوية ع: م٤ = مساويةً ع: م٥ .

(٢٧) ز ط ه] ز ط ط ه: م٣ ، م٥ .

And again, let angle KMO be equal to angle AHT, line MO being equal to line HT, and we draw around line KL and point O a segment of a circle, KOL; then angle O, contained by KOL, is equal to angle T, and angle S is [also] equal to angle T, contained by ETZ. Therefore, AB is seen as greater than EZ; and again, because angle T, contained by ETZ, is greater than angle G, contained by GTD, then EZ is seen as greater than GD. And that is what we wished to demonstrate.

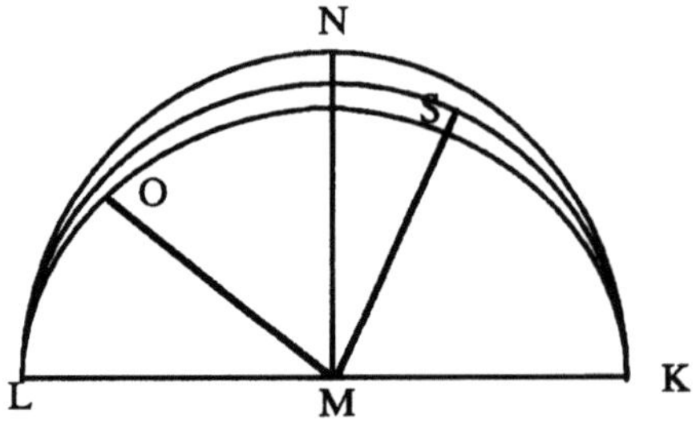

FIGURE 41.2

A	B	G	D	E	W	Z	H	T	I	K	L	M	N	S	O	F
ا	ب	ج	د	ه	و	ز	ح	ط	ى	ك	ل	م	ن	س	ع	ف

وأيضاً فليكن زاوية كمع مساويةً لزاوية احط
ويكون خطّ مع مساوياً لخطّ حط
وندير على خطّ كل وعلى نقطة ع قطعة دائرة عليها كعل
فتكون زاوية ع التى يحيط بها كعل مساويةً لزاوية ط
وزاوية س مساويةً لزاوية ط التى يحيط بها هطز
فيكون يرى اب أعظم من هز
وأيضاً لأنّ زاوية ط التى يحيط بها هطز
أعظم من زاوية جـ التى يحيط بها جـطد
يكون هز يرى أعظم من جد
وذلك ما اردنا أنْ نبيّن ٥٥

---

(٢٨) احط] احح ط: م٣ .
(٢٨) كمع] كمس: م٢= كممن: م٣= كمو: م٤= كمن: م٥ .
(٢٩) مساوياً لخطّ] مساو خطّ: م٤ .
(٣٠) على نقطة ع] نقطة ع: م٥ .
(٣١) فتكون] فيكون: م٢ .
(٣١) ع التى يحيط بها] ــ م٥ .
(٣١) ط] + التى يحيط بها اطب: م١= التى يحيط اطب م٣ .
(٣٣) يرى اب] يرى م١= اب يرى: م٥ .
(٣٤) هطز] هز: م٢ ، م٤ ، م٥ .
(٣٥) جـ] ط: م٢ ، م٣ ، م٤ ، م٥ .
(٣٥) جـطد] جد: م٢ ، م٤ ، م٥ .
(٣٦) هز] ــ م٢ ، م٤ ، م٥ .
(٣٧) أن نبيّن] بيانه: م٥ .

42

Wheels of chariots are seen at times as circular and at other times as distorted.

Let there be a wheel, ABGD, and we draw in it two diameters, AB, GD, intersecting at right angles at point E; and let the eye be on a plane parallel to the plane of the circle. Then if the line drawn from the eye's position to the circle's center is perpendicular to the circle's surface, or else equal to half its diameter but not perpendicular, then the diameters are seen as equal; therefore it is demonstrated that the wheel is seen as circular; but if the line drawn from the eye's position to the circle's center is not perpendicular to the circle's surface and not equal to half its diameter, then the diameters are seen as unequal; therefore it is demonstrated that the wheel is seen in this case as distorted. And that is what we wished to demonstrate.

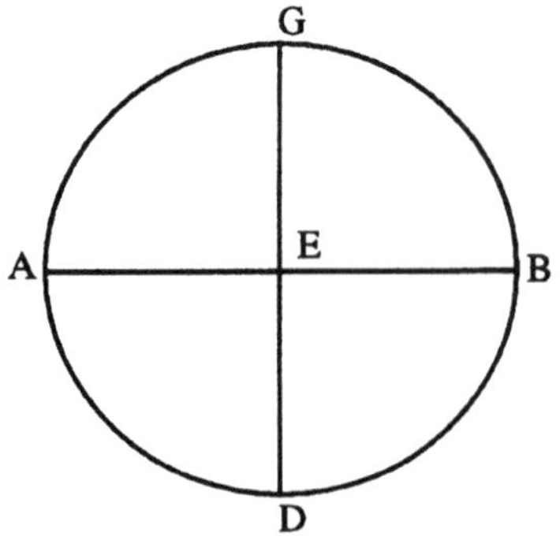

FIGURE 42

A	B	G	D	E	W	Z	H	T	I	K	L	M	N	S	O	F
ا	ب	ج	د	ه	و	ز	ح	ط	ى	ك	ل	م	ن	س	ع	ف

مب
بكرات العجل ترى مرّةً مستديرةً ومرّةً معوجّةً
فليكن بكرة عليها ا ب ج د ونخرج فيها قطرين
عليهما ا ب ج د يتقاطعان على زوايا قائمة على نقطة ه
وليكن البصر على سطح موازٍ لسطح الدائرة
فإنْ كان الخطّ الذى يُخرَج من موضع البصر إلى مركز الدائرة
عموداً على سطح الدائرة
أو مساوياً لنصف قطرها ولا يكون عموداً
فإنّ اقطارها ترى متساويةً
فتبيّن أنّ البكرة ترى مستديرةً
وإنْ لم يكن الخطّ الذى يُخرَج من البصر إلى مركز الدائرة
عموداً على سطح الدائرة
ولا مساوياً لنصف قطر الدائرة
فانّ الأقطار ترى مختلفةً
فتبيّن أنّ البكرة ترى كذلك معوجّةً
وذلك ما أردنا أنْ نبيّن ە°

----

(٢) ترى مرّة مستديرة ومرّة معوجّةً] ترى مرة معوجة ومرة مستديرة: م١= مرة ترى مستديرة ومرة ترى معوجة:م٢ ، م٤ ، م٥= مرة يرى معوجة ومرة مستديرة: م٣ .

(٣) فليكن] فلتكن: م٥ .

(٣-٤) ونخرج ...ا ب ج د] هش م١ .

(٥) مواز] موازى: م٣ .

(٦) كان] _ م٢ . (٦) إلى] على: م٥ .

(٧) عموداً] عمود: م٢ .

(٨) أو مساوياً] ومساوياً: م٢ .

(٩) أقطار] الاقطار: م١ .

(١٠) فتبيّن] فبين: م٥ .

(٩، ١٤) ترى] يرى: م٣ .

(١٢) الدائرة] _ م٤ ، م٥ .

(١٥) ترى] + كذلك: فس: م١ .

(١٥) ترى كذلك] ليرى كذلك: م٣= ترى: م٥ .

(١٦) أن نبيّن] _ م٢= وذلك ما اردنا بيانه: م٥ .

43

There is a place for the eye [such that], once it is fixed in it
while a viewed [object] is being moved,
then the viewed [object] is seen in different positions
always as the same [in size].
For let the eye's position be point A, and the object, magnitude BG,
and let ray[s] AB, AG fall;
and we draw around ABG a circle, ABG.
Then I say that when the eye is fixed in its position,
and the magnitude to be viewed is moved, it is always seen as the same [in size].
For let the magnitude be moved from its position, and let it be GD,
and we join ray AD.
Now because BG is equal to GD, arc BG is equal to arc DG.
And since two equal angles fall upon the two equal arcs GB and DG,
namely, the two angles DAG and BAG,
then angle BAG is equal to angle DAG,

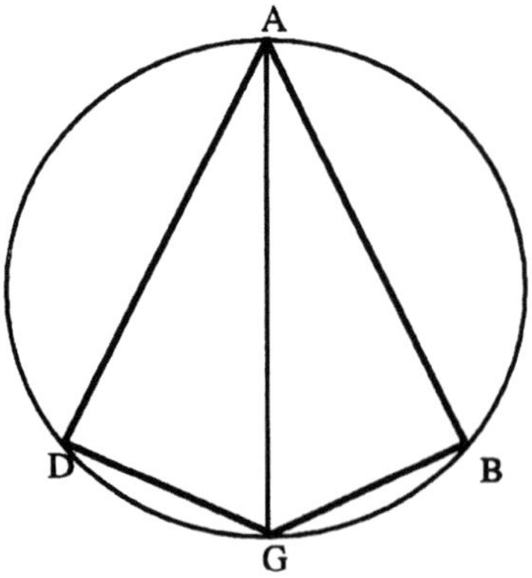

FIGURE 43

A	B	G	D	E	W	Z	H	T	I	K	L	M	N	S	O	F
ا	ب	ج	د	ه	و	ز	ح	ط	ى	ك	ل	م	ن	س	ع	ف

مج

للبصر موضعٌ إذا هو ثبَت فيه
وكان المنظور إليه منتقلاً
يرى المنظور إليه فى مواضع مختلفةٍ
متساوياً أبداً
فليكن موضع البصر نقطة ا والمنظور إليه قدر ب ج
وليقع شعاع ا ب ا ج
وندير على ا ب ج دائرةً عليها ا ب ج
فأقول إنّه إذا ثبَت البصر فى موضعه
وانتقل العظَم المنظور إليه يكون يرى متساوياً أبداً
فلينتقل العَظَم من موضعه ويكون عليه ج د
ونصل شعاَع ا د
فلأنّ ب ج مساو ل ج د يكون قوس ب ج مساويةً لقوس د ج
ولأنّه وقعت زاويتاَن على قوسى ج ب ج د المتساويتين
وهما زاوية د ا ج و زاوية ب ا ج
تكون زاوية ب ا ج مساويةً لزاوية د ا ج

---

(٢) للبصر ... ثبت فيه] كذلك للبصر ... ثبت فيه: م١ .

(٥) متساوياً] متساوية: م٣ .

(٧) وليقع] + عليه: م٥ .

(٨) دائرة عليها ا ب ج] هش م٢ .

(٩) موضعه] موضع: م٢ .

(١٠) متساوياً أبداً] ابداً متساوياً: م٥ . (١٠) يكون] ـ م٥ .

(١١) فلينتقل] فينتقل: م٤= فننقل: م٥ .

(١٣) مساو ل جد] مساوى لج: م٣ . (١٣) يكون] تكون: م٥ .

(١٣) مساوية لقوس جد] مساوية لقوس دج: م١ ، م٣ ، م٥= مساو لقوس جد: م٢ .

(١٤) ولأنّه وقعت زاويتان على قوسى جب جد المتساويتين] وقعت زاويتين على قوس جب جد المتساويتين: م٢ ، م٤= فلانه وقعت على قوسى جب جد المتساويتين زاويتان: م١ ، م٣= قد وقعت زاويتان على قوس جب جد المتساويتين: م٥ .

(١٥) زاوية ... لزاوية] زاويتا ... لزاويتا: م٢ . (١٦) ب ا ج] هش م٢ .

therefore GB is seen as equal to GD in this way;
then again, let the position of the eye be point B
and of the viewed [object], AG;
then we move the eye's position and it becomes point D,
and we fix the viewed [object].
Now if AG is the circle's diameter,
then angles ABG and ADG are right, and AG is seen as equal [from B and D],
but if it is not the circle's diameter,
but the two rays AB, BG are equal to the two rays AD, DG,
and AG is common, [then] angle ADG is equal to angle ABG;
therefore it is clear that the viewed [object] is seen as equal [from B and D].
And that is what we wished to demonstrate.

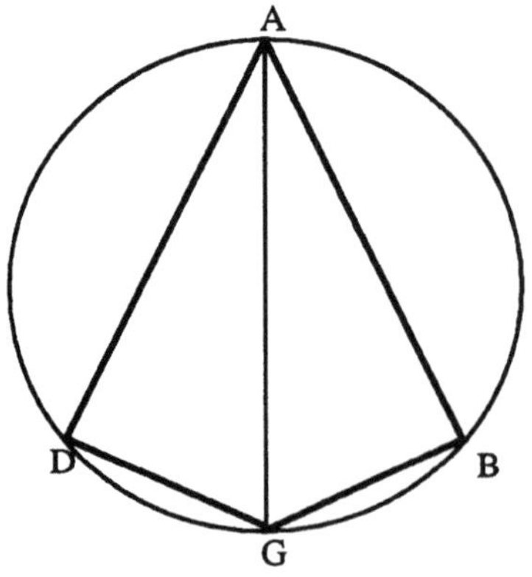

FIGURE 43

A	B	G	D	E	W	Z	H	T	I	K	L	M	N	S	O	F
ا	ب	ج	د	ه	و	ز	ح	ط	ى	ك	ل	م	ن	س	ع	ف

فيرى جـ ب كذلك مساوياً لـ جـ د
وليكن أيضاً موضع البصر نَقطة ب
والمنظور إليه ا جـ
ثمّ ننقُل موضع البصر ويصير نقطة د
وتثبت المنظور إليه
فإنْ كان ا جـ قطر الدائرة
فإنّ زاوية ا ب جـ وزاوية ا د جـ قائمتان ويرى ا جـ متساوياً
وإنْ لم يكن قطر الدائرة
ولكنّ كان شعاعا ا ب ب جـ مساويين لشعاعى ا د د جـ
وكان ا جـ مشتركاً تكون زاوية ا د جـ مساوية لزاوية ا ب جـ
فبيّن أنّ المنظور إليه يكون يرى متساوياً
وذلك ما اردنا أنْ نبيّن ٥°

---

(١٧) فيرى] ويرى: م١ .

(١٧) مساوياً لـ جد] مساو لـ جد: م١= مساوياً لـ جـ: م٣ .

(١٨) وليكن] وَلكنّ: م١ ، م٢ ، م٣ ، م٤ .

(١٨) أيضاً موضع البصر] يكون ايضاً موضع البصر: م٢ ، م٤= موضع البصر ايضاً: م٥ .

(٢٠) موضع] ـ م٢ ، م٥ .

(٢٠) يصير] نصيره: م٢ ، م٤ ، م٥ .

(٢٢) ا جـ] اد: م٢ ، م٤ ، م٥ .

(٢٢، ٢٤) قطر الدائرة] قطراً: م٢ ، م٤ ، م٥ .

(٢٣) فإنّ زاوية] وزاوية: م٣ .

(٢٣) ا ب جـ وزاوية ا د جـ قائمتان] ا د جـ...ا ب جـ قائمتين: م٢ ، م٣ ، م٤ ، م٥ .

(٢٣) متساوياً] متساوى: م٣ .

(٢٥) شعاعا ...لشعاعى] شعاعى ...لشعاعى: م٢ ، م٣ ، م٤ ، م٥ .

(٢٦) ا جـ] ـ م٣ .

(٢٦) مساويين] متساويين: م٢ ، م٤ .

(٢٨) وذلك ما أردنا أن نبيّن] ـ م٢= وذلك ما اردنا بيانه: م٥ .

44

When there is a raised magnitude being perpendicular to the plane underneath, and then the eye comes to be [placed] on a point on this plane, and the viewed [object] is moved on a circle, then the viewed [object] is always seen as the same [in size]. For let there be a magnitude to be viewed that is AB in height and perpendicular to the plane underneath, and let the eye's position be point G; and we join GB, and we draw a circle, BD, on center G and with radius GB. Then I say that if magnitude AB is rotated and moved around the circle's circumference, then from the eye's position, point G, it is always seen as the same [in size]. Now because AB is perpendicular, it produces a right angle with BG,

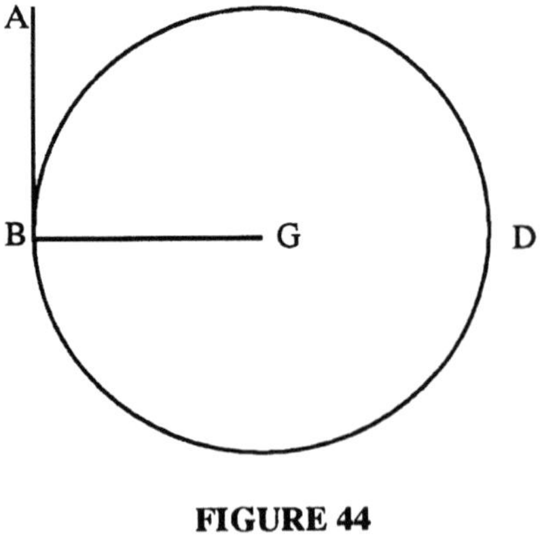

FIGURE 44

A	B	G	D	E	W	Z	H	T	I	K	L	M	N	S	O	F
ا	ب	ج	د	ه	و	ز	ح	ط	ى	ك	ل	م	ن	س	ع	ف

مد

إذا كان عظَمٌ مرتفعٌ
وكان قائماً على السطح السفلى
ثمّ صير البصر على نقطة من السطح
ونُقل المنظور إليه على استدارة
فإنّ المنظور إليه يرى أبداً متساوياً
فليكن عظَم منظور إليه وهو ا ب فى ارتفاع
وقائم علَى السطح السفلى
وليكن موضع البصر نقطة جـ ونصل جـ ب
وعلى مركز جـ وببعد جـ ب ندير دائرةً عليها ب د
فأقول إنّه إنْ ادير عظَم ا ب
ونُقل على محيط الدائرة
يكوَن يرى من موضع البصر نقطة جـ أبداً متساوياً
فلأنّ ا ب قائم يحدث مع ب جـ زاويةً قائمةً

---

(٢) كان] ـ م٥ .
(٤) نقطة] نقط : م٢ .
(٥-٦) على استدارة فإنّ المنظور إليه] ـ م٤ .
(٦) يرى أبداً متساوياً] ابداً يرى متساوياً : م١ ، م٣ = يرى متساوياً ابداً : م٥ .
(٧) عظم] عظماً : م٣ .
(٧) منظور إليه وهو ا ب] منظور اليه عليه ب : م١ = منظورا اليه عليه اب : م٣ .
(٩) وليكن] وليكن على : م٤ .
(٩) موضع] على موضع : م٤ .
(٩) ونصل] ولنصل : م٣ .
(١٠) وببعد جـ ب ندير دائرة عليها ب د] ـ م٤ .
(١٠) ببعد] بعد : م١ . (١٠) ب د] جـ : م٢ ، م٥ .
(١٢) على] عن : م٣ .
(١٣) يكون] ـ م٢ ، م٤ .
(١٣) البصر] بصر : م١ ، م٣ ، م٤ ، م٥ .
(١٣) نقطة جـ] ـ م٢ = جـ : م٥ .
(١٤) فلأنّ] فان : م٤ . (١٤) قائم] + قائم : م٥ .
(١٤) زاويه] دائره : م١ .

and for every ray drawn from center G to magnitude AB
its [formed] angles become equal,
so the viewed [object] is seen as the same [in each case];
and in this way again, if a line is erected from point G at right angles,
and the eye is placed on it,
and [the line] is parallel to the magnitude being seen and [then] moved,
then it is always seen as the same [in size].
And that is what we wished to demonstrate.

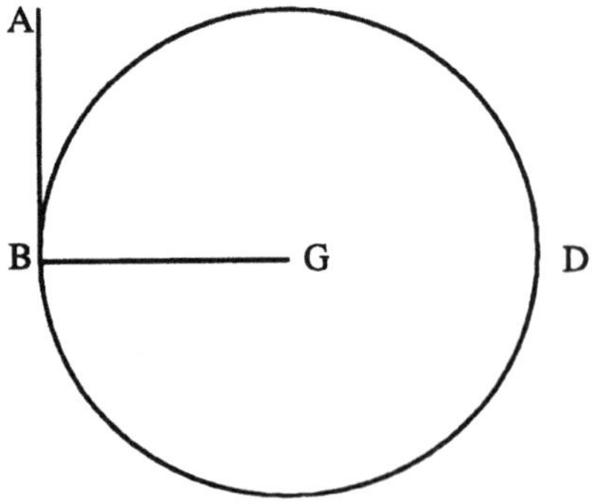

FIGURE 44

A	B	G	D	E	W	Z	H	T	I	K	L	M	N	S	O	F
ا	ب	ج	د	ه	و	ز	ح	ط	ى	ك	ل	م	ن	س	ع	ف

فكلّ الشعاع الذى يُخرَج من مركز جـ إلى عِظَم آب
يصير زواياه متساويةً
فيكون المنظور إليه يرى متساوياً
وكذلك أيضاً إنْ أُقيم من نقطة جـ خطّ على زوايا قائمة
ووُضِع البصر عليه
وكانَ موازياً للعظَم الذى يرى وينقل العِظَم
فإنّه يرى متساوَياً أبداً
وذلك ما أردنا أنْ نبيّن ٥٥

----

(١٦) يصير] تصير: م٥ .
(١٧) يرى] ـ م٥ .
(١٨) خطّ] خطاً: م١ ، م٣ .
(١٨) قائمة] فس م١ .
(١٨) زوايا] زاوية: م١ .
(١٨-٢١) وكذلك ... متساوياً أبداً]
وذلك ما اردنا بيانه وكذلك ...طابداً متساوياً: م٥ .
(٢٠) ينقل] نقل: م٣ .
(٢١) فإنّه يرى] فانه يكون يرى: م٣ .
(٢٢) وذلك ما أردنا أن نبيّن] ـ م٢ ، م٥ .

45

We draw a circle, AD,
and we mark point D on its circumference,
and from it we erect line DZ not at right angles to the circle,
the eye's position being point E.
Then I say that if line DZ is moved on the circle's circumference
then it is seen at times as equal [to],
at times as greater [than], and at other times as smaller [than itself].
Now line DZ is either equal to half the circle's diameter
or greater than it, or smaller than it.
So let it be greater, and we draw from point E, which is the center,
a line parallel to line DZ, namely EG, and let EG be equal to DZ;
and we draw from point G to the plane a perpendicular, GH ,
and let it fall on the plane at point H, and we join line EH;
then we extend it rectilinearly
and let it fall on the circle's circumference at point A;
and we draw from point A a line parallel to line GE, namely AB,
and let line AB be equal to line DZ.

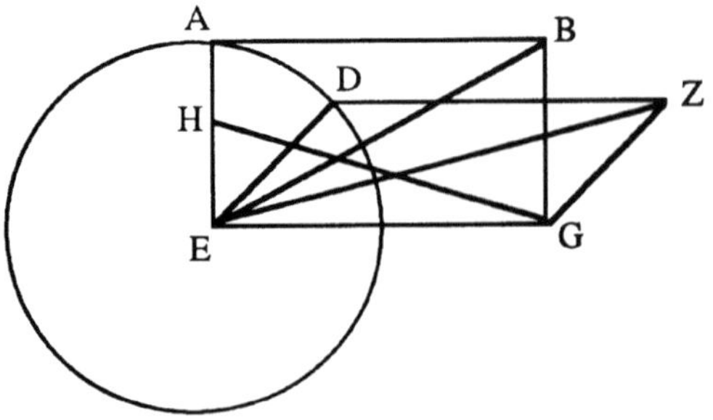

FIGURE 45.1

A	B	G	D	E	W	Z	H	T	I	K	L	M	N	S	O	F
ا	ب	ج	د	ه	و	ز	ح	ط	ى	ك	ل	م	ن	س	ع	ف

مه

ونخطّ دائرةً عليها اد
ونتعلّم على محيطها نقطة د
ونقيم منها خطّ د ز على غير زوايا قائمة عند الدائرة
ويكون موضع البصر نقطة ه
فأقول إنّ خطّ د ز إذا انتقل على محيط الدائرة
فمرّةً يرى متساوياً
ومرّةً أعظم ومرّةً أصغر
وخطّ د ز إمّا أنْ يكون مساوياً لنصف قطر الدائرة
وإمّا أعظم منه وإمّا أصغر منه
فليكن أعظم ونخرج من نقطة ه التى هى مركز
خطّاً موازياً لخطّ د ز عليه ه ج وليكن ه ج مساوياً لِ د ز
ونخرج من نقطة ج إلى السطح عموداً عليه ج ح
وليقع على السطح على نقطة ح ونصل خطّ ه ح
ثمّ نخرجه على استقامة
وليقع على محيط الدائرة على نقطة ا
ونخرج من نقطة ا خطّاً موازياً لخطّ ج ه عليه ا ب
وليكن خطّ ا ب مساوياً لخطّ د ز

---

(٢) ونخطّ] فنخط: م٢، م٥ .
(٦) انتقل] ـ م٣ .
(٧) يرى متساوياً] ترى متساوية: م٢ .
(١٠) منه] ـ م٥ .
(١١) فليكن اعظم] فليكن اولاً اعظم: م٥ .
(١١) ه] ـ م١، م٣ .
(١١) هى مركز] هى المركز: م٢، م٥= فى المركز: م٣ .
(١٢) د ز] زد: م٥ .
(١٣) من نقطة ج] د: م٣ .
(١٤) خطّ] ـ م٥ .
(١٦) وليقع على محيط الدائرة] الى محيط الدائرة: م٢، ؟م٤، م٥ .
(١٨) خطّ ا ب] اب: م٣ .

Then I say that AB is seen as smaller than all the lines that are moved on the circle's circumference. For let us join lines ED, EZ, EB, BG, GZ. Now we have [already] demonstrated in proposition thirty six that angle GEA is smaller than all the angles produced from the lines that pass through E and line EG. And because GE is parallel to AB and equal to it EA is also equal and parallel to GB; thus, AEGB is a parallelogram, and as a result GEDZ is a parallelogram; [further], because it is required that we demonstrate that magnitude AB is <seen as> smaller than magnitude DZ, [then] it is required that we demonstrate that angle AEB is smaller than angle DEZ; and since it has been demonstrated that angle AEG is smaller than all angles produced by lines that pass through point E and line EG, then angle GEA is less than angle GED.

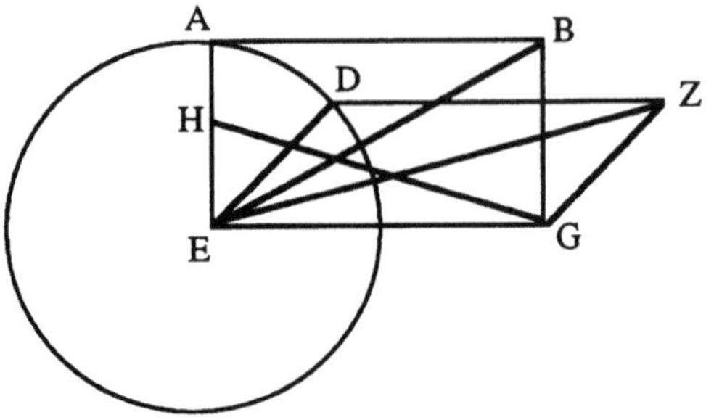

FIGURE 45.1

A	B	G	D	E	W	Z	H	T	I	K	L	M	N	S	O	F
ا	ب	ج	د	ه	و	ز	ح	ط	ى	ك	ل	م	ن	س	ع	ف

فأقول إنّ اب يرى أصغر
من جميع الخطوط التي ينتقل على محيط الدائرة
فلنصل خطوط ەد ەز بج ەب جز
وقد بيّنا فى الشكل السادس والثلثين
أنّ زاوية جەا أصغر من جميع الزوايا الحادثات من الخطوط
التي هي على نقطة ە ومن خطّ ەج
فلأنّ جە يوازى اب ومساو له
فَ ەا اذاً مساو ل جب ومواز له
فَ اەجب متوازِى الأضلاع
ومن أجْل ذلك يكون جەدز متوازى الاضلاع
ولأنّه ينبغى أنْ نبيّن أنّ
عظَم اب <يرى> أصغر من عِظَم دز
ينبغى أنْ نبيّن أنّ
زاوية اەب أصغر من زاوية دەز
فلأنّه قد استبان أنّ زاوية اەج أصغر من جميع
الزوايا الحادثات من الخطوط التي تمُرّ على نقطة ە ومن خطّ ەج
تكون زاوية جەا أصغر من زاوية جەد

----

(١٩) يرى] ـ م٤ .

(٢١) فلنصل] فنصل: م٢ ، م٣ ، م٤ .

(٢١) ەد ...جز] ەد ەز بج ەز جز: م١= ەد ەز بج زج ەب: م٢ ، م٤ ، م٥ .

(٢٢) السادس والثلثين] ٣٦ : م٥ .

(٢٣) جەا] جە: م١= دەا: م٢ ، م٤ ، م٥= جەەا: م٣ .

(٢٤) خطّ ەج] ەج: م٥ .

(٢٥) فلأنّ جەا] فلان خطّ جە: م٢ .

(٢٥) مساو له] يساويه: م٥ .

(٢٦) مواز] موازى: م١ ، م٣ .

(٢٧، ٣٣) فـ] و: م٣ .

(٢٨) ومن أجل...الأضلاع] ـ م٢ ، م٤ ، م٥ .

(٢٩) لأنّه ينبغى] لا ينبغى: م٥ .

(٣١) ينبغى] وينبغى: م١ ، م٤ .

(٣١) ينبغى أن نبيّن أنّ] ينبعى ان يكون: م٢ ، م٤ .

So let KOL be equal to the semicircle, its center being at N,
and we extend line LK rectilinearly to point S,
and let line NS be equal to each of the lines GE, AB, DZ;
and let angle SNM be equal to angle GEA,
and angle SNO equal to angle GED;
and we draw line MF equal and parallel to line NS,
and we join FS;
then FN is a parallelogram with equal angles, and similar to EB.
Again, we draw from point O line OQ equal and parallel to line NS,
and we join QS;
then parallelogram QN becomes
equal to parallelogram EZ and similar to it.
Then we join diagonals FN, QN.

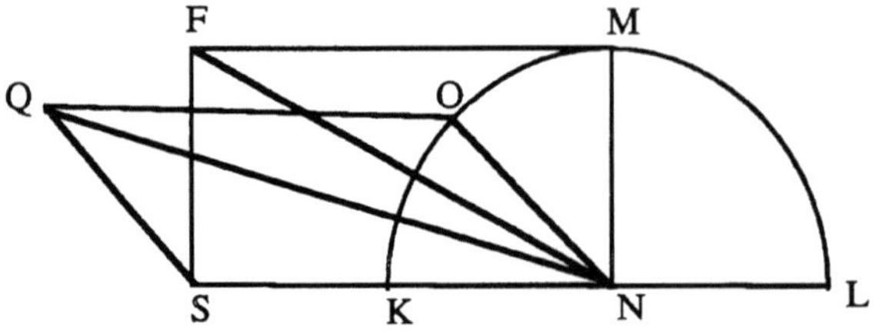

FIGURE 45.2

A	B	G	D	E	W	Z	H	T	I	K	L	M	N	S	O	F
ا	ب	ج	د	ه	و	ز	ح	ط	ى	ك	ل	م	ن	س	ع	ف

فليكن كعل مساو لنصف الدائرة ويكون مركزه نقطة ن
ونخرج خطّ لك على استقامة إلى نقطة س
وليكن خطّ ن س مساوياً لكلّ واحد من خطوط جه اب دز
وليكن زاوية س ن م مساويةً لزاوية جه ا
وزاوية س ن ع مساويةً لزاوية جه د
ونخرج خطّ م ف مساوياً وموازياً لخطّ ن س
ونصل ف س
فَ ن ف متوازى الأضلاع ومتساوى الزوايا ويشبه ه ب
وأيضاً نخرج من نقطة ع خطّ ع ق مساوياً وموازياً لخطّ ن س
ونصل ق س
فيصير إذاً ق ن المتوازى الأضلاع
مساوياً لـ ه ز المتوازى الأضلاع وشبهه
فنصل الأقطار ف ن ق ن

(٣٦- ٣٧) كعل ... لك] كمع ل ... زك: م٥ .

(٣٦، ٣٨) مساو] مساوى: م٣= مساوياً: م٤، م٥ .

(٣٦-٣٧) ن ونخرج خطّ لك] ز ... لك: م٢= ن ... ز ك: م٥ .

(٣٨) وليكن خطّ ن س مساوياً] مساو: م١ .

(٣٨) ن س] مس: م٥ .

(٣٩) جه ا] + وزاوية س ل ع مساوية لزاوية جه ا: م١ .

(٤٣) ومساوى ... ه ب] ومساو وشبيه بـ ه ب: م٢ .

(٤٣) مساوى الزوايا ويشبه] متساوى ويشبه: م٤= متساوى وشبيه به: م٥ .

(٤٤) ع ق ... ن س] ع و ... ن س: م٢، م٤= ع و ... ف س: م٥ .

(٤٥) ق س] و س: م٢، م٤= ف س: م٥ .

(٤٦) فيصير إذاً ق ن] ف ع س اذا: م٢، م٤، م٥= اذا: م٣ .

(٤٧) مساوياً لـ ه ز] مساو لـ ه د: م٢= مساوى بـ ه ز ا: م٣= مساوياً لـ ه ز: م٤= .
مساوياً لـ ه ب: م٥ .

(٤٧) شبهه] يشبه به: م٢، م٤= يشبهه به: م٣= وشبيه به: م٥ .

(٤٨) ف ن ق ن] ف ن ع س: م١= ف ن و ن: م٢، م٤، م٥= ق ن: م٣

Now angle SNF is smaller than angle SNQ;
but angle SNF is equal to angle AEB,
and angle SNQ is equal to angle DEZ; 51
therefore, angle AEB is smaller than angle DEZ;
and in this way magnitude AB is seen as smaller than magnitude DZ.
And that is what we wished to demonstrate. 54

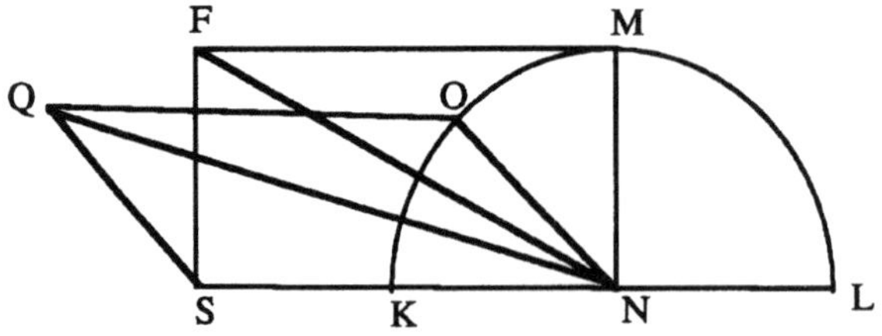

FIGURE 45.2

A	B	G	D	E	W	Z	H	T	I	K	L	M	N	S	O	F
ا	ب	ج	د	ه	و	ز	ح	ط	ى	ك	ل	م	ن	س	ع	ف

فتكون زاوية س ن ف أصغر من زاوية س ن ق
ولكنّ زاوية س ن ف مساوية لزاوية ا ه ب
وزاوية س ن ق مساوية لزاوية د ه ز ٥١
فزاوية ا ه ب أصغر من زاوية د ه ز
فكذلك يكون قدر ا ب يرى أصغر من قدر د ز
وذلك ما أردنا أنْ نبيّن ٥ ٥٤

---

(٤٩) س ن ف أصغر من زاوية س ن ق] زاوية س ن ف اصغر من س ن و: م٢ = زاوية س ن ف اصغر من زاوية س ن و: م٤، م٥ .

(٥٠-٥١) س ن ف ... س ن ق] س ن و ... س ن ف: م٢، م٤، م٥ .

(٥٢) د ه ز] د ه ب: م٣ .

(٥٤) أن نبيّن] بيانه: م٥ .

46

And we set line DZ equal to half the diameter,
and we hold everything else in the state we mentioned
in the illustration that preceded.
So let the semicircle equal to the semicircle be TKLM,
and [let] its center be point N,
and because KN is equal to the semidiameter, KN is equal to TN;
and let angle TNK be equal to angle GEA;
so we draw LS parallel to TN and equal to it, and we join ST;
so let angle TNL be equal to angle GED,
and we draw line KO equal and parallel to line TN, and we join OT.
Now each of TK, TL is a parallelogram,
and they are equal to [parallelograms] EZ, EB and similar to them;

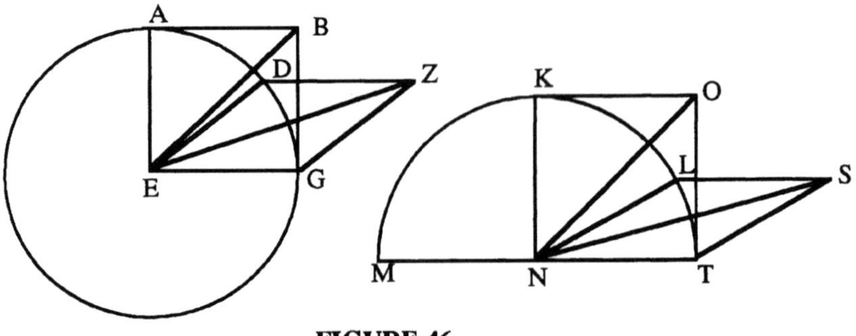

FIGURE 46

A	B	G	D	E	W	Z	H	T	I	K	L	M	N	S	O	F
ا	ب	ج	د	ه	و	ز	ح	ط	ى	ك	ل	م	ن	س	ع	ف

مو

ونجعَل خطّ د ز مساوياً لنصف القطر
وتثبت كلّ ما سوى ذلك على حال ما ذكرنا
فى الصورة التى تقدّمت
وليكن نصف الدائرة المساوى لنصف الدائرة عليها ط ك ل م
ومركزها نقطة ن
ولأنّ ك ن مساو لنصف القطر يكون ك ن مساوياً لـ ط ن
وليكن زاوية ط ن ك مساويةً لزاوية ج ه ا
ونخرج ل س موازياً لـ ط ن ومساوياً له ونصل س ط
وليكن زاوية ط ن ل مساويةً لزاوية ج ه د
ونخرج خطّ ك ع مساوياً وموازياً لخطّ ط ن ونصل ع ط
فكلّ واحد من ط ك ط ل متوازى الأضلاع
وهى مساوية لـ ه ز ه ب ويشبه بها

(٢، ٩) مساوياً] متساوياً: م١ .

(٢) قطر] قطر الدائره: م٥ .

(٢) نثبت] ثبت: م٢ .

(٤) الصورة التى تقدّمت] الصورة الاولى: م٥ .

(٥) نصف الدائرة المساوى] النصف الدائرة المتساوية: م١ ، م٣= نصف الدائرة المساوية: م٢= لنصف الدائرة المساوية: م٤ .

(٥) عليها] التى عليها: م١ ، م٣ .

(٧) ولأنّ] فلان: م٢ ، م٣ ، م٤ ، م٥ .

(٧) ك ن مساوياً] د ز مساو: م٢= ك ز مساوى: م٣= د ز مساوياً: م٤ ، م٥ .

(٧) القطر] قطر: م٣ .

(٨) وليكن زاوية ط ن ك مساويةً لزاوية ج ه ا] ولكنّ زاوية ط ن ن ك مساوية لزاوية جه ه ا: م٣= وليكن زاوية ط ن ك ...لزاوية ج ه د: م٥ .

(٩) موازياً لـ ط ن ومساوياً له] مساوى لـ طن وموازياً له : م٣= موازياً لـ ن ط ومساوياً: م٥ .

(١١) وليكن ...لزاوية] ولكنّ زاوية م ن ك مساوى لزاوية: م٣

(١٢) ط ل] فس م١= ع ز: م٢ ، م٥= مربعى ط ل: م٣ .

(١٣) يشبه بها] شبيهه بهما: م٢ ، م٤ ، م٥= يشبها بهما: م٣ .

and in this way, angle TNL is equal to angle GED, and angle TNK is equal to GEA; but angle GED is smaller than angle GEA; therefore, angle TNL is smaller than angle TNK. Then we draw diagonals SN, ON so that angle TNS is smaller than angle TNO; but angle TNS is equal to angle AEB, and angle TNO is equal to angle DEZ, so angle EAB is then smaller than angle DEZ; therefore, magnitude AB is seen as smaller than magnitude DZ. And that is what we wished to demonstrate.

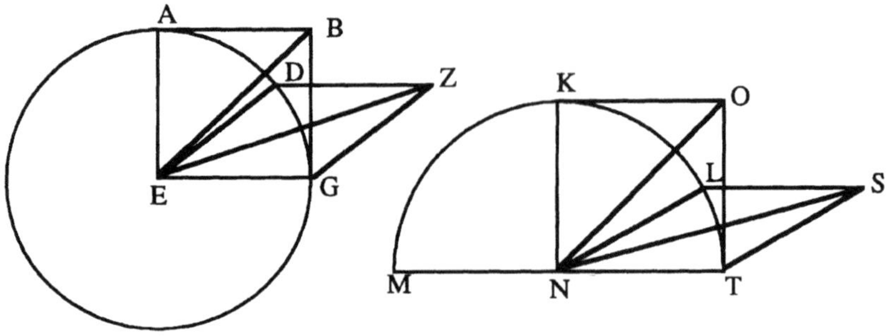

FIGURE 46

A	B	G	D	E	W	Z	H	T	I	K	L	M	N	S	O	F
ا	ب	ج	د	ه	و	ز	ح	ط	ى	ك	ل	م	ن	س	ع	ف

فكذلك تكون زاوية ط ن ل مساويةً لزاوية جـ ه د
وزاوية ط ن ك مساويةً لزاوية جـ ه ا
ولكنّ زاوية جـ ه د تكون أصغر من زاوية جـ ه ا
فزاوية ط ن ل أصغر من زاوية ط ن ك
ونخرج اقطار س ن ع ن
فزاويةَ ط ن س أصغر من زاوية ط ن ع
ولكنّ زاوية ط ن س مساوية لزاوية ا ه ب
وزاوية ط ن ع مساوية لزاوية د ه ز
فزاوية ا ه ب إذا تكون أصغر من زاوية د ه ز
فيرى قدر ا ب أصغر من قدر د ز
وذلك ما أردنا أنْ نبيّن ه°

---

(١٤) تكون] ـ م٢ .
(١٤-١٥) ط ن ل... جـ ه د...ط ن ك... جـ ه ا] ط ن ل ... جـ ه ا ...ط ن ك ... جـ ه ا : م٥ .
(١٦) لكنّ] ـ م٥ .
(١٦) جـ ه د ...جـ ه ا] جـ ه ا ...جـ ه د : م١ = جـ ه ا ...جـ ه ب : م٢ ، ـ م٣ .
(١٦) جـ ه د] جـ ه ا : م٤ .
(١٦-١٧) من زاوية جـ ه ا ...من زاوية] ـ م٤ .
(١٧) أصغر] اذا اصغر: م٣ .
(١٧) ط ن ل ...ط ن ك] ط ن ك ...ط ن ل: م٣ ، م٥ .
(١٧) فزاوية ... ط ن ك] + فزاوية ط ن ل اذا اصغر من زاوية جـ ه د فزاوية ط ن ك اذا اصغر من زاوية جـ ه د: م١ = فزاوية ط ن ل: م٤ .
(١٨) ونخرج] فنخرج: م١ .
(١٩-٢٠) ط ن ع ولكنّ زاوية] ـ م٤ .
(٢٢) إذا تكون] ـ م٢ ، م٤ .
(٢٢-٢٣) زاوية د ه ز...أصغر من] ـ م٢ ، م٤ .
(٢٣) قدر] هش م٢ .
(٢٣) فيرى قدر ا ب] فقدر ا ب يرى: م٥ .
(٢٤) وذلك ما أردنا أن نبيّن] ـ م٢ = وذلك ما اردنا بيانه: م٥ .

47

And we set line DZ less than half the diameter,
and we hold everything else in the state we mentioned in what preceded.
So let the semicircle that is equal to the semicircle be TKLM,
its center being point N;
and from line TN, we cut off a line, SN, equal to DZ.
and let angle TNL be equal to angle GEA,
and angle TNK be equal to angle GED;
and from point L we draw a line parallel to line SN and equal to it,
namely, LO, and we join OS;
and we draw from point K a line equal and parallel to line SN,
namely, KF, and we join FS.
Then each of KS, SL is a parallelogram,
and NF is equal to EB and similar to it,
and SK is equal and parallel to GD.

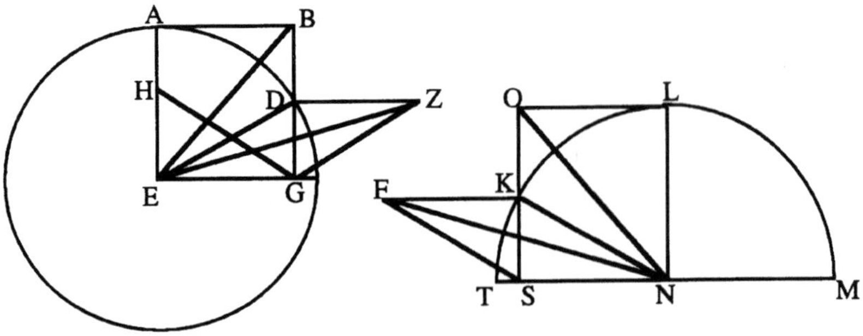

FIGURE 47

A	B	G	D	E	W	Z	H	T	I	K	L	M	N	S	O	F
ا	ب	ج	د	ه	و	ز	ح	ط	ى	ك	ل	م	ن	س	ع	ف

مز

ونجعَل خطّ د ز أصغر من نصف القطر
ونثبت كل ما سوى ذلك على حال ما ذكرنا فيما تقدّم
وليكَن نصف الدائرة المساوى لنصف الدائرة عليها ط ك ل م
ويكون مركزها نقطة ن
وننقُص من خطّ ط ن خطاً مساوياً لخطّ د ز عليه س ن
وليكن زاوية ط ن ل مساويةً لزاوية ج ه ا
وزاوية ط ن ك مساويةً لزاوية ج ه د
ونخرِج من نقطة ل خطاً موازياً لخطّ س ن ومساوياً له
عليه ل ع ونصل ع س
ونخرِج من نقطة ك خطاً مساوياً وموازياً لخطّ س ع
عليه ك ف ونصل ف س
فكلّ واحد من ك س س ل متوازى الأضلاع
ون ف مساو لِ ه ب ويشبه به
وس ك مساوٍ وموازٍ لِ ج د

---

(٢) خطّ د ز] د ز: م٥ . (٣) نثبت كل] ثبت كل: م٢ = نثبت: م٥ .
(٤، ٧) وليكن] ولتكن: م٤ ، م٥ .
(٤) نصف ...المساوى] النصف ...المساوية: م١ ، م٣ = النصف ...المساوى: م٤ ، م٥ .
(٤) عليها] الذى عليه: م١ = عليه: م٣ .
(٥) ويكون مركزها نقطة ن] ويكون مركزه نقطة ن: م١ = مركزه نقطة ن: م٣ = مركز نقطة ن: م٤ = ومركزها نقطة ز: م٥ .
(٦) د ز عليه س ن] ك ن عليه س ن: م٣ = د ن عليه س: م٤ .
(٧) ط ن ل] ط ن ه: م١ .
(٩) موازياً ...ومساوياً له] موازياً...ومساوياً: م١ = مساوياً لخطّ س ن: م٢ ، م٤ .
(١٠) عليه ل ع] عليه ل ف: م٥ .
(١١-١٢) ونخرج من نقطة ك ...لخطّ س ع] – م٢ ، م٤ ، م٥ .
(١١) س ع] س: م٣ .
(١٢-١٤) ك ف ... ن ف] ل ف...س ك = م٢ ، م٤ ، م٥ .
(١٤) يشبه به] شبهه: م١ = شبيه: م٢ = وشبيهه به: م٣ ، م٤ .
(١٥) مساو وموازٍ لِ ج د] مساو وموازى لِ ه = م١ = مساو وموازٍ لِ ه د: م٢ ، م٤ = مساوى وموازى لِ ه ز: م٣ = مساو لِ ه ز: م٥ .

Thus, in this way, angle TNK is equal to angle GED,
and angle TNL is equal to angle GEA,
and angle GEA is greater than angle GED; 18
therefore, angle TNL is greater than angle TNK.
So let us join ON, FN.
Then angle SNO is smaller than angle SNF; 21
but angle SNO is equal to angle AEB,
and angle SNF is equal to angle DEZ,
hence AEB is smaller than angle DEZ, 24
therefore, magnitufe AB is seen as smaller than magnitude DZ.
And that is what we wished to demonstrate.

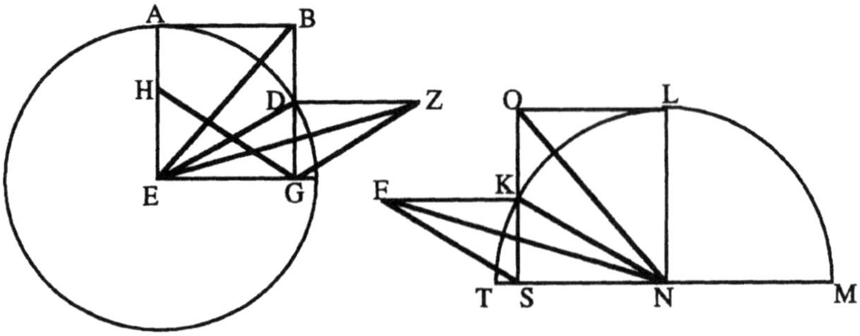

FIGURE 47

A	B	G	D	E	W	Z	H	T	I	K	L	M	N	S	O	F
ا	ب	ج	د	ه	و	ز	ح	ط	ى	ك	ل	م	ن	س	ع	ف

فكذلك تكون زاويه ط ن ك مساويةً لزاوية جـ ه د
وزاوية ط ن ل مساويةً لزاوية جـ ه ا
وزاوية جـ ه ا أعظم من زاوية جـ ه د
فزاوية ط ن ل أعظم من زاوية ط ن ك
فلنصل ع ن ف ن
فزاوية س ن ع أصغر من زاوية س ن ف
ولكنّ زاوية س ن ع مساوية لزاوية ا ه ب
وزاوية س ن ف مساوية لزاوية د ه ز
فزاوية ا ه ب أصغر من زاوية د ه ز
فعظَم ا ب يرى أصغر من عِظَم د ز
وذَلك ما أردنا أنْ نبيّن ٥°

---

(١٦) جـ ه د] جـ ه ا: م١، م٥ .
(١٧-١٨) جـ ه ا ... جـ ه ا ... جـ ه د] جـ ه د ... جـ ه د ... جـ ه ك: م١ = جـ ه ك ... جـ ه ك ... جـ ه ا= م٢، م٤= جـ ه د ... جـ ه ا ... جـ ه د: م٥ .
(١٩) ط ن ك] ك ن ل: م٤ .
(١٩-٢٠) س ن ع ... س ن ف ... س ن ع] س ع ن ... س ف ن ... س ف ن: م١، م٢، م٣، م٤= س ن ع ... س ن ف ... س ن ع: م٥ .
(٢٠) لكنّ] _ م٢، م٤، م٥ .
(٢٠-٢٢) ا ه ب ... لزاوية] _ م٤ .
(٢٢) لزاوية ا ه ب] لزاوية د ه ز وزاوية ا ه ب مساوية لزاوية س ن ف: م٥ .
(٢٣) فزاوية ا ه ب أصغر من زاوية د ه ز] فزاوية د ه ز اصغر من زاوية ا ه ب: م٣، م٥= فزاوية ا ه ب= م٢ .
(٢٤) د ه ز] + وعظم د ز يرى من زاوية د ه ز: هش م٢= وعظم ا ب يرى من زاوية ا ه ب وعظم د ز يرى من زاوية د ه ز: م٣ .
(٢٥) يرى ... د ز] يرى هكذا تحت فكل واحد من ك س س ل متوازى الاضلاع ومساويان ويشبهان به ه ب فكذلك تكون زاوية ط ن ك مساوية لزاوية جـ ه ا وزاوية ط ن ك لزاوية جـ ه د وزاوية جـ ه ا اعظم من زاوية جـ ه د فزاوية ط ن ك اعظم من زاوية ط ن ل فنصل ع ن ف ن فزاوية س ن ف اصغر من س ن ع ولكنّ زاوية س ن ف مساوية لزاوية ا ه ب وزاوية س ن ع مساوى لزاوية د ه ز فزاوية ا ه ب اصغر من زاوية ا ه ب وعظم د ز يرى من زاوية د ه ز فعظم ا ب يرى اصغر من عظم د ز وعظم د ز يرى من زاوية د ه ز فعظم ا ب يرى اصغر من زاوية ا ه من عظم د ز: م١= فعظم ا ب اصغر من عظم د ز: م٢، م٤، م٥ .
(٢٦) أن نبيّن] بيانه: م٥ .

48

There is a place to be found for the eye [such that]
when the visible object is fixed in it while the eye is being moved,
the viewed [object] is always seen as the same [in size].
For let the magnitude of the visible [object] be AB,
and [let] the eye's position be point G;
and we draw from the eye ray[s] GA, GB,
and we inscribe around triangle GAB a circle, AGB.
Then I say that if the viewed [object] is fixed while the eye is moved
[on the circle], it is seen [from each position] as the same [in size].
So let the eye be moved and we make it reach the position of point D,
and we draw ray[s] DA, DB.
Now because angle E and angle Z are in the same segment,
angle E is equal to angle Z;
and for that [reason], the viewed [object] is [always] seen as the same [in size].
And that is what we wished to demonstrate.

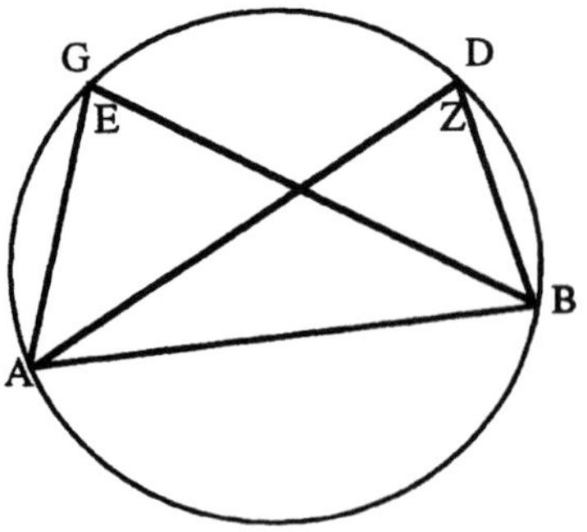

FIGURE 48

A	B	G	D	E	W	Z	H	T	I	K	L	M	N	S	O	F
ا	ب	ج	د	ه	و	ز	ح	ط	ى	ك	ل	م	ن	س	ع	ف

مح

قد يوجد للبصر موضعٌ
إذا ثبَت فيه الشيء المبصر وانتقل البصر
يرى المنظور إليه متساوياً أبداً
فليكن قدر المبصر عليه اب
وموضع البصر نقطة ج
ونخرِج من البصر شعاع جا جب
وندير على مثلث جاب دائرةً عليها اجب
فأقول إنّ المنظور إليه إذا ثبَت وانتقل البصر
يرى متساوياً
فلينتقل البصر و نصيره فى موضع نقطة د
ونخرِج شعاع دا دب
فلأنّ فى قطعة واحدة زاوية ه وزاوية ز
تكون زاوية ه مساويةً لزاوية ز
فيكون لذلك المنظور إليه يرى متساوياً
وذلك ما أردنا أنْ نبيّن ٥°

---

(٣) موضع] موضعاً: م٣ .
(٣، ٩) ثبت] اثبت: م١ ، م٣ ، م٥ .
(٣) انتقل البصر] انتقل البصرى: م١= ــ م٤ .
(٥) قدر المبصر] القدر المبصر: م١ ، م٣= المبصر: م٥ .
(٧، ١١) شعاع] شعاعى: م٢ ، م٤ .
(٨) جاب دائرةً عليها] جا دائرةً عليها: م٢= اجب دائرةً: م٥ .
(٩) ثبت و] ــ م٢= اثبت و: م٥ .
(١٠) د] ــ م٥ .
(١١) ونخرج] ولنخرج: م٢ ، م٤ .
(١١) شعاع] ــ م٣= شعاعى: م٤ ، م٥ .
(١٣) تكون] وتكون: م٤ .
(١٤) يرى] ــ م٢ ، م٣ ، م٤ ، م٥ .
(١٤) ز] د: م٥ .
(١٥) وذلك ما أردنا أن نبيّن] ــ م٢= وذلك ما اردنا بيانه: م٥ .

49

When the viewed [object] is perpendicular to the assumed plane, that is, raised on it at right angles, and the eye is moved along the circle's circumference, [then] the viewed [object] is always seen as the same [in size]. For let the viewed [object] be AB raised on the assumed plane, and [let] the position of the eye be G, and we inscribe a circle around center B, with radius BG namely GD. Then I say that if the eye is moved from the position of point G on the circle's circumference, AB is [always] seen as the same [in size]. And this is clear because every ray drawn from point G to AB is drawn at equal angles, because angle B is a right angle; therefore, the viewed [object] is [always] seen as the same [in size]. And that is what we wished to demonstrate.

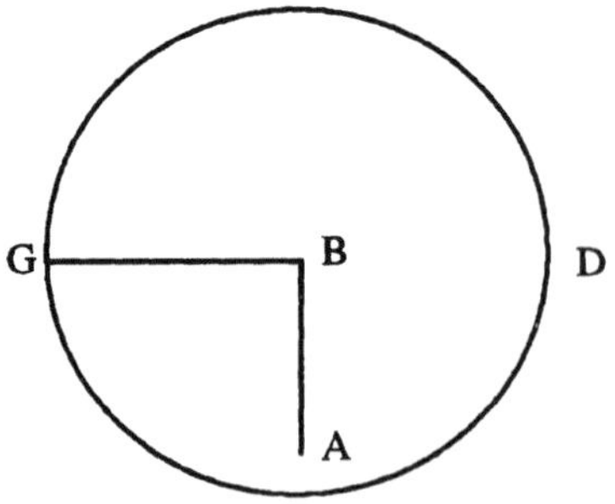

FIGURE 49

A	B	G	D	E	W	Z	H	T	I	K	L	M	N	S	O	F
ا	ب	ج	د	ه	و	ز	ح	ط	ى	ك	ل	م	ن	س	ع	ف

مط

إذا كان المنظور إليه قائماً على السطح الموضوع أعنى عالى عليه على زاوية قائمة وانتقل البصر على محيط الدائرة يكون المنظور إليه يرى أبداً متساوياً فليكن المنظور إليه عليه اب عالياً على السطح الموضوع وموضع البصر نقطة ج وعلى مركز ب وببعد ب ج ندير دائرةً عليها ج د فأقول إنّه إنْ انتقل البصر من موضع نقطة ج على محيط الدائرة يكون اب يرى أبداً متساوياً وذلك بيّنٌ لأنّ كلّ شعاع الذى يُخرَج من نقطة ج الى اب يُخرَج على زوايا متساوية لأنّ زاوية ب قائمة فيكون المنظور إليه يرى متساوياً وذلك ما أردنا أنْ نبيّن ٥٥

---

(٢) إذا] واذا: م١، م٣ .

(٣) عالى... زاويه] عال... زاويه: م٤= عالى... زوايا: م٣= عالياً... زوايا= م٥ .

(٤) الدائرة] دائرة: م٥ .

(٦) المنظور إليه] المبصر: م٥ .

(٧) عالياً] ـ م٢، م٤، م٥= عالى: م٣ .

(٧) الموضوع] الموضع: م١ .

(٩) وعلى] وندير على: م٥ .

(١١) البصر] هش م٢ .

(١٢) اب يرى أبداً] يرى ابداً: م٤= ا ب يرى: م٥ .

(١٣) شعاع] الشعاع: م٢، م٣، م٥ .

(١٤-١٥) من نقطة ج إلى اب يخرج] هش م٢ .

(١٧) وذلك ما أردنا أن نبيّن] ـ م٢= وذلك ما اردنا بيانه: م٥ .

50

There is a place to be found [such that] when the viewed object is fixed in it, and the eye is being moved on the same line, then the viewed [object] is seen [from each position] as different [in size].

For let the viewed [object] be AB, and let line GD be to the side of it; and we extend BA rectilinearly to point G, and we take GE as a mean proportional of the two lines BG, GA, and let the eye's position be point E; then we move it and make it come to the position of point D on that line.

Then I say that from points E [and] D, the viewed object is seen as unequal.

So let us join lines AE, EB, BD, and we inscribe a segment of the circle around triangle AEB, namely AEB, and we join AD and BZ.

Now because angle AZB is greater than angle ADB, and angle AZB is equal to angle AEB, because the two [angles] are in the same segment,

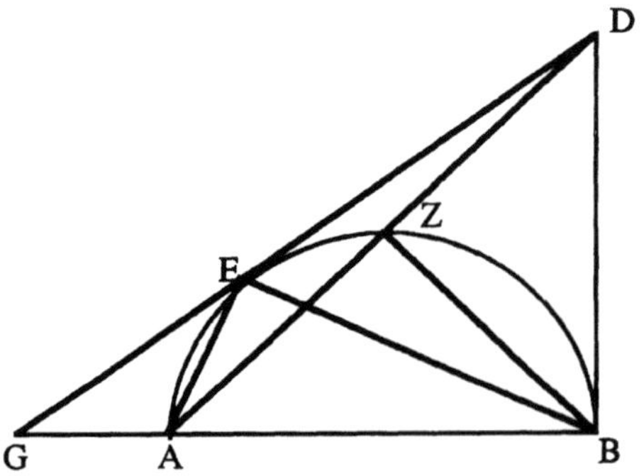

FIGURE 50

A	B	G	D	E	W	Z	H	T	I	K	L	M	N	S	O	F
ا	ب	ج	د	ه	و	ز	ح	ط	ى	ك	ل	م	ن	س	ع	ف

ن

قد يوجد موضعٌ إذا ثبَت فيه المنظور إليه
وانتقل البصر على خطّ واحد
يكون المنظور إليه يرىً مختلفاً
فليكن المنظور إليه عليه اب
وليكن خطّ جد جانباً منه
ونخرج با على استقامة الى نقطة ج
ونأخذ جه بين خطّى بج جا على نسبة
وليكن موضع البصر نقطة ه
ثمّ ننقُله ونصيره على ذلك الخطّ فى موضع نقطة د
فأقول إنّ المنظور إليه من موضعى ه د يرى مختلفاً
فلنصل خطوط اه هب بد
وندير على مثلث اهب قطعة دائرة عليها اهب
ونصل اد و بز
فلأنّ زاوية ازب أعظم من زاوية ادب
وزاوية ازب مساوية لزاوية اهب
من أجْل أنّهما فى قطعةٍ واحدةٍ

----

(٢) ثبت] اثبت: م١، م٣، م٥ .
(٥) إليه عليه] اليه: م١= م٣، ـ م٤، م٥ . (٥) اب] اد: م٢ .
(٨) بين خطى] بين خطين: م١، ـ م٣ .
(٩) موضع] ـ م٢، م٤، م٥ .
(١٠-١٢) موضع...فلنصل] ـ م٢، م٤، م٥ .
(١٢) خطوط] الخطوط: م٣ . (١٢) اه] اب: م٢، م٥= ا د: م٤ .
(١٣) عليها] ـ م٣ .
(١٤) اد و بز] ادابزباه فى دائرة: م٢= ادابزباهد فى دائرة: م٣= ادزب: م٥ .
(١٦) ازب] ادب: م٢، م٤ . (١٦) لزاويه] فس م١ .
(١٦) اهب] ازب: م١ .
(١٧) من أجل أنّهما فى قطعةٍ واحدةٍ] من انهما فى قطعة واحدة: م٣= لانهما فى قطعة واحدة من دايرة: م٥ .

then angle AEB is greater than angle ADB; but AB is seen under angle AEB and under angle ADB; therefore, it is seen [from these positions] as different [in size] while the eye is being moved on the same line. And that is what we wished to demonstrate.

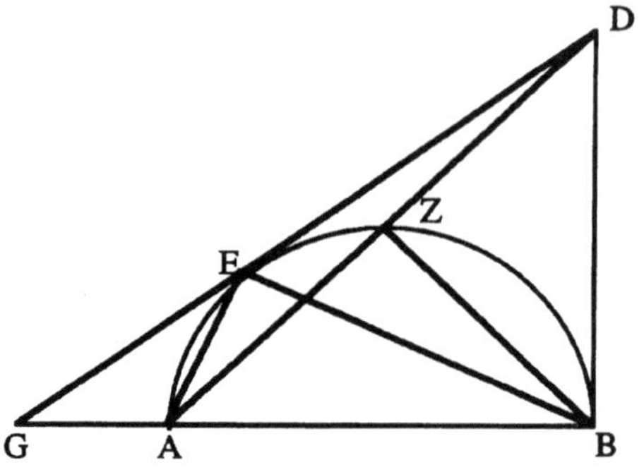

FIGURE 50

A	B	G	D	E	W	Z	H	T	I	K	L	M	N	S	O	F
ا	ب	ج	د	ه	و	ز	ح	ط	ى	ك	ل	م	ن	س	ع	ف

تكون زاوية اهب أعظم من زاوية ادب
ولكنّ اب يرى من زاوية ادب ومن زاوية اهب
فهو إذاً يرى مختلفاً
والبصر منتقل على خطٍّ واحدٍ
وذلك ما أردنا أنْ نبيّنَ ٥٥

---

(١٨) تكون زاويه] فزاويه: م٥ .

(١٩) ولكنّ ... اهب] ولكن اب يرى من زاوية اهب اعظم: م١ .

(١٩) ولكنّ اب يرى من زاوية ادب] هش م٢ .

(٢٠) إذاً يرى] يرى اذاً: م٢ ، م٣ ، م٤ .

(٢٢) وذلك ما أردنا أن نبيّن] _ م٢ = وذلك ما اردنا بيانه: م٥ .

51

Let the magnitude of the viewed [object] be AB, and GD is parallel to AB; and we cut AB in half at point E, and from point E we erect line EZ at right angles to AB, [such that] the eye's position becomes point Z; and we join lines ZA, ZB, and we inscribe around triangle AZB a segment of a circle, namely AZB; and we draw GA, BD parallel to EZ, and we move the eye and make it reach the position of point D, and we draw the two rays DA, DB. Then I say that what is seen from the positions Z [and] D is seen as unequal. So we join BH. Now because angle AZB is equal to angle AHB, and angle AHB is greater than angle ADB, then angle AZB is greater than angle ADB; but AB is seen from angle AZB and angle ADB, thus the same viewed [object] is seen as unequal from positions D [and] Z;

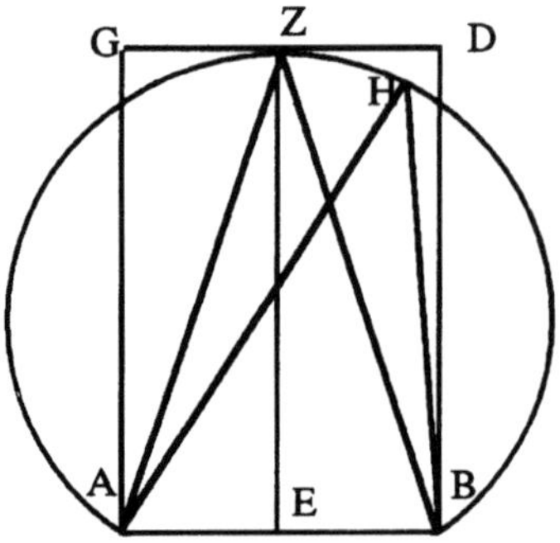

FIGURE 51

A	B	G	D	E	W	Z	H	T	I	K	L	M	N	S	O	F
ا	ب	ج	د	ه	و	ز	ح	ط	ى	ك	ل	م	ن	س	ع	ف

نا

وليكن قدر المنظور إليه عليه اب
ويكون جد يوازى اب ونقطع خطّ اب بنصفين على نقطة ه
ونقيم من نقطة ه خطّ ه ز على زوايا قائمة من اب
ويكون موضع البصر نقطة ز ونصل خطوط زا زب
وندير على مثلث ازب قطعة دائرة عليها ازب
ونخرج جا بد موازيين لـ ه ز
وننقُل البصر ونصيره فى موضع نقطة د
ونخرج شعاعى دا دب
فاقول إنّ الذى يرى من موضعى ز د يكون مختلفاً
ونصل بح
فلأنّ زاوية ازب مساوية لزاوية احب
وزاوية احب أعظم من زاوية ادب
تكون زاوية ازب أعظم من زاوية ادب
ولكنّ اب يرى من زاوية ازب ومن زاوية ادب
فيرى لذلك الشىء الواحد المنظور إليه مختلفاً من موضعى د ز

---

(٢) وليكن] فليكن: م١ .
(٣) يوازى اب ونقطع خط اب] موازى‌اب ونقطع اب: م٢ ، م٣ = يوازى اب ونقطع اب: م٥ ، ــ م٤ .
(٤) من اب] عند اب: م٢ ، م٤ ، م٥ .
(٥) زا زب] زا دب: م٢ ، م٤ .
(٩) شعاعى دا] شعاعين دا: م١= شعاعى زا: م٢ ، م٤ .
(١٠) من] فى: م١ . (١١) بح] زب: م١= دب: م٣ .
(١٢-١٣) ازب... احب... احب.. ادب] ازب...احد...احد...ادب: م١= ازب...اجد...اجد...ادب: م٤ . (١٣) اعظم] اصغر: م٢ ، م٤ .
(١٤) تكون‌زاوية ازب ...ادب] هش م١= تكون زاوية ادب...ازب= م٢ .
(١٥) ازب] فس م٢ .
(١٦) فيرى] ويرى: م٢ ، م٤ .
(١٦) من موضعى د ز] + من موضعى دز: م٣ من موضعى ز د= م٥ .
(١٦) لذلك] ذلك: م٢ ، م٤ ، ــ م٥ .

so because DZ is equal to ZG,
the viewed [object] is seen from the position of point G 18
smaller than what is seen from the position of point Z;
while from the position of points G and D it is seen as equal.
And that is what we wished to demonstrate. 21

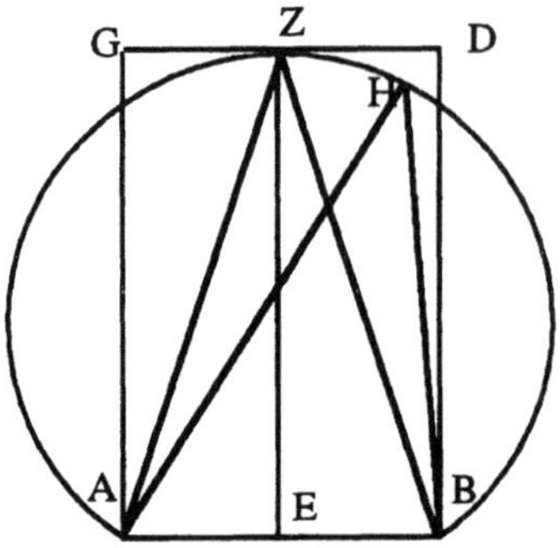

FIGURE 51

A	B	G	D	E	W	Z	H	T	I	K	L	M	N	S	O	F
ا	ب	ج	د	ه	و	ز	ح	ط	ى	ك	ل	م	ن	س	ع	ف

فلأنّ د ز مساوٍ لـ ز جـ
يكون يرى المنظوَر إليه من موضعِ نقطة جـ
أصغر ممّا يرى من موضعِ نقطة ز
ويرى من موضع نقطة جـ ونقطة د متساوياً
وذلك ما أردنا أنْ نبيّن °°

----

(١٧) فلأنّ] ــ م٣ .
(١٧) د ز مساو لـ زجـ] دزا مساو لـ زجـ ا: م٢= مساوى لـ زجـ: م٣= دز مساو لـ دجـ: م٤= مساو لـ زجـ: م٥ .
(١٨) يكون] ــ م٢ .
(١٨) المنظور إليه] المنظور: م١ .
(٢٠) نقطة د] نقد د: م٥ .
(٢٠) متساوياً] متساوياً له: م١= مساوى له: م٣= سوا: م٥ ، ــ م٢ ، م٤ .
(٢١) أردنا أن نبيّن] اردناه: م٢= اردنا بيانه: م٥ .

52

There is a common place to be found for the eye [such that] once it is on it,
equal magnitudes are seen [from there] as unequal.
For let the two magnitudes AB, BG be equal,
and we draw from point B line BD at right angles.
Then I say that if the eye is placed on any position on line BD,
then AB, BG are seen as equal.
So we move the eye and we set it on the position of point E.
Then I say from the position of point E they are seen as unequal.
So let us draw ray[s] AE, EB, EG,
and we inscribe a circle around triangle AGE, namely EADG,
and we extend line BD rectilinearly as line BZ
and line EB rectilinearly as line BH.
Now because AB is equal to BG, and GE is greater than AE,
then AB is seen greater than BG,

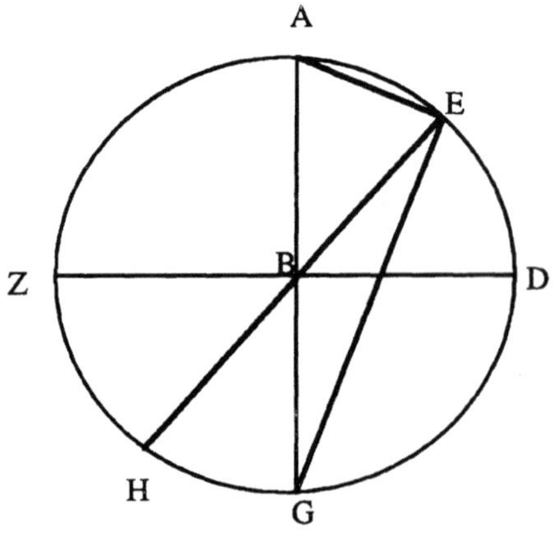

FIGURE 52

A	B	G	D	E	W	Z	H	T	I	K	L	M	N	S	O	F
ا	ب	ج	د	ه	و	ز	ح	ط	ى	ك	ل	م	ن	س	ع	ف

نب

قد يوجد للبصر موضعٌ مشتركٌ إذا كان البصر عليه
يكون الأقدار المتساوية ترى مختلفةً
فليكن قدرا اب بج متساويين
ونخرج من نقطة ب خطّ بد على زوايا قائمة
فاقول إنّه إذا وُضع البصر على اىّ موضع كان من خطّ بد
يكون اب بج ترى متساوية
وننقُل البصر ونجعَله فى موضع نقطة ه
فأقول إنّهما يريان من موضع نقطة ه مختلفين
فلنخرِج شعاع اه هب هج
ونديرُ على مثلّث اجه دائرةً عليها هادج
ونخرِج على خطّ بد خط بز على استقامة
وعلى هب خطّ بح على استقامة
فلأنّ اب مساو ل بج وَ جه أعظم من اه
يكون اب يرى أعظم من بج

---

(٢) موضع مشترك] موضعًا مشتركًا: م٣ .
(٢) للبصر] _ م٤ ، م٥ .
(٣) يكون] تكون: م٤ ، م٥ .
(٣) ترى] يرى: م٣ .
(٤) قدرا] قدرى: م١ ، م٢ ، م٣ ، م٤ .
(٤) متساويين] يرى متساويان: م١= متساوية: م٢ ، م٤ .
(٥) زوايا] زاوية: م٢ . (٦) وضع] وقع: م٢ .
(٧) ترى] _ م٣= يرى: م٥ .
(٧) متساوية] متساويان: م١= متساوياً: م٥ .
(٨-٩) نقطة ه... من موضع] هش م٢ .
(٩) مختلفين] مختلفان: م٣ .
(١٠) فلنخرج] نخرج: م٢ .
(١٠) شعاع] شعاعى: م٢ ، م٥ ، _ م٣ .
(١١) اجه...هادج] اهج...ادج: م٥ .
(١٢) خطّ بد] بد: م٢ ، م٣ ، م٤ ، م٥ .
(١٢) بز] بزى: م١= بد: م٥ .

so if the eye is placed at [any point on] ZB, it is seen as equal, and if the eye is placed at BH, it is seen as unequal; and in this way, in all parts of the circle except the position from which a line is drawn at right angles, the magnitudes are seen as [being] unequal, even when the eye is placed outside of the circle. And that is what we wished to demonstrate.

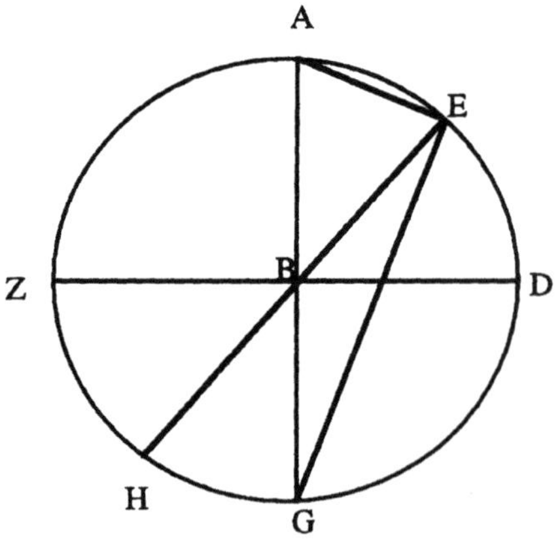

FIGURE 52

A	B	G	D	E	W	Z	H	T	I	K	L	M	N	S	O	F
ا	ب	ج	د	ه	و	ز	ح	ط	ى	ك	ل	م	ن	س	ع	ف

فإنْ وُضِع البصر على زب تكون ترى متساويةً
وإنْ وُضِع البصر على بح رؤيت مختلفةً
وكذلك فى جميع أجزاء الدائرة ما خلا
الموضع الذى يُخرَج إليه الخطّ على زوايا قائمة
فإنّ الأقدار ترى مختلفةً
وإنْ وُضع البصر خارجاً من الدائرة
وذلك ماَ أردنا أنْ نبيّن ٥

---

(١٦) فإنّ] وان: م٢، م٣، م٥.

(١٧) تكون ترى] يكون يرى: م٥.

(١٨) فى] ــ م١، م٣.

(١٨) اجزاء] اجزا: م١، م٣.

(٢٠) فإنّ] وان: م٢.

(٢٠) ترى مختلفةً] يكون متساوية: م٢ = تكون متساوية: م٤، م٥.

(٢٢) وذلك ما اردنا أن نبيّن] ــ م٢ = وذلك ما اردنا بيانه: م٥.

53

Let there be two equal magnitudes, AB, GD
being at right angles to the assumed plane.
Then I say that there is to be found a place [such that] once the eye is in it,
then magnitudes AB, GD are seen sometimes as equal, sometimes as unequal.
So let us join point B to point D by line BD,
and we cut it in half at point E,
and from point E we erect line EZ at right angles.
Then I say that if the eye is on line EZ,
magnitudes AB, GD are seen as equal.
For let the eye's position on line EZ be position Z,
and we draw ray[s] ZB, ZA, ZG, ZD;
then side ZB is equal to side ZD, and AB is equal to GD;
as a result, its position is the same as what we set forth before,
so sides AB, BZ are equal to sides GD, DZ;

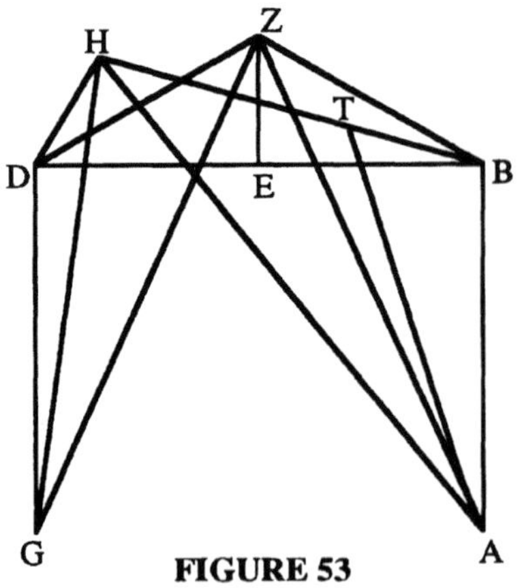

FIGURE 53

A	B	G	D	E	W	Z	H	T	I	K	L	M	N	S	O	F
ا	ب	ج	د	ه	و	ز	ح	ط	ي	ك	ل	م	ن	س	ع	ف

نج

ليكن قدران متساويان عليهما اب جد
ويكونان قائمتين على السطح الموضوع
فأقول إنّه قد يوجد موضعٌ إذا كان البصر فيه
يكون قدرا اب جد يريان مرّةً متساويين ومرّةً مختلفين
فلنصل نقطة ب بنقطة د بخطّ ب د
ونقطعه بنصفين على نقطة ه
ونقيم من نقطة ه خطّ ه ز على زوايا قائمة
فأقول إنّه إنْ كان البصر على خط ه ز
يكون قدرا اب جد يريان متساويين
فليكن موضع البصر على خطّ ه ز فى موضع ز
ونخرج شعاع ز ب زا زج زد
فضلع ز ب يكون مساوياً لضلع ز د و اب مساو لِ جد
فإنّه لذلك كان وُضْعه فيما قدّمنا
فضلعا اب ب ز مساويين لضلعى جد د ز

---

(٢) قدران متساويان] قدرين متساويين: م٣ ، م٤= قدرين مختلفين متساويين: م٥ .
(٣) ويكونان قائمتين] ويكون قائمتين: م٢= ويكونان قايمان: م٥ .
(٣) على السطح الموضوع] على السطح الموضع: م٢= على على السطح الموضع: م٤ .
= على السطح المفروض: م٥ .
(٤) موضع] موضعا: م١ ، م٣ .
(٥) قدرا] قدرى: م٥ .
(٦-٧) نقطة ب بنقطة د خط ب د ونقطعه بنصفين على] ـ م٣ .
(٧) بنصفين] بقطعتين: م٢ ، م٤ .
(١٠) قدرى] قدرا: م٥ .
(١١) على خطّ ه ز فى موضع ز] نقطه ز على خط ه ز: م٢ ، م٤ ، م٥ .
(١٣) زد] ز ب د: م١ .
(١٣) فضلع] قطع: م٣ .
(١٣) يكون متساوياً] مساو: م٥ .
(١٤) وضعه] ضعه: م٢ ، م٤= نصفه: م٥ .
(١٤) لذلك] كذلك: م٥ .
(١٥) فضلعا] فضلعى: م٥ .
(١٥) ب ز] ـ م١= ب ج: م٢ ، م٤ .

and they enclose equal angles,
so base AZ is equal to base ZG,
and angles on bases with equal sides subtending them are equal;
so angle BZA is equal to angle DZG,
therefore AB [and] GD are then seen as equal.
Then I say that it is [also] possible that they are seen as unequal.
So we move the eye to point H,
and we draw ray[s] HB, HA, HG, HD,
BH being greater than HD;
and from line HB we cut off a line equal to line HD, which is line BT,
and we join AT
so that angle BTA is equal to angle GHD.
Now the exterior angle BTA is greater than the interior angle BHA,
so angle GHD is greater than angle BHA;
therefore, magnitude GD is seen as greater than magnitude AB.
And that is what we wished to demonstrate.

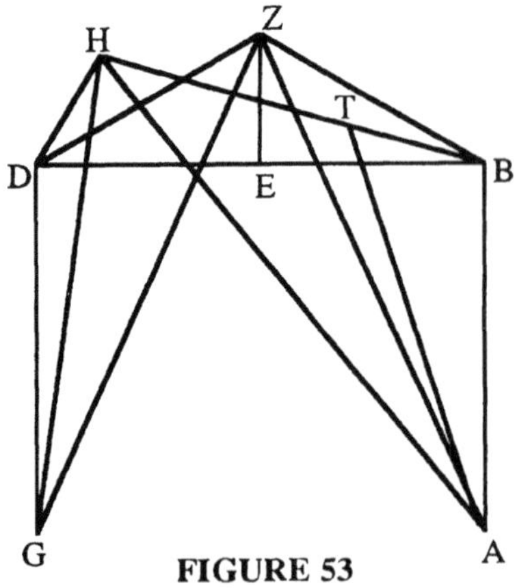

FIGURE 53

A	B	G	D	E	W	Z	H	T	I	K	L	M	N	S	O	F
ا	ب	ج	د	ه	و	ز	ح	ط	ى	ك	ل	م	ن	س	ع	ف

ويحيط بزوايا متساوية
فتكون قاعدة ا ز مساويةً لقاعدة ز ج
والزوايا التي على القواعد التي توترها الأضلاع المتساوية متساوية
فزاوية ب ز ا مساوية لزاوية د ز ج
فَ ا ب ج د إذاً يريان متساويين
فأقول إنّه يمكن أنْ يريا مختلفين
فننقل البصر إلى نقطة ح
ونخرج شعاع ح ب ح ا ح ج ح د
فيكون ب ح أعظم من ح د
ونَفصل من خطّ ح ب خطّاً مساوياً لخطّ ح د وهو خطّ ب ط
ونصلَ ا ط
فزاوية ب ط ا مساوية لزاوية ج ح د
ولكنّ زاوية ب ط ا الخارجة أعظم من زاوية ب ح ا الداخلة
فزاوية ج ح د أعظم من زاوية ب ح ا
فيرى قدر ج د أعظم من قدر ا ب
وذلك ما أردنا أنْ نبيّن ٥°

----

(١٦) متساوية] + قائمة: م٢ ، م٤ ، م٥ .
(١٧) فتكون قاعدة] لتكون قاعدة: م٢ ، م٤= + فتكون قاعدة: م٥ .
(١٧) ا ز ... ز ج] ا د ... ب ج: م١= ا ب ... ز ج: م٢ ، م٤= ا ب ... ج د: م٥ .
(٢٠) إذاً] اذا: م١ ، م٣ .
(٢١) فأقول] واقول: م٢ ، م٣ ، م٤ ، م٥ .
(٢١) يريا] يريان: م٢ ، م٣ ، م٥ .
(٢٢) البصر] نقطه البصر: م٢ .
(٢٥) ح ب] ه ب: م٥ .
(٢٦) ونصل ا ط] ـ م٢ ، م٥ .
(٢٨) ولكنّ زاوية ب ط ا ...الداخلة] ولكن زاوية ب ح ا الخارجة اعظم من الداخلة فزاوية ج ح د اعظم من زاوية ب ح ا الخارجة اعظم من الداخلة: م٣ .
(٢٨-٢٩) الخارجة ... ج ح د] ـ م٢ ، م٤ ، م٥ .
(٢٩) فزاوية ج ح د] ـ م٢ ، م٤ ، م٥ .
(٣١) وذلك ما أردنا أن نبيّن] ـ م٢= وذلك ما اردنا بيانه: م٥ .

54

There is a place to be found from which it is possible to see
unequal magnitudes as equal.
For let there be two unequal magnitudes, AB, BG,
AB being greater than BG.
And we seek to find a single position from which they are seen as equal.
So we inscribe around line BG a segment of circle
greater than a semicircle, BDG,
and we draw around AB a segment of a circle
similar to segment BDG, namely BDA.
Then segment BDA is greater,
because line AB is greater than line BG;
and we draw two lines from point D
wherever one of the segments cuts the other, to line ABG,
and these are lines DA, DG.
Now because segment BDG is similar to segment ADB,
angle ADB is equal to angle GDB;

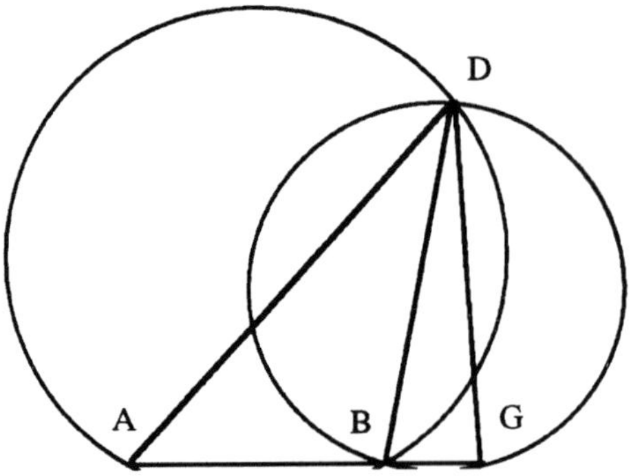

FIGURE 54

A	B	G	D	E	W	Z	H	T	I	K	L	M	N	S	O	F
ا	ب	ج	د	ه	و	ز	ح	ط	ى	ك	ل	م	ن	س	ع	ف

نَد

قد يوجد موضعٌ يمكن أنْ يرى منه الأقدار المختلفة متساويةً فليكن قدران عليهما ا ب ب ج مختلفين ويكون ا ب أعظم من ب ج ونطلب وجود موضع واحد يريان منه متساويين فندير على ب ج قطعةَ دائرةٍ أعظم من نصف دائرة عليهاً ب د ج ونديرعلى ا ب قطعة دائرة شبيهة بقطعة ب د ج عليها ب د ا فقطعة ب د ا أعظم من أجْل أنّ خطّ ا ب أعظم من خطّ ب ج ونخرجِ خطوطاً من موضع نقطة د حيثُ يقطَع أحدى القطعتين الأخرى إلى خطّ ا ب ج وهيِ خطوط د ا د ج فلأنّ قطعة ب د ج شبيهة بقطعة ا د ب تكون زاوية ا د ب مساويةً لزاوية ج د ب

---

(٢) موضع] موضعا: م٣ . (٢) أن يرى] ايرى: م١= ان ترى: م٤ .
(٤) قدران ... ا ب ب ج] قدرين ... ا ب ب ج: م١، م٣، م٤= قدرين ... ا ب ج د: م٥ .
(٥) من ب ج] ـ م١، م٣، م٤، م٥ .
(٦) وجود موضع واحد يريان منه] وجودهما من موضع واحد يريان منه: م١= وجود موضع يريان منه: م٢، م٤= وجودهما من موضعين يريان منه: ٣ .
(٨) أعظم ... ب د ج] عليها ب د ج: م١، م٣= اعظم ... ب ج د: م٥ .
(٩) وندير على ا ب قطعة دائرة] وعلى اب قطعة: م٢، م٤، م٥ .
(١٠) ب د ا] اب د: م٢، م٤، م٥ .
(١١) فقطعة ... اعظم] فقطعة ... ترى اعظم: م١= يرى اعظم: م٣ .
(١٣) نخرج خطوطا] نخرج خطوط: م١= نصل خطوطا: م٢، م٤، م٥ .
(١٤) القطعتين] العظمين: م١ .
(١٥) وهى خطوط د ا د ج] وهى دا دب: م١= خطوط دا دج: م٢، م٣، م٤= وهىخطوط دا دج بد= م٥ . (١٦) ا د ب] ب د ا: م٢، م٤، م٥ .

therefore it has been demonstrated that magnitude AB is seen 18
equal to magnitude BG, from the position of point D.
And that is what we wished to demonstrate.

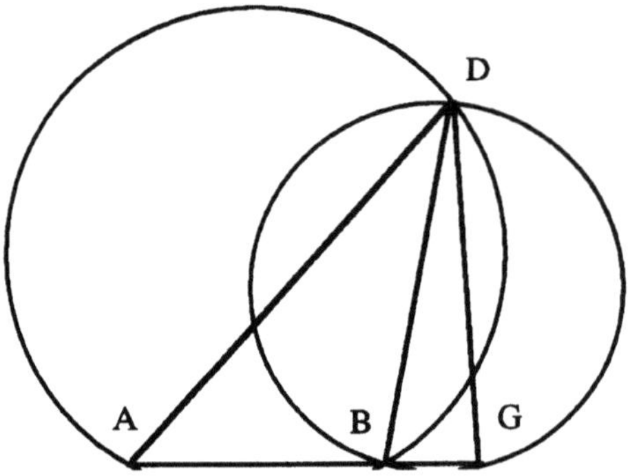

FIGURE 54

A	B	G	D	E	W	Z	H	T	I	K	L	M	N	S	O	F
ا	ب	ج	د	ه	و	ز	ح	ط	ى	ك	ل	م	ن	س	ع	ف

فتبيّن أنّ قدر اب يرى مساوياً لقدر بج من موضع نقطة د وذلك ما أردنا أنْ نبيّن ه

(١٨-١٩) يرى مساوياً] يريا مساوياً: م٣ .

(١٩) نقطة د] نقطة ز: م٢ ، م٤ .

(٢٠) أن نبيّن] بيانه: م٥ .

55

There is a place to be found [such that] once the eye is in it
unequal magnitudes taken together are seen as equal
to the amount seen of each one of them when seen separately.
For let there be two magnitudes, namely AB, BG,
AB being greater than BG;
and we inscribe around AB a semicircle, ADB,
and around BG a semicircle, BEG,
and around the whole of ABG a semicircle, AZG;
and we cut arc AG in half at point Z,
and we join ZA, ZG, and we also join DB, BE.
Now because angle ADB is a right angle,
and angle BEG is also a right angle;
then angle BEG is equal to angle ADB,
so AB [and] BG are seen as equal;
but angle AZG is a right angle, and AG is seen from it,
therefore, AG is seen as equal to each of AB, BG.
And that is what we wished to demonstrate.

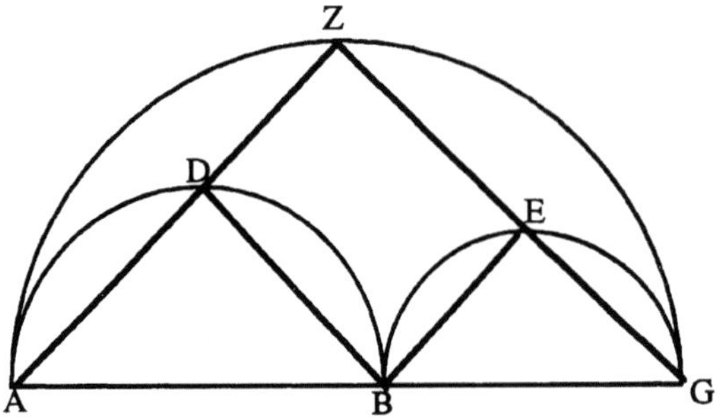

FIGURE 55

A	B	G	D	E	W	Z	H	T	I	K	L	M	N	S	O	F
ا	ب	ج	د	ه	و	ز	ح	ط	ى	ك	ل	م	ن	س	ع	ف

نه

قد يوجد موضعٌ إذا كان فيه البصر
يكون الأقدار المختلفة إذا اجتمعت ترى فيه مساويةً
لقدر ما يرى من كلّ واحد من المختلفة إذا رأى على حِدَته
فليكن قدران عليهما اب بج
ويكون اب أعظم من بج
ونديرعلى اب نصف دائرة عليها ادب
وعلى بج نصف دائرة عليها به‍ج
وعلى جميع ابج نصف دائرةً عليها ازج
ونقطع قوس اج بنصفين على نقطه ز
ونصل زا زج وأيضاً نصل دب به
فلأنّ زاوية ادب قائمة
وزاوية به‍ج أيضاً قائمة
فزاوية به‍ج مساوية لزاوية ادب
فَ اب بج يريان متساويان
ولكنّ زاوية ازج قائمة ويرى منها اج
فَ اج يرى مساوياً لكلّ واحد من اب بج
وذلك ما أردنا أنْ نبيّن ٥°

---

(٢) موضع] موضعاً: م٣ . (٣) مساويةً] مساوية : م٢ .

(٣) يكون ...اجتمعت ترى فيه] يكون ... ترى: م٢= تكون ...جمعت ترى فيه: م٥ . (٤) رأى] رى: م١ ، م٢ ، م٤ ، م٥ .

(٥) قدران] قدرين: م٣ ، م٤ ، م٥ .

(٦) من بج] ـ م٢ ، م٣ ، م٤ ، م٥ .

(٩-١١) وعلى جميع ابج...ازج ونصل زا زج] ـ م٢ ، م٤= وعلى اج نصف الدائرة عليها ازب ونصل از...زج: م٥ .

(١١) وأيضاً نصل] ـ م٥ .

(١٥) فَ اب بج ...متساويان] ب ج...متساويين: م٣ .

(١٦) ويرى] فيرى: م٥ .

(١٧) ازج] ادج: م٤ .

(١٨) وذلك ما أردنا أن نبيّن] ـ م٢= وذلك ما اردنا بيانه: م٥ .

56

How we find positions for the eye from which a magnitude is seen as its half, its quarter, or according to any part by which the angle is divided. Let the magnitude of the viewed [object] be AB; then I say that for magnitude AB there is a position such that, when the eye is in it, it is seen at times in its entirety, at times as its half, and at other times as its quarter, or [else] according to any supposed amount. So we inscribe a circle around line AB, namely AEB, such that AB is not a diameter; while the circle's center is point G the eye's position is upon it, and we join lines GA, GB; so AB is seen from the position of point G.

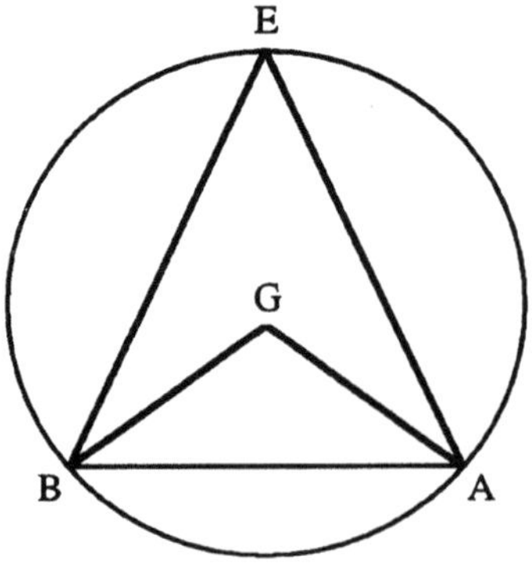

FIGURE 56

A	B	G	D	E	W	Z	H	T	I	K	L	M	N	S	O	F
ا	ب	ج	د	ه	و	ز	ح	ط	ى	ك	ل	م	ن	س	ع	ف

نو

كيف نجد مواضع للبصر يرى منها القدر
على نصفه وعلى ربعه وعلى أيّ جزء تنقسم به الزاوية
فليكن قدر المنظور إليه عليه اب
فأقول إنّ قدر اب له موضعٌ
إذا كان فيه البصر
يرى مرّةً كلّه ومرّةً مثل نصفه ومرّةً مثل ربعه
وعلى كلّ قدر مفروض
فندير على خطّ اب دائرةً عليها اهب
ولا يكون اب قطراً
ويكون مركز الدائرة نقطة جـ ويكون عليه نقطة البصر
ونصل خطوط جا جب
فَ اب يرى من موضع نقطة جـ

---

(٢) كيف نجد] فنجد : م٤ .
(٢) للبصر] البصر: م٢ .
(٣) جزء] جز: م٥ .
(٣) الزاوية] الزوايا : م٢ .
(٢) مواضع] مواضعاً : م٣ .
(٤) عليه] ـ م٢ ، م٤ ، م٥ .
(٥) موضع] مواضع: م٣ ، م٥ .
(٦) فيه] فيها : م٣ ، م٤ ، م٥ .
(٧) مثل ربعه] على ربعه: م٢ ، م٤ ، م٥ .
(٩) اهب] ا ه ج: م٥ .
(١٠) ولا يكون] فلا يكون: م٢ ، م٤ .
(١٠) اب قطراً] خطّ اب قطراً: م٢= خطّ اب قطر: م٥ .
(١١) عليه نقطة البصر] عليها البصر: م٢ ، م٤ ، م٥ .
(١٣) نقطه] ـ م٢ ، م٤ ، م٥ .

Then we move the eye to the circumference of the circle at point E, and we join ray[s] AE, EB. Now because angle AEB is equal to half of angle AGB, then magnitude AB is seen from the position of point G twice [as large] as what is seen from the position of point E; and in the same way it is seen as equal to its quarter, if the [center] angle is four times the [corner] angles; and according to whatever amount it is divided, it is seen in that way . And that is what we wished to demonstrate.

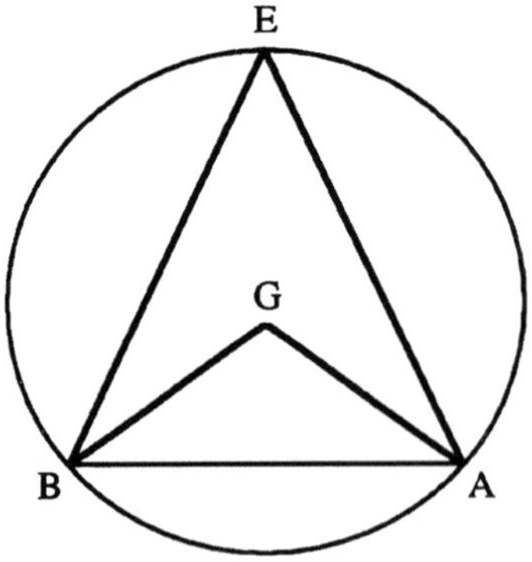

FIGURE 56

A	B	G	D	E	W	Z	H	T	I	K	L	M	N	S	O	F
ا	ب	ج	د	ه	و	ز	ح	ط	ى	ك	ل	م	ن	س	ع	ف

ونَنقُل البصر إلى محيط الدائرة على نقطة ه
ونصل شعاع ا ه ه ب
فلأنّ زاوية ا ه ب مثل نصف زاوية ا ج ب
فقدر ا ب يرى من موضع نقطة ج
مثلى ما يرى من موضع نقطة ه
وعلى مثل ذلك يرى مثل ربعه
إذا كانت الزاوية أربعة أمثال الزاوية
وعلى أيّ قدر قسمت يرى على مثل ذلك
وذلك ما أردنا أنْ نبيّن ٥°٥

----

(١٤) ه] ــ م٥ .

(١٥) ا ه ب مثل نصف زاوية ا ج ب] اجب مثلا اهب: م١= اجب مثلا زاوية اهب: م٣ .

(١٧) فقدر] يكون قدر: م٢ ، م٤ ، م٥ .

(١٧-١٨) موضع نقطة ج ...من موضع] هش م١= موضع ج ...من موضع: م٢ ، م٤ ، م٥= موضع ج ... من: م٣ .

(١٨) مثلى ما يرى] مثلا ما يرى: م٣ .

(١٨) ربعه] اربعه: م٥ .

(١٩) إذا كانت الزاوية أربعة] ــ م٣ .

(٢٠) يرى] ترى: م٤ ، م٥ .

(٢٠) على] ــ م٣ .

(٢١) وذلك ما اردنا أن نبيّن] ــ م٢= وذلك ما اردنا بيانه: م٥ .

57

When objects of equal [speed of] motion approach the eye on the same line,
the one to the rear is thought to be ahead;
and once their [manner of] seeing is changed,
the one ahead is thought to be following,
and the one following [is thought to be] ahead.
For let us move magnitudes AB, GD, EZ with equal speed,
and let them be on the same [vertical] line, namely BZ,
and [let] the position of the eye be point H; and we join ray[s] HZ, HD, HB
so that the ray HB is higher than ray HD and HD is higher than ray HZ.
Then it is thought that AB is overtaking GD and GD overtaking EZ
so that AB overtakes [the others] while GD is behind [it], and EZ follows.
Then we set parallel lines TK, LM, NS to the right of line NT,
and let ray[s] HT, HL, HN fall;

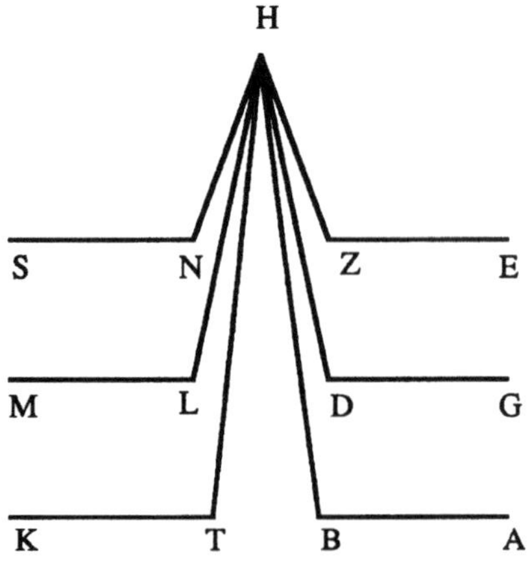

FIGURE 57

A	B	G	D	E	W	Z	H	T	I	K	L	M	N	S	O	F
ا	ب	ج	د	ه	و	ز	ح	ط	ى	ك	ل	م	ن	س	ع	ف

نز

الأشياء المتساوية الحركة التى تكون على خطٍّ واحدٍ إذا دنت إلى البصر
فإنّ آخرها يُظَنّ أنّه متقدّم
فإذا هى اختلف رؤيتها
يكون المتقدّم يُظَنّ أنّه لاحق
واللاحق متقدّم
فلنحرّك أقداراً حركةً متساوية السرعة عليها اب جد هز
وتكون على خطّ واحدٍ وهو بز
وموضع البصر نقطة ح ونصل شعاع حز حد حب
فشعاع حب أرفع من شعاع حد وشعاع حد أرفع من شعاع حز
فيُظَنّ أنّ اب سابق لجد وَ جد سابق لِ هز
فيسبق إذاً اب ويتاخّر جد ويلحق هز
ونجعَلها متوازية متيامنة على خطّ نط عليها طك لم نس
وليقع شعاع حط حل حن

---

(٢) تكون] – م٥ .
(٢) إلى] من: م٢ ، م٤ ، م٥ .
(٢) دنت] ادنيت: م١ ، م٣ ، م٥ .
(٤) اختلف] اختلفت: م٣ .
(٥) يُظنّ أنّه لاحق] يظن لاحقاً: م١ ، م٣ .
(٥) المتقدّم] المقدم: م٢ ، م٣ .
(٦) متقدّم] متقدماً: م١ .
(٧) فلنحرّك] فلنتحرك: م٥ .
(٧) أقداراً] فلنحرك اقدار: م٢ ، م٣ ، م٤ ، م٥ .
(١١) لـ هز] لـ ه: م١ .
(١٢) فَيسبق اذا] فنسبق اذًا: م٢= فتسبق اذًا: م٤ .
(١٣) نط] بط: م٤ . (١٤) وليقع] + وليقع: م٥ .
(١٤) حط حل حن] جط جل جز: م٤ .
(١٤) حن] + فلانّ اب يظنّ أنّه سابق لِ جد وَ جد سابق لِ ه ز: م١ ، م٢ ، م٣ ، م٤ ، م٥ .

Now let AB come to the position TK, GD to position LM, EZ to position NS, and [so] point B to point K, A to T, G to L, and D to M. And it was [then] thought that AB is overtaking GD and GD overtaking EZ; but [now] HN is overtaking HL and HL overtaking HT, and therefore NS is [thought to be] overtaking LM and LM [overtaking] TK; so the one in the rear is thought to be overtaking [the others], and the one thought to be behind is in this case the one overtaking [the others]. And that is what we wished to demonstrate.

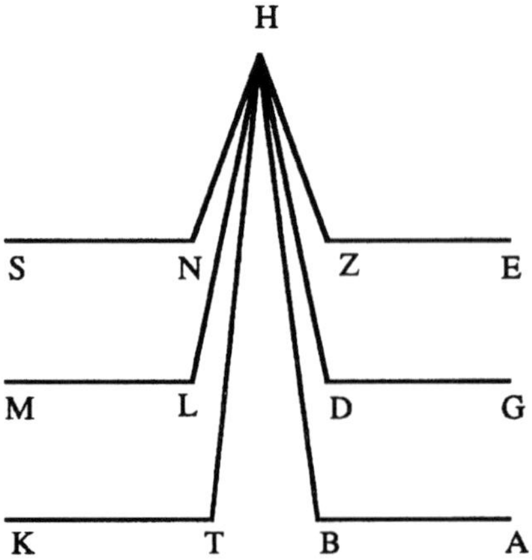

FIGURE 57

A	B	G	D	E	W	Z	H	T	I	K	L	M	N	S	O	F
ا	ب	ج	د	ه	و	ز	ح	ط	ى	ك	ل	م	ن	س	ع	ف

وليكن اب صار إلى موضع طك
و جد إلى موضع لم و هز إلى موضع نس
ونقطة ب إلى نقطة ك و ا إلى ط و ج إلى ل و د إلى م
ويُظَنّ أنّ اب سابق لـ جد و جد سابق لـ هز
ولكنّ حن سابق لـ حل و حل سابق لـ حط
فـ نس إذاً سابق لـ لم و لم لـ طك
فيُظَنّ أنّ الآخر سابق
وفإنّ الذى كان يُظَنّ أنّه المتأخّر فذاك هو السابق
وذلك ما اردنا أنْ نبيّن ٥٥

---

(١٥) وليكن] ولكن: م٢ ، م٣ ، م٤ ، م٥ .

(١٦) إلى موضع لم] فى موضع لم: م٥ .

(١٧) ط] موضع ط: م٣ .

(١٨) و جد سابق لـ هز] ـ م١ ، م٢ ، م٣ ، م٤ ، م٥ .

(١٩) حن سابق لـ حل] حز سابق لـ حد: م٢ ، م٥= ج ن سابق لـ ج د= م٤ .

(٢٠) إذا] اذاً: م٢ .

(٢١) فيظنّ] + اذا: م٤ ، م٥ .

(٢٢) فإنّ الذى] وان الذى: م٢= والذى: م٣ ، م٤ ، م٥ .

(٢٣) وذلك ما أردنا أن نبيّن] ـ م٢= وذلك ما اردنا بيانه: م٥ .

58

When magnitudes are moving with unequal [speed of] motion,
and the eye is moving with a motion equal [in speed] to one of them,
then it is thought that some of the ones moving are fixed in their place,
and some of them [moving] ahead, and some moving in the opposite direction.
For let us move three points, A, B, G, with unequal [speed of] motion,
and let the eye be point D, its motion being equal in speed
to the speed of the motion of point B.
Then I say that it is thought that point B is [staying] fixed,
while point G is [moving] ahead and point A moving in the opposite direction.
So let ray[s] DA, DB, DG fall.
Now because A, B, [and] G are moving with a motion of unequal [speed],
and point B has a motion equal [in speed] to that of the eye's point, namely D,
since point D is not [staying] fixed, then it is thought that point B is fixed,
because they are on the same line;
And as point B is [becoming] distant from point G, point G is [moving] ahead,

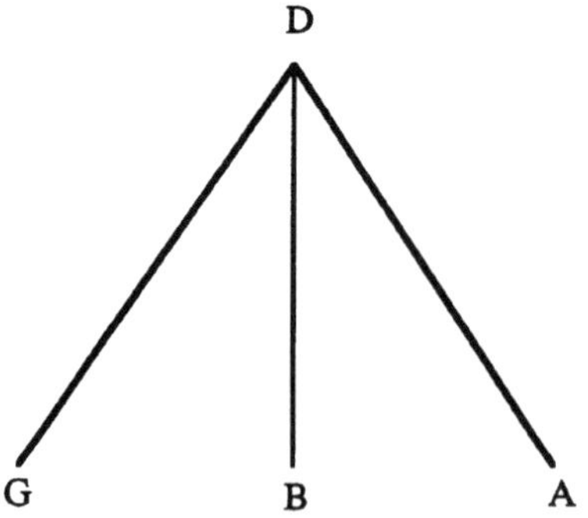

FIGURE 58

A	B	G	D	E	W	Z	H	T	I	K	L	M	N	S	O	F
ا	ب	ج	د	ه	و	ز	ح	ط	ى	ك	ل	م	ن	س	ع	ف

نح

إذا كانت أقدار متحركة حركةً مختلفةً
والبصر متحرّك حركةً مساويةً لحركة أحدها
فإنّه يُظَنّ أنّ بعض متحرّكة ثابت فى مكانه
وبعضها متقدّم وبعضها متحرّك إلى خلاف ذلك
فلنحرّك ثلثة نقط حركةً مختلفةً عليها ا ب ج
وليكن البصر نقطة د وحركته مساوية السرعة
لسرعة حركة نقطة ب
فأقول إنّ نقطة ب يُظَنّ أنّها ثابتة
ونقطة ج متقدّمة ونقطة ا متحرّكة إلى خلاف ذلك
فليقع شعاع دا دب دج
فلأنّ ا ب ج متحرّك حركةً مختلفةً
ونقطة ب مساوية الحركة لنقطة البصر التى هى نقطة د
فإذا لم تثبت نقطة د تكون نقطة ب يُظَنّ أنّها ثابتة
من أجْل أنّها على خطّ واحد
فلأنّ نقطة ب متباعدةً عن نقطة ج ونقطة ج متقدّمة

---

(٢) كانت] كان: م٢ .
(٣) متحرك] متحركة: م٣ ، م٥ .
(٣) لحركة أحدها] لحركة احدهما: م١= بحركة اخرها: م٢= بحركة احدها: م٣ لحركة اخرها: م٤ .
(٤) فإنّه] + فانه: م١ . (٤) مكانه] مكانها: م٣ .
(٦) ثلثة ... ا ب ج] ثلث نقط عليها اب ج: م٢ ، م٤= ثلاث نقط عليها ا ب ج: م٥ .
(٧) السرعة] _ م٥ .
(٨) لسرعة حركة نقطة ب] بحركة ب: م٢= لحركة نقطة ب: م٤ ، م٥ .
(١٠) نقطة ا] نقطة ج: م٥ .
(١٢) فلأنّ اب ج] _ م٣ . (١٣) نقطة د] د: م٥ .
(١٤-١٧) نقطة د ...لم تثبت] _ م٣ .
(١٤) نقطة ب] ب: م٢ ، م٥ . (١٤) أنّها] _ م١ ، م٣ ، م٥ .
(١٥) أنّها] انهما: م٢ ، م٥ .
(١٦) عن نقطه ج] عن ج: م٢ .

and since point D is not fixed, it is thought that point G is [moving] ahead; but point A is [also becoming] distant from point B, <and so> since point D is not fixed, point A is seen as moving in the opposite direction, because of the slowness of its motion. And that is what we wished to demonstrate.

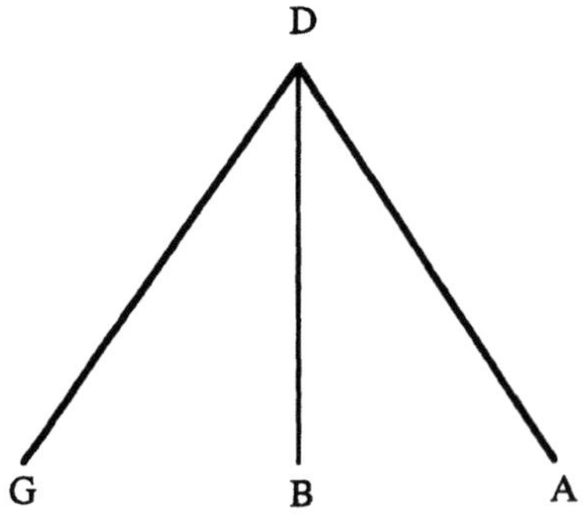

FIGURE 58

A	B	G	D	E	W	Z	H	T	I	K	L	M	N	S	O	F
ا	ب	ج	د	ه	و	ز	ح	ط	ى	ك	ل	م	ن	س	ع	ف

فإذا لم تثبت نقطة د تكون نقطة جـ يُظَنّ أنّها متقدّمة
ولكنّ نقطة آ متباعدة عن نقطة ب
<فَ> تكون نقطة آ إذا كانت نقطة د غير ثابتة
ترى متحرّكةً إلى خلاف ذلك من أجْل إبطاء الحركة
وذلك ما أردنا أنْ نبيّن ٥°

---

(١٧) تكون نقطة جـ] تكون ب : م٥ .
(١٧) يظنّ أنها] فس م١ .
(١٨) ولكنّ . . . متباعدة] ولأن . . . متباعد : م٣ = فلانّ . . . متباعدة : م٥ .
(٢٠) ترى] ــ م٢ ، م٤ ، م٥ .
(٢٠) ذلك] ــ م١ ، م٣ ، م٤ ، م٥ .
(٢١) وذلك ما أردنا أن نبيّن] ــ م٢ = وذلك ما اردنا بيانه : م٥ .

59

When the eye is approaching a body,
then it is thought that this body is rising and enlarging.
For let the body of the viewed [object] be AB,
and [let] the eye's position be point G;
and we draw ray[s] GA, GB [such that] AB is enclosed by GA, GB;
then the eye comes closer to it, its position becoming point D,
and we draw ray[s] DA, DB.
Now because angle ADB is greater than angle AGB,
then when AB is viewed from the position of point D,
it is thought that it is greater than when it is viewed from point G.
And that is what we wished to demonstrate.

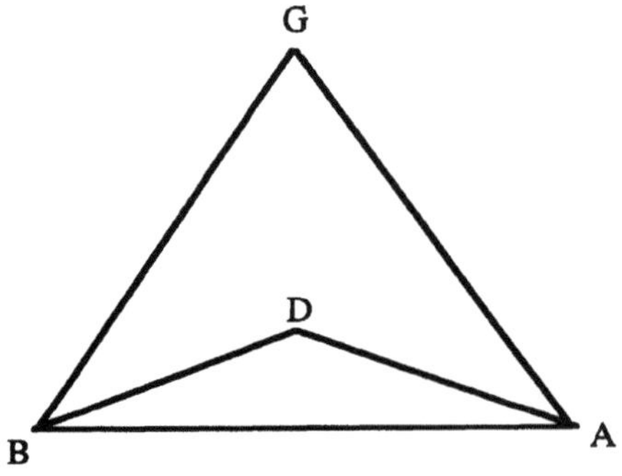

FIGURE 59

A	B	G	D	E	W	Z	H	T	I	K	L	M	N	S	O	F
ا	ب	ج	د	ه	و	ز	ح	ط	ى	ك	ل	م	ن	س	ع	ف

نط

إذا كان البصر يدنو إلى جسم
فإنّ ذلك الجسم يُظَنّ أنّه ينشو و يزيد
فليكن جسم منظور إليه عليه ا ب
وموضع البصر نقطة ج
ونخرِج شعاع ج ا ج ب فَ ا ب يحيط به ج ا ج ب
ويصير البصر أقرب اليه ويكون موضعه نقطة د
ونخرِج شعاع د ا د ب
فلأنّ زاوية ا د ب أعظم من زاوية ا ج ب
فيكون ا ب إذا نُظر إليه من موضع نقطة د
يُظَنّ أنّه أعظم منهَ إذا نُظر إليه من نقطة ج
وذلك ما أردنا أنْ نبيّن ٥٥

---

(٢) يدنو] يدنوا: م٣، م٥ .
(٣) ينشو ويزيد] ينمو او يزيد: م١ = ينشوا ويزيد: م٥ .
(٤) جسم منظور إليه] جسماً منظوراً اليه: م١، م٣ .
(٦) يحيط به ج ا ج ب ويصير] يحيط د ه ج ح ا ج ب ونصير: م٥ .
(٧) موضعه] موضع: م١، م٣ = موضع البصر: م٥ .
(٨) د ا د ب] داب: م٣ .
(١٠) فيكون] يكون: م١، م٣، م٥ .
(١٠) إذا ... ج] واذا نظر من موضع نقطة د: م١ .
(١١) إذا] واذا: م١، م٥ .
(١٢) وذلك ما أردنا أن نبيّن] ـ م٢ = وذلك ما اردنا بيانه: م٥ .

60

[Of] magnitudes with equal [speed of] motion, the most distant one is thought to have the slowest motion. For let two points A [and] B be moving rectilinearly to the side of G, the eye being point Z; and we draw ray[s] ZA, ZB, ZE, ZD. Then I say that point A, which is the more distant one, is thought to have a slower motion than point B. Now because angle BZE is greater than angle AZD, BE is seen as greater than AD; and because AD, BE are moving with an equal [speed of] motion, H [and] D are seen by ray ZD. If we now extend ray ZE rectilinearly, when it is moved with an equal speed along ray ZEG, it falls behind point A. Therefore, [of] objects moving with equal speed, the more distant one seems have a slower motion, and that is what we wished to demonstrate.

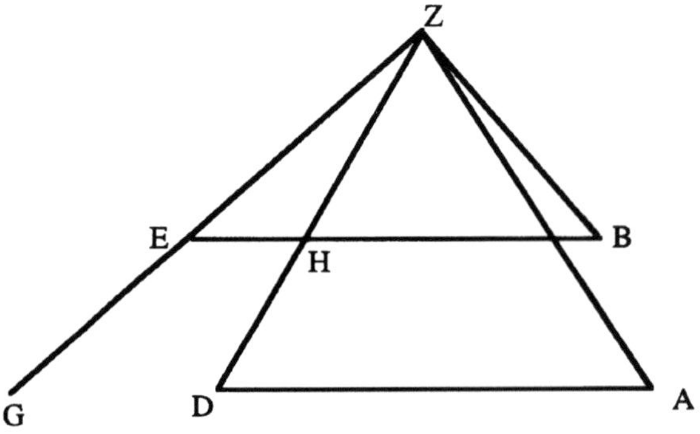

FIGURE 60

A	B	G	D	E	W	Z	H	T	I	K	L	M	N	S	O	F
ا	ب	ج	د	ه	و	ز	ح	ط	ى	ك	ل	م	ن	س	ع	ف

س

الأقدار المتساوية الحركة
فإنّ أبعدها يُظَنّ أنّه أبطأ حركة
فليتحرّك نقطتا ا ب على استقامة إلى ناحية ج
ويكون البصر نقطة ز ونخرج شعاع زا زب زه زد
فاقول إنّ نقطة ا التى هى أبعد
يُظَنّ أنّها أبطأ حركةً من حركة نقطة ب
فلأنّ زاوية ب ز ه أعظم من زاوية ا ز د
يرى ب ه أعظم من ا د
فلانّ ا د ب ه يتحركان بحركةٍ متساويةٍ
يرى ح د بشعاع ز د
فإنْ نحن أُخرجنا شعاع زه على استقامة
يكون إذا يتحرّك بسرعة متساوية على شعاع ز ه ج يتأخّر نقطة ا
فالأشياء المتحرّكة بسرعةٍ متساويةٍ
يكون أبعدها يُظَنّ أنّه أبطأ حركةً
وذلك ما أردنا أنْ نبيّن ٥٥

(١-١٦) الأقدار... نبيّن] _ م٤ .
(٣) أنّه] فس م١ .
(٤) نقطتا ...ج] نقطتى: م١ ، م٣= نقطتا...نقطة ج: م٥ . (٤) ناحيه] فس م٥ .
(٥) زا زب زه زد] زا زه زب زد: م٥ .
(٧) نقطة ب] نقطة حركة ب: م٣ . (٨) ب ز ه ...ا ز د] ب ز ه...ا ز ه: م٣ .
(٨-٩) فلانّ...ا د] فلان از زد يحيطان بزاوية اصغر من الزاوية التى يحيط بها زب زه يكون ب ه يرى اعظم من ا د: م٢= فلان ا ز زد يحيطا بزاوية اصغر من الزاوية التى يحيط بهاب ز زه يكون ب ه يرى اعظم من ا د: م٥ .
(١٠) ب ه] د ه: م١ .
(١٠) فلانّ ...يتحركان] فلان ...يحركا: م١ ، م٢= ولان ...يحركا: م٣ .
(١١) يرى ح د بشعاع] يرى ج د شعاع: م٥ . (١٢) فا...زه] _ م٢ .
(١٣) يتحرّك] تحركت: م٣ ، م٥ . (١٣) ز ه ج] زح ج: م٢ .
(١٦) وذلك ما أردنا أن نبيّن] _ م٢= وذلك ما اردنا بيانه: م٥ .

61

When the eye is moving,
distant objects are thought to be different [in speed] from those closer.
For let A [and] G be two viewed [objects] on rectilinear [extensions] AB, GD,
and [let] the eye's position be E, and we draw ray[s] EG, ED, EA.
Then I say point A that is farther is thought to be different [in speed from G].
So let us extend ED until it falls on the viewed [object],
and let it be EB;
Now because angle GEB is greater than angle AEB,
AB is seen as smaller than GD;
point A is then [thought to be] different [in speed from point G].
And that is what we wished to demonstrate.

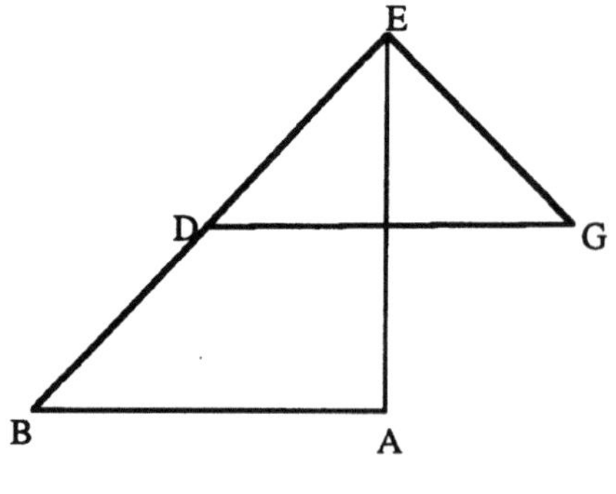

FIGURE 61

A	B	G	D	E	W	Z	H	T	I	K	L	M	N	S	O	F
ا	ب	ج	د	ه	و	ز	ح	ط	ی	ك	ل	م	ن	س	ع	ف

سا

إذا كان البصر متحركاً

تكون الأشياء البعيدة يُظَنّ أنّها مختلفة عمّا هو أقرب منها
فليكن ا جـ منظورا إليهما ويكونا على استقامة ا ب جـ د
وموضع البصر نقطة ه ونخرِج شعاع ه جـ ه د ه ا
فأقول إنّ نقطة ا التى هى أبعد يُظَنّ أنّها مختلفة
فلنخرج ه د حتّى يقع على المنظور إليه
وليكن ه ب
فلأنّ زاوية جـ ه ب أعظم من زاوية ا ه ب
يكون ا ب يرى أصغر من جـ د
فنقطة ا إذاً مختلفة
وذلك ما أردنا أنْ نبيّن ٥٥

(٢-٤) إذا كان البصر متحركاً تكون االأشياء البعيدة يظنّ أنّها مختلفة عمّا هو أقرب منها فليكن ا جـ منظورا إليهما] ـ م٤ .

(٤) ا جـ منظورا إليهما] اب منظور اليهما: م٢ ، م٥ .

(٤) ويكونا] ويكونان: م١ .

(٥) ه جـ ه د ه ا] ه جـ د ه ا: م١= ه جـ ه د .

(٧) فلنخرج] فنخرج: م١ ، م٣ .

(٨) وليكن] فليكن: م١ .

(١٠) جـ د] + وهو متساوية= م١ .

(١١) إذا] اذاً: م٢ .

(١٢) أن نبيّن] بيانه: م٥ .

62

Magnitudes that grow [in size] are thought to become closer to the eye. For let the magnitude of the viewed [object] be AB, and [let] the eye be point G; and we draw ray[s] GA, GB, and let [line] AB grow [in size] and come to the status of [line] BD, and we draw ray GD. Now because angle BGD is greater than angle BGA, BD is greater than BA; and objects that are seen under large angles are thought to be closer to the eye. And that is what we wished to demonstrate.

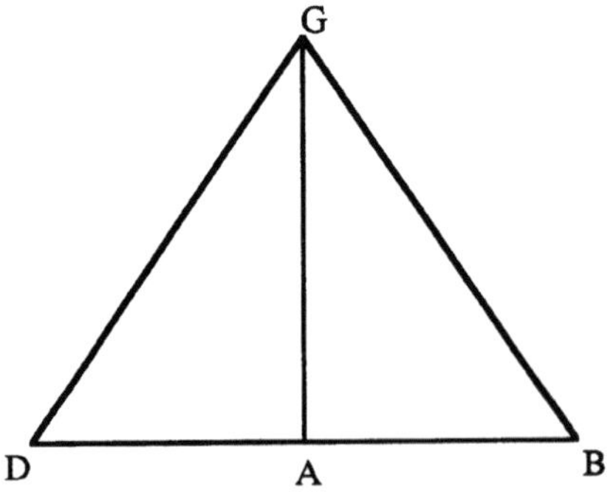

FIGURE 62

A	B	G	D	E	W	Z	H	T	I	K	L	M	N	S	O	F
ا	ب	ج	د	ه	و	ز	ح	ط	ي	ك	ل	م	ن	س	ع	ف

سب
الأقدار التى تنمو يُظَنّ أنّها تقرب إلى البصر
فليكن المنظور إليه عليه اب
والبصر نقطة ج ونخرِج شعاع جا جب
ولينمو اب ويصير بمنزلة بد
ونخرِج شعاع جد
فلأنّ زاوية بجد أعظم من زاوية بجا
يكون بد أعظم من با
والأشياء التى ترى من زوايا عظام
يُظَنّ أنّها أقرب إلى البصر
وذلك ما أردنا أنْ نبيّن ه‍ه

---

(٢) تنمو] تنموا: م١، م٣، م٥ .
(٣) فليكن ... اب] فليكن قدرا منظورا إليه عليه اب: م١، م٣ .
(٤) نقطة] ـ م٥ .
(٥) ولينمو] ولينموا: م٥ .
(٥) اب] با: م٢، م٣، م٤، م٥ .
(٥) ويصير] او يصير: م١= ويصيره: م٢ .
(٧) بجد] بجد ا: م١ .
(٨) يكون ... با] ـ م١، م٣ .
(٩) عظام] عظاما: م١.
(١٠) يظن] يظن ترى: م١ .
(١٠) أقرب] فس م١ .
(١٠) وذلك ما أردنا أن نبيّن] ـ م٢= وذلك ما اردنا بيانه: م٥ .

63

[When] unequally distant objects, with nonaligned ends and middle,
are [placed] on a straight line,
then such a figure is seen at times as concave, and at other times as convex.
For let the viewed [objects] be A, B, G,
and points D, E, Z not in line;
and let the eye be at point H, and let ray[s] HB, HA, HG fall,
and we join lines BA, AG.
Now if we change the view from the position of point H
and make it come to point B, and A <and> G,
then BAG is seen as concave
because the points are not on a straight line;
then the eye becomes point T,
and let ray[s] TD, TE, TZ fall.

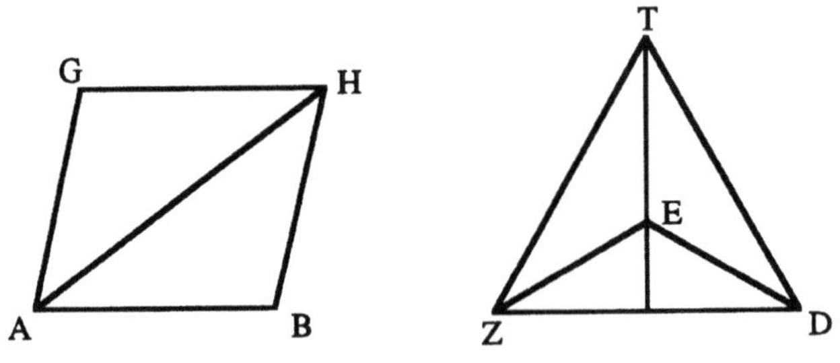

FIGURE 63

A	B	G	D	E	W	Z	H	T	I	K	L	M	N	S	O	F
ا	ب	ج	د	ه	و	ز	ح	ط	ى	ك	ل	م	ن	س	ع	ف

سجـ

الأشياء المختلفة البُعد وأطرافها على غير مرتّبةٍ عند الوسط
على خطّ مستقيم
فإنّ ذلكً الشكلً مرّةً يرى غائراً ومرّةً يرى متحدّباً
فليكن أشياء منظور إليها و هى ا ب جـ
ونقط د ه ز وهى على غير مرتّبة
وليكنْ البصر على نقطة ح وليقعْ شعاع ح ب ح ا ح جـ
ونصل خطوط ب ا ا جـ
فإذا بدّلنا بالنظر من موضع نقطة ح
وصرنا إلى نقطة ب و ا ‹و› جـ
يكون ب ا جـ يرى غائراً
من أجْل أنّ النقط ليست على خطٍّ مستقيمٍ
ثمّ يصير البصر نقطة ط
وليقع شعاع ط د ط ه ط ز

----

(٢) أطرافها] اطرافيها: م٣ .
(٤) مرّةً يرى غائراً] مرّةً يرى غاير: م٤= يرى مرّةً غايراً: م٥ .
(٤) متحدّباً] منحدواً: م٥ .
(٥) فليكن أشياء منظور] فليكن اشياء منظورا: م٣= فلتكن اشياء منظور: م٤ .
(٥) ا ب جـ] خط ا ب جـ: م٢= خط اب اجـ: م٥ .
(٦) نقط] نقطة: م٣ ، م٤ . (٧) على] – م٢ ، م٣ ، م٤ ، م٥ .
(٧) ح ب ح ا ح جـ] ح ب ح د: م٣= جب جا جح: م٤= جا جب جح: م٥ .
(٩) بدّلنا] تس م٢= بدنا: م٢= بدانا: م٣ ، م٤ ، م٥ .
(١٠-١١) وَ ا ‹وَ› جـ يكون ب ا جـ يرى غائراً] وَ ا جـ يكون اب جـ يرى غايراً: م١=
و ا وجـ يكون ا جـ ه يرى غائراً: م٢= وَ ا وجـ يكون اب جـ يرى غايراً: م٣ ، م٥ .
(١٠) صاعداً] هش م١ .
(١٢) النقط] النقطة: م٤ ، م٥ .
(١٣) يصير] نصير: م٥ .
(١٣-١٥) ط وليقع شعاع ط د ط ه ط ز...نقطة ط] ط وليقع شعاعا
ط ز ط د ... نقطة ط: هش م٢= م٥= ط وليقع شعاع ط ز ط د ...نقطة ط: م٤ .

Now if we change the view from the position of point T, then it is seen once at D and once at Z, and DEZ is seen as convex, because the points that are in front of it are not seen. And that is what we wished to demonstrate.

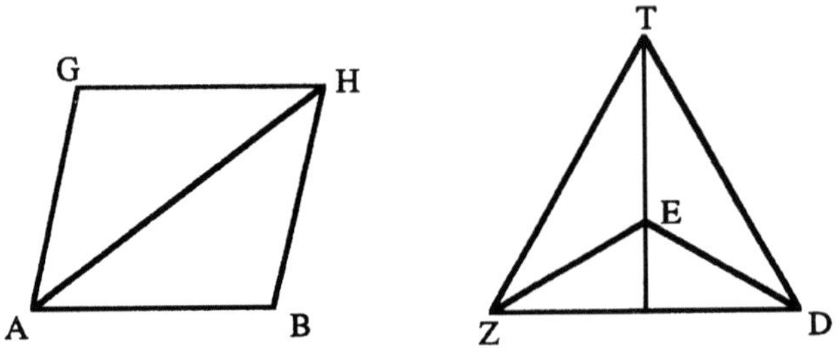

FIGURE 63

A	B	G	D	E	W	Z	H	T	I	K	L	M	N	S	O	F
ا	ب	ج	د	ه	و	ز	ح	ط	ى	ك	ل	م	ن	س	ع	ف

فإذا بدّلنا بالنظر من موضع نقطة ط يرى مرّةً إلى د ومرّةً إلى ز ويكون د ه ز يرى متحدّباً من أجْل أنّ المواضع التى بين يديه ليس يراه وذلك ما اردنا أنْ نبيّن ٥°

----

(١٥) بدّلنا] بدنا: م٢ = بدانا: م٣، م٤، م٥ .

(١٧) د ه ز] ه د ز: م٥ .

(١٧) متحدّباً] منحدواً: م٥ .

(١٨) من أجل] فمن اجل: م٢ .

(١٨) المواضع التى ...يراه] الموضع الذى بين يديه ليس يرانيه: م١ = الموضع الذى بين يديه ليس يراه: م٣ .

(١٩) وذلك ما أردنا أن نبيّن] – م٢ = وذلك ما اردنا بيانه: م٥ .

64

When there is a quadrilateral figure, and a line is erected at right angles from the position of the intersection of its diameters, and the eye is placed on it, then the sides of the quadrilateral figure are seen as equal, and its diameters are seen as equal.

For let there be a quadrilateral figure, namely ABGD, and we draw the diameters, and they are AG, BD; and we erect from their intersection, that is point E, a line upward at right angles to the assumed plane, namely [line] EZ, the eye's position being point Z, and let ray[s] ZA, ZB, ZG, ZD fall.

Now because DE is equal to EG, and EZ is common between them and the angles are right, then base ZG is equal to base ZD;

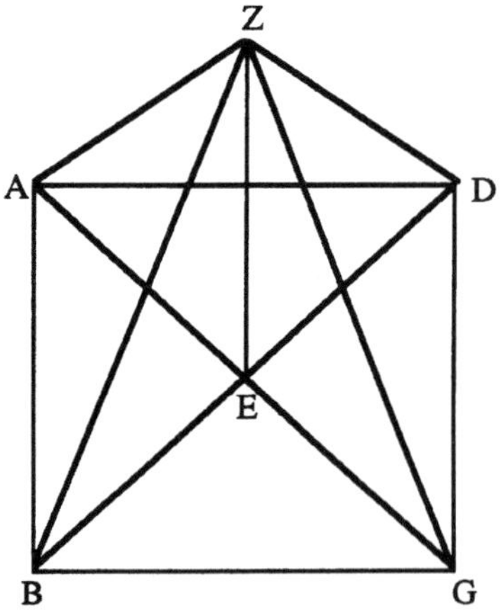

FIGURE 64

A	B	G	D	E	W	Z	H	T	I	K	L	M	N	S	O	F
ا	ب	ج	د	ه	و	ز	ح	ط	ى	ك	ل	م	ن	س	ع	ف

سد

إذا كانت مربّعة وأُقيم خطّ على زوايا قائمةٍ
مِن موضع تقاطع قطريها
ووُضِع البصر عليه
فإنّ أَضلاع المربّعة ترى متساويةً
وأقطارها ترى متساويةً
فليكن مربّعة عليها ا ب جـ د
ونخرِج قطريها وهما ا جـ ب د
ونقيم من موضع تقاطعهما وهو نقطة ه خطّاً فى الهواء
على زوايا قائمة من السطح الموضوع عليه ه ز
ويكون البصر على نقطة ز
وليقع شعاع ز ا ز ب ز جـ ز د
فلأنّ د ه مساوٍ لـ ه جـ
و ه ز مشترك بينهما وزوايا قائمة
تكون قاعدة ز جـ مساوية لقاعدة ز د

---

(٢) خطّ] على خط: م٢، م٤.
(٣) تقاطع] _ م٢.
(٦) وأقطارها ترى متساوية] _ م٢، م٥= واقطارها يرى متساوية: م٣.
(٧) فليكن] فلتكن: م٤= م٥.
(٨) وهما] _ م١= _ م٣.
(٩) تقاطعهما] تقاطعها: م١.
(٩) الهواء] الهوى: م٥.
(١٠) الموضوع] هش م٢= الموضع: م٢، م٤.
(١١) البصر على] البصر عليه على: م٢، م٤.
(١٢) زجـ زد] زد زجـ: م١.
(١٣، ٢٢) مساو] مساوى: م٣.
(١٤) وزوايا] على زوايا: م١، م٢، م٤، م٥.
(١٤-١٥) بينهما وزوايا قايمة تكون قاعدة زجـ متساوية] _ م٣.
(١٥) زجـ...زد] زب: م١=ب جـ...زد: م٢= زجـ...زب: م٣= م٥.

and angles at bases that are enclosed by equal lines
are equal,
so angle EZG is equal to angle EZD,
[and] therefore EG is seen equal to ED;
and in the same way, angle AZE is also equal to angle BZE,
so AG is seen equal to BD;
[again] because GZ is equal to ZD, and AZ equal to ZB,
and AB equal to GD, then three of them are equal to the three [of these],
and the angles likewise;
therefore AB is seen equal to GD;
and in the same way, the remaining sides are seen as equal.
And that is what we wished to demonstrate.

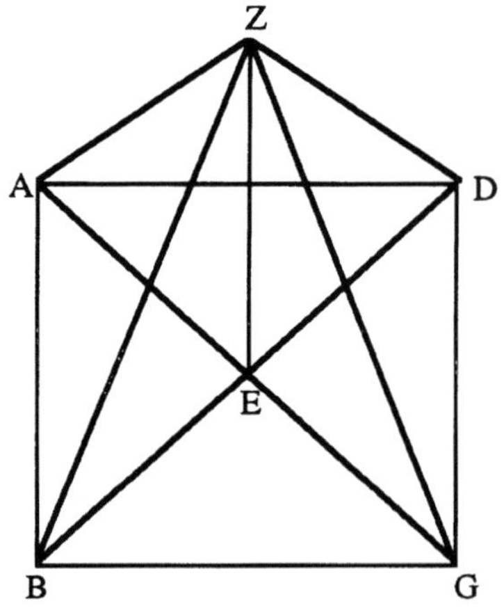

FIGURE 64

A	B	G	D	E	W	Z	H	T	I	K	L	M	N	S	O	F
ا	ب	ج	د	ه	و	ز	ح	ط	ى	ك	ل	م	ن	س	ع	ف

والزوايا التى على القواعد التى يحيط بها خطوطٌ متساويةٌ فهى متساويةٌ
فزاوية ه ز ج مساوية لزاوية ه ز د
فَ ه ج يرى مساوياً لِ ه د
وكذلك أيضاً زاوية ا ز ه مساوية لزاوية ب ز ه
فَ ا ج يرى مساوياً لِ ب د
وأيضاً لأنّ ج ز مساوٍ لِ ز د وَ ا ز مساوٍ لِ ز ب
وَ ا ب مساوٍ لِ ج د فالثلثة تكون مساويةً لثلثة
وعلى مثل ذلكَ تكون الزوايا
فَ ا ب يرى مساوياً لِ ج د
وكذلك أضلاع الباقيةَ ترى متساويةً
وذلك ما أردنا أنْ نبيّن ٥٥

---

(١٦) على القواعد التى] ــ م١= عند القواعد التى: م٥ .

(١٦) بها] فس م١ .

(١٨) ه ز د] ه ز ب: م١ ، م٣ .

(١٩) ه د] ه ب: م١ ، م٣ .

(٢٠ ، ٢٣) متساوية] ــ م٢ . (٢٠) ب ز ه] د ز ه: م١ .

(٢٢) لِ ز د وَ ا ز مساوٍ] ــ م١ ، م٣ .

(٢٣) وَا ب ... لثلثة] وا د ... لثلاثة: م٥ .

(٢٥) مساوياً] مساوٍ: م٢ .

(٢٦) ترى متساوية]+ الاضلاع الباقية: م١ ، ــ م٢ ، م٤ ، م٥ .

(٢٧) أن نبيّن] + تم كتاب اقليدس فى اختلاف المناظر والشعاعات وهو اربعة وستون شكلاً والحمدلله ربّ العالمين والصلوة على سيد الانبيا محمد وآله اجمعين: م١= تم كتاب المناظر لاوقليدس الصورى وهذا بحمد والمنة كتبه ابن قاهر قونوى وكان اصله سقيماً مغلقا بتراً (جداً؟): م٢ (+ بلخ: هش م٢)= كمل كتاب اوقليدس فى اختلاف المناظر وهو اربعة ستون شكلاً والحمدلله رب العالمين: م٣= تم بعون الله؟ وتوفيقه ...الثالث والعشرون من ذى القعده من شهور وسنه احدى وخمسين وسبعمئة للهجره وصلى الله على سيدنا محمد خاتم النبيين واشرف المرسلين وعلى اصحابه المنتخبين وسلّم سليماً كثيراً ويسمّى هذا الكتاب بالمناظر: م٤= بيانه + تم كتاب المناظر لاوقليدس برحمه هلىل بن سرحون والحمدالله رب العالمين والصلاة على محمّد وعلى آله وسلم تسليماً: م٥ .

<Correction of Optics>

Abū Yūsuf Ya'qūb ibn Isḥāq al-Kindī said in the "Correction of the Book of Optics" in proposition 60 [that] [of] magnitudes with equal [speed of] motion, such as [magnitudes] AB, GD [moving] on parallel lines such as [lines] BT, DI, [and] cutting the perpendicular such as EW issued from the eye at right angles, the one closer to the eye, which is BA, is seen at times faster than the farther one, when it is directed toward the perpendicular, at times equal to it, when its beginning [part]s end at the perpendicular, and at other times slower [than it], when it crosses the perpendicular. Then he lengthened his claim with a demonstration, which is clear from this figure, besides being one of the necessary [consequences] of proposition 57.

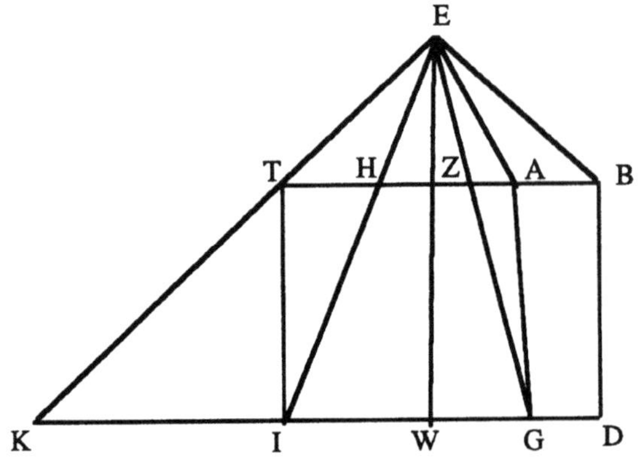

FIGURE 1

A	B	G	D	E	W	Z	H	T	I	K	L	M	N	S	O	F
ا	ب	ج	د	ه	و	ز	ح	ط	ى	ك	ل	م	ن	س	ع	ف

<إصلاح المناظر>

قال أبو يوسف يعقوب بن إسحق الكندى
فى إصلاح كتاب المناظر فى شكل س
الأقدار المتساوية الحركة كَ اب جد
على خطوط متوازية كَ ب ط دى
قاطعة للعمودِ الخارجِ من البصرِ كَ ه و على قوائم
يرى أقربُها من البصر وهُو ب ا
تارةً أسرع مِن الأبعد إذا كان متوجّهاً إلى العمود
ومرّةً مساو له إذا انتهت أوائلها إلى العمود
وأُخرى أبطأ إذا جاوز العمود
ثمّ طوّل فى إبانة دعواه
وهو ظاهرٌ من هذا الشكل
مع أنّه من لوازم شكل نز

---

(١) إصلاح المناظر] ـ ص ١ ، ص ٢ ، ص ٣ ، ص ٤ ، ص ٥ ، ص ٦ ، ص ٧ = من المناظر: ص ٨ = ومن افادات الخواجه: ص ٩ .

(٢) أبو يوسف يعقوب بن إسحق] ابو يوسف يعقوب بن اسحق بن: ص ٩ .

(٣) فى إصلاح] فى اصلاحه: ص ١ ، ص ٢ ، ص ٤ ، ص ٦ ، ص ٧ ، ـ ص ٢ .

(٣) كتاب المناظر] لهذا الكتاب: ص ١ ، ص ٢ ، ص ٤ ، ص ٥ ، ص ٦ ، ص ٧ ، ـ ص ٣ = فى اصلاح هذا الكتاب: ص ٩ .

(٣) س] سه : ص ١ ، ص ٢ ، ص ٣ ، ص ٤ ، ص ٥ ، ص ٨ = سد: ص ٩ .

(٤) المتساوية الحركة] المتساوية الحركات: ص ٩ .

(٦) ه و] هز: ص ٩ .

(٧) ب ا] ا: ص ٣ = با: ص ٤ ، ص ٨ .

(٨) إذا] اذاذا: ص ٥ .

(٩) ومرّةً مساو لَه إذا انتهت أوائلها إلى العمود] ـ ص ٦ .

(٩) مساو له] مساويا له: ص ٤ ، ص ٩ .

(٩) أوائلها] باوائلها: ص ٨ .

(١١) طوّل فى] ـ ص ٩ .

(١٣) مع] ـ ص ٣ .

He [also] said in [proposition] 61 [that] [for] all marks on a straight line like A [and] B, when the eye, which is G, is moving in the direction of the path of the line upon which these [marks] are, namely AB, then they are seen in different orders, that is, the one closer to the eye, like A, is seen at times ahead of B, the farther one, and at times together with it on a straight line, and at other times behind it. And that is also clear from this figure.

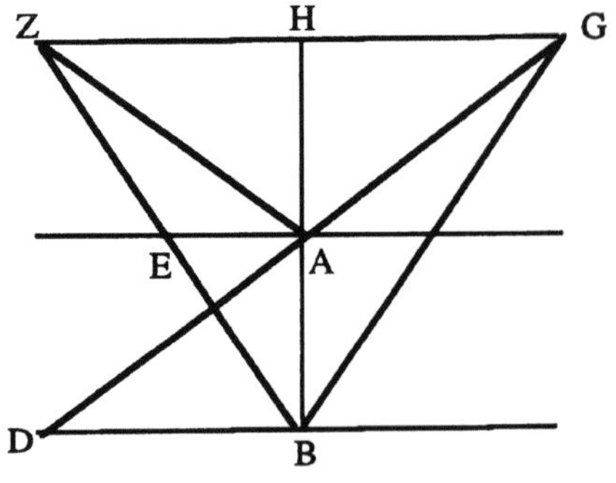

FIGURE 2

A	B	G	D	E	W	Z	H	T	I	K	L	M	N	S	O	F
ا	ب	ج	د	ه	و	ز	ح	ط	ى	ك	ل	م	ن	س	ع	ف

وقال في سا
كلّ العلامات التي علي خطّ مستقيمٍ كَ آ بَ
إذا كان البصر وهو جَ متحَرّكاً
نحو سمت الخطّ التي هي عليه يعني آب
فإنّها يرى مختلفة الترتيب
أعني أنّ الأقرب من البصر كَ آ يرى تارةً متقدّماً على بَ الأبعد
ومرّةً معه على خط مستقيمٍ وأُخرى متأخّراً عنه
وهُو أيضاً ظاهرٌ مِن هذا الشكل

(١٤-١٥) في سا كل] سا كل: ص٨ = فما كلّ: ص٩ .

(١٨) يرى] ترى: ص٥ .

(١٩) كَ آ] كما: ص٢ ، ص٨ = _ ص٤ .

(٢٠) متأخّراً] متاا: ص٨ .

(٢١) أيضاً] فس: ص٩ .

(٢١) هذاالشكل] الشكل: ص١ ، ص٢ ، ص٣ ، ص٤ ، ص٦ ، ص٧ =
+ والله اعلم: ص٩ .

Appendices

Appendix A

Manuscript Plates

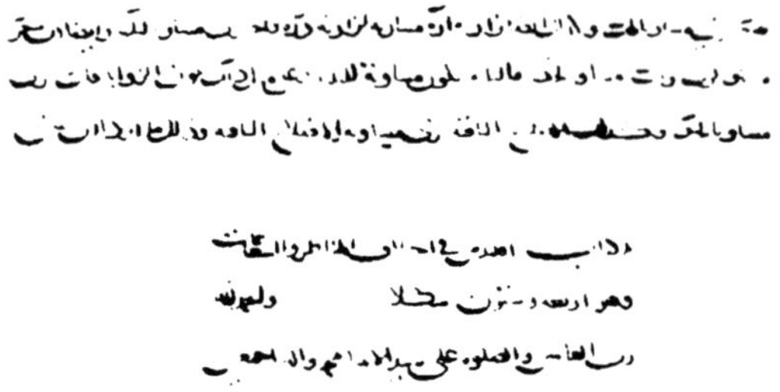

Manuscript Plate I: Courtesy of the Topkapı Museum New Library

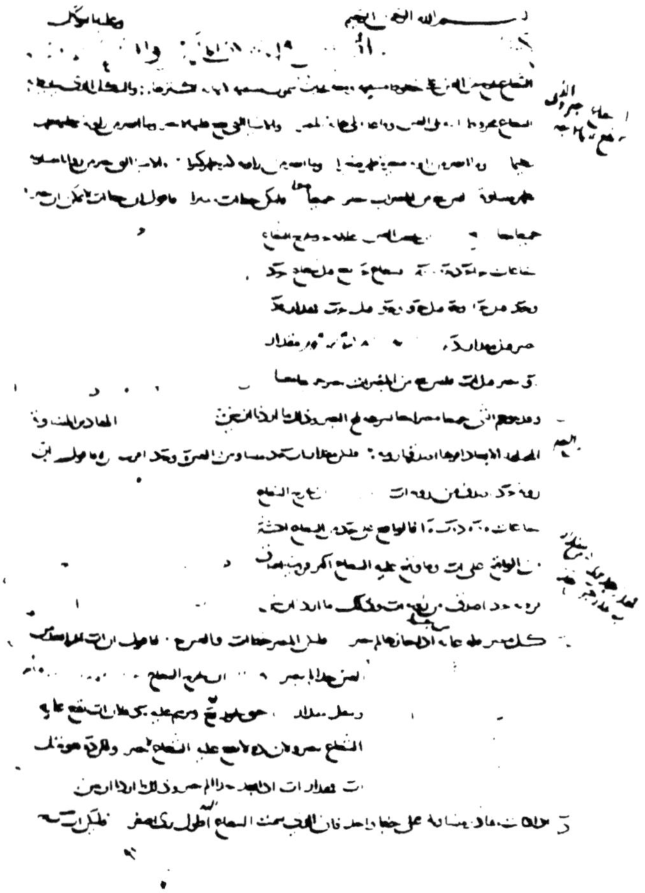

Istanbul: Topkapı Sarayı, Ahmet III 3464, 4: 59b–74a, 625H. m1/م١

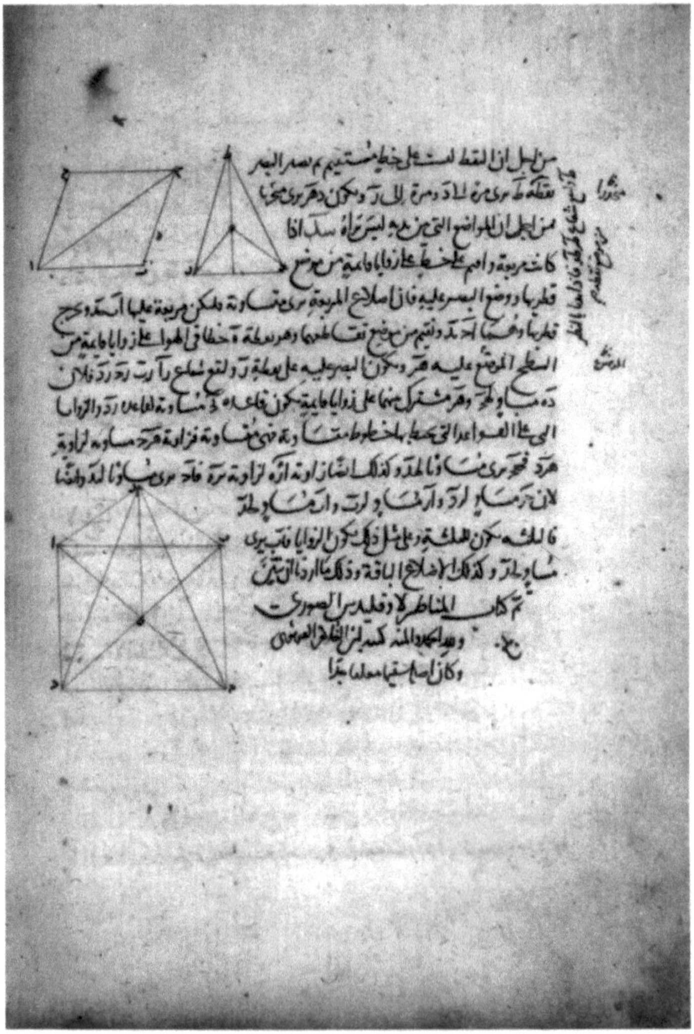

Manuscript Plate II: Courtesy of Leiden University Library

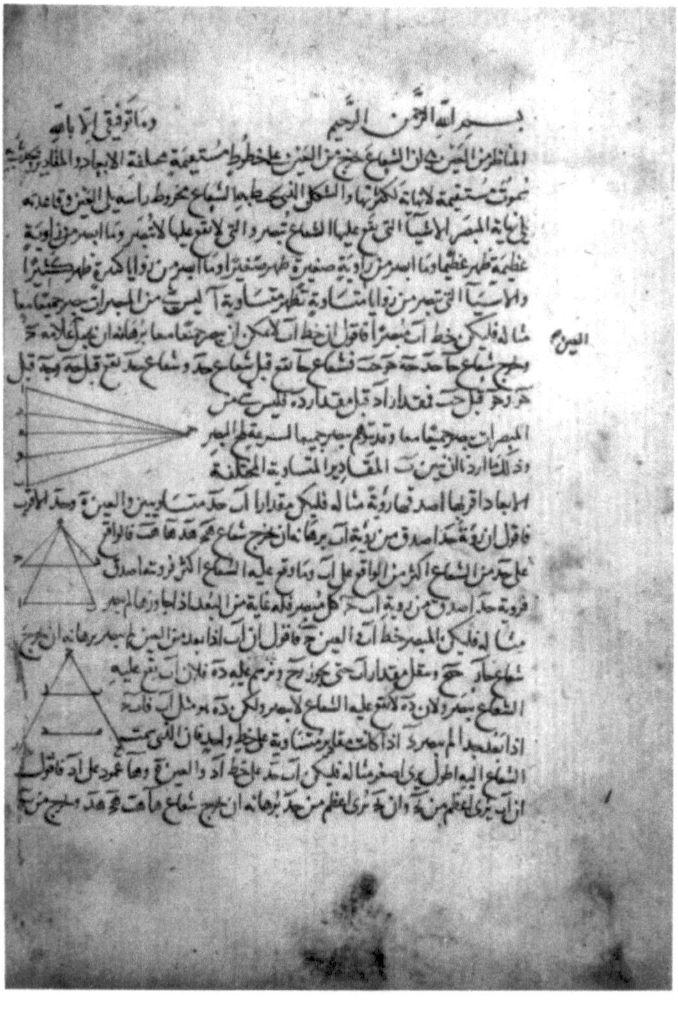

بسم الله الرحمن الرحيم وما توفيقي إلا بالله

Leiden: Universiteitsbibliotheek, Or. 133, 5: 81–105, 692H. m2/٢م

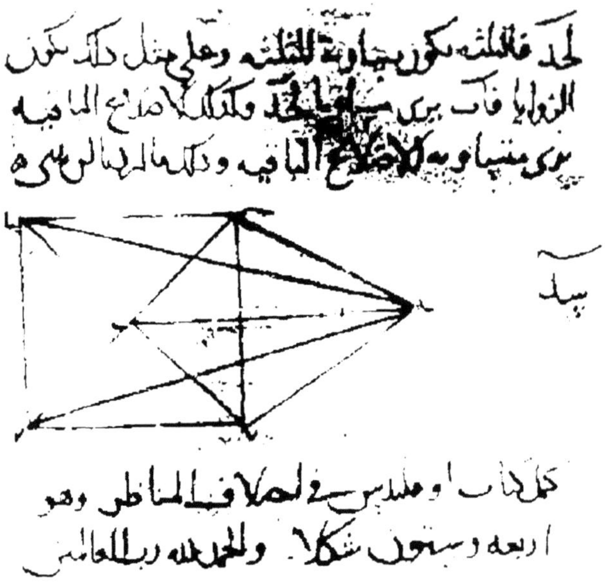
كمل كتاب اوقليدس في اختلاف المناظر وهو
اربعة وستون شكلا والحمد لله رب العالمين

Manuscript Plate III: Courtesy of the Egyptian National Library

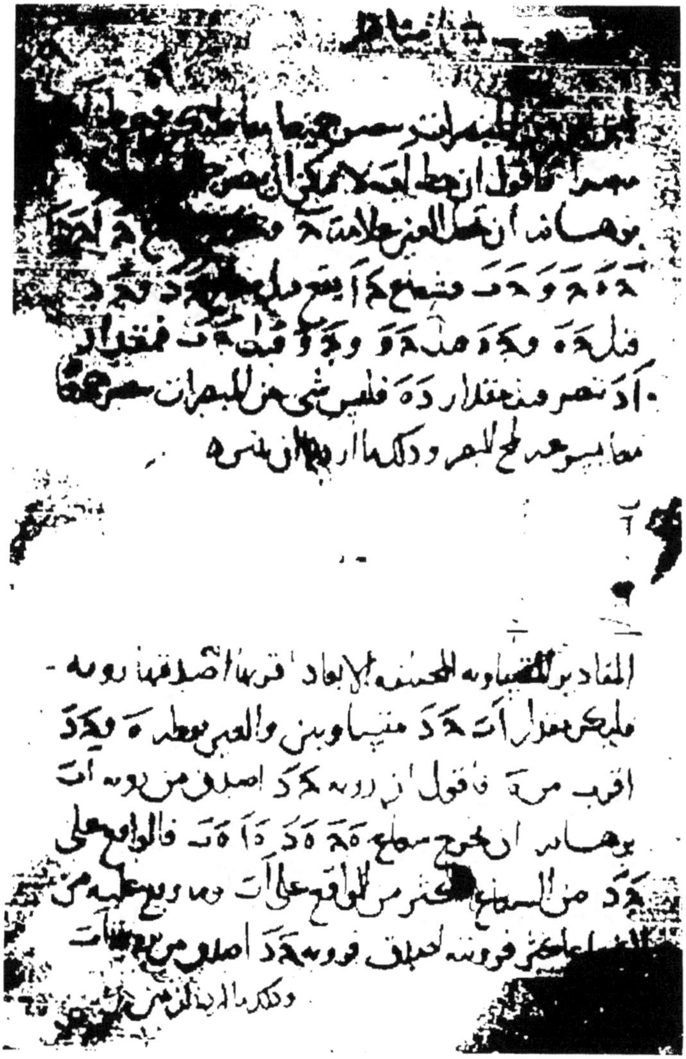

Cairo: Dār al-Kutub al-Miṣriyya, Dr 260, 2. 59a–92a, 600H. m3/٣م

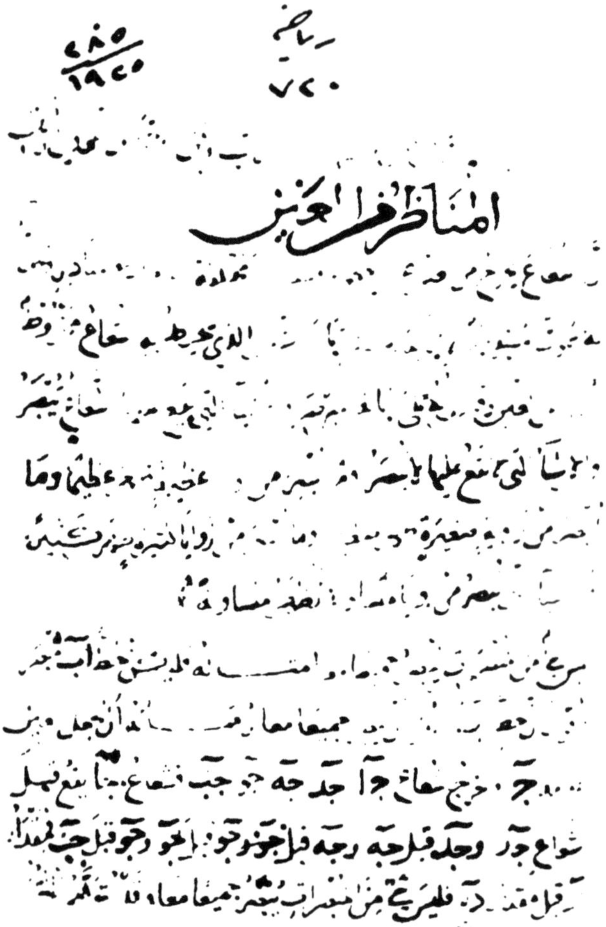

Manuscript Plate IV: Courtesy of the Egyptian National Library

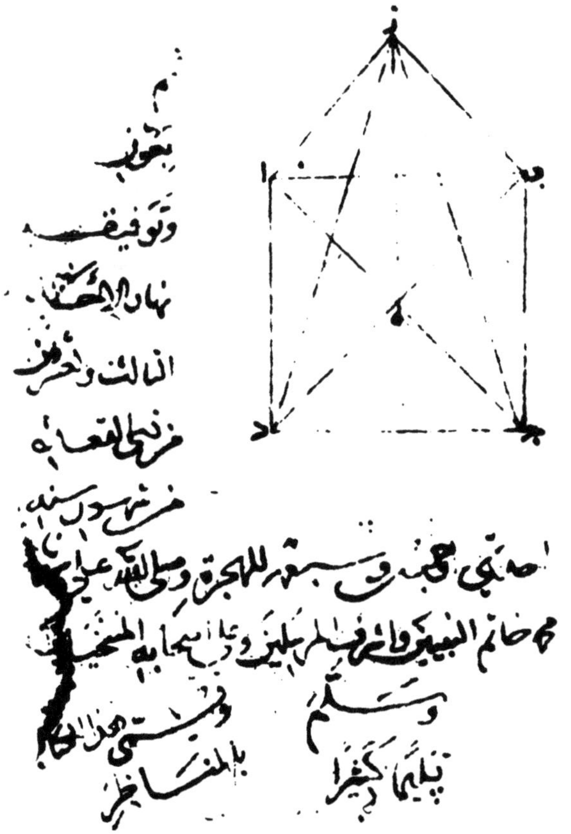

Cairo: Dār al-Kutub al-Miṣriyya, Dr 720: 2–77, 751H. m4/٤م

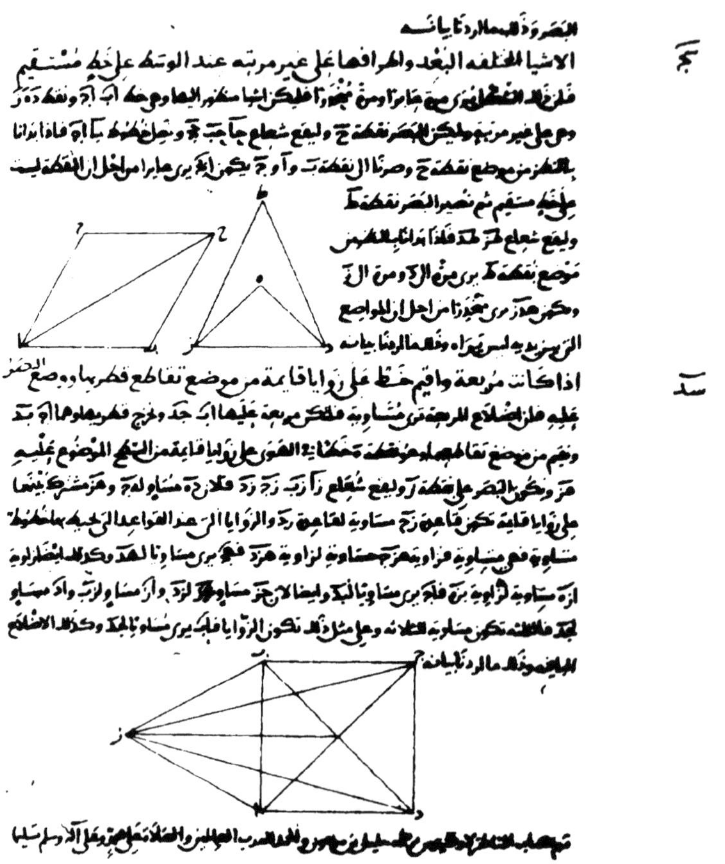
الاشيا المختلفه البعد واطرافها على غير مرتبه عند الوسط على خط مستقيم

اذا كانت مربعة واقيم خط على زوايا قايمة من موضع تقاطع قطريها ووضع البصر

Manuscript Plate V: Courtesy of Basel University Library

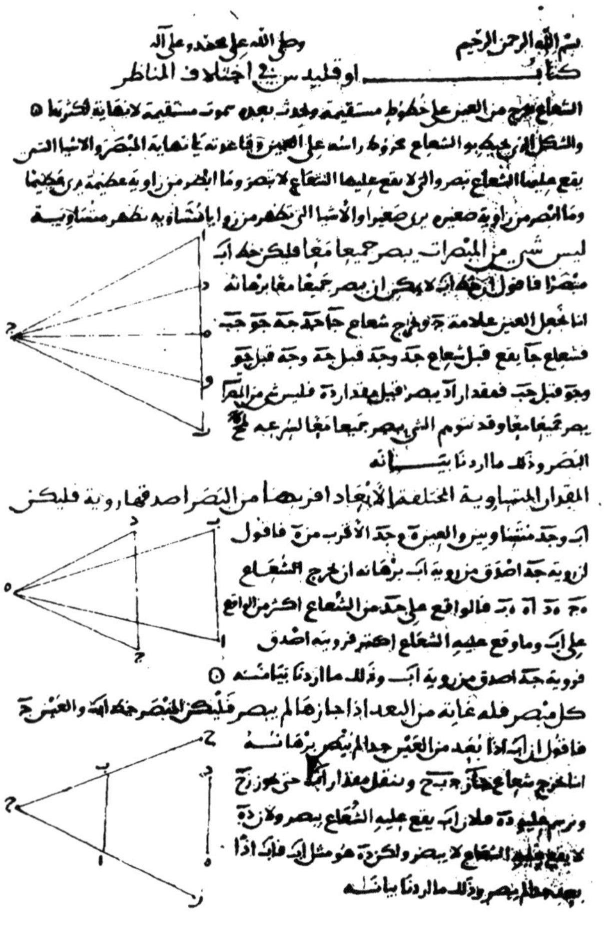
بسم الله الرحمن الرحيم وصلى الله على محمد وعلى آله

كتاب أوقليدس في اختلاف المناظر

الشعاع يخرج من العين على خطوط مستقيمة وتحدث بعده سموت مستقيمة لانهاية لكثرتها

والشكل الذي يحيط به الشعاع مخروط رأسه على العين وقاعدته في نهاية المبصر والأشياء التي

يقع عليها الشعاع تبصر والتي لا يقع عليها الشعاع لا تبصر وما أبصر من زاوية عظيمة يرى عظيما

وما أبصر من زاوية صغيرة يرى صغيرا والأشياء التي تظهر من زوايا أشباه تظهر متساوية

آ

ب

ج

Basel: Universitätsbibliothek Microfilm, 2: 20b–32b, ca. 680H m5/٥م

Appendix B

Supplementary Passages

Appendix B Supplementary Passages

From al-Ya‘qūbī, *Ta'rīkh*, ed. Houtsma, v. 1, p. 139:

ولأقليدس هذا كتاب فى المناظر واختلافها من مخارج العيون والشعاع
يقول فيه
[١] إنّ الشعاع تخرج من العين على خطوط مستقيمة
وتحدث بعد سموت لا نهاية لكثرتها
[٢] فإنّ الأشياء التى يقع عليها الشعاع تُبصَر
والتى لا يقع عليها الشعاع لا تُبصَر
ويمثّل فى ذلك أشكالاً مختلفة يبيّن بها مخرج النظر
وكيف يختلف عدّة الأشكال التى يبيّن بها ذلك
وهى أربعة وستون شكلاً

From [Pseudo-] Euclid, *De speculis, Kitāb al-Mir'āh li-Uqlīdis*
Florence: Biblioteca Medicea Laurenziana Or. 152:

كتاب المراءى [sic] لأوقليدس :
[١] العين ينبثّ من ناظرها قوّة نوريّة
تؤثّر فيما لاقت من الجوّ أجمع ضياء صنوبريّاً
زُجّه أعنى مستحدّه عند الناظر نفسه
وكلّما بَعُدت اتّسعت قاعدته
[٢] ويكون الشكل الذى يحيط به ذلك الضياء مخروطاً أسطوانيّاً
مستحدّه عند الناظر ونهايته يلى المنظور إليه
[٣] فما وقع عليه الضياء ‹أدركه البصر›
وما لم يقع عليه لم يدركه البصر

From *Kitāb al-Manāẓir wa al-marāyā al-muḥriqa* (Aḥmad ibn ʿĪsā)
Istanbūl: Süleymaniye, Ragip Paşa 934, ḥ1, fols. 5b–6b:

كتاب فيه المناظر والمرايا المحرقه (أحمد بن عيسى):
قالت الفلاسفة وأقليدس منهم ومعهم إنّ
[١] العين ينبثّ من ناظرها قوّة نوريّة
تؤثّر فيما لاقت من الجوّ المضيء أعنى الهواء إذا كان مضيئاً
أجمع ضياء شكله صنوبرياً[sic] شبيه بالزُجّ
وزُجّه أعنى مستحدّه عند الناظر نفسه وكلّما بَعُد اتّسعت قاعدته
[٢] فيكون الشكل الذى يحيط به ذلك الضياء مخروطاً أسطوانياً
مستحدّه عند الناظر ونهايته تماسّ المنظور إليه
[٣] فما وقع عليه ذلك الشعاع النورىّ
وهو الضياء المنبثّ من البصر رآه البصر وأدركه
وما لم يقع عليه ذلك الشعاع لم يره البصر و لم يدركه
فيصير ذلك النور الشعاعىّ على المخروط الخارج من البصر
المنبثّ فى الهوى المضيء كالعضو للحىّ الناظر
فيحسّ به الناظر المرتفعة عنه الموانع كلّما لامسه
ووقع عليه من الأجرام من مستحدّه الذى هو زُجّه
وهو مخرج النور من البصر يخرج على زاوية
[٤-١] فما كانت نهايتى [sic] ذلك النور الخارج من البصر
المماسّ لنهايتى الجرم المنظور إليه يحيطان من الناظر إلى المنظور إليه
بزاوية عظيمة رُ‹ؤ›ى ذلك الجرم المنظور إليه عظيماً
[٤-٢] وما كانت نهايتى ذلك النور الخارج عن البصر
المماسّ لنهاية الجرم المنظور إليه يحيطان من الناظر إلى المنظور إليه
بزاوية ‹صغيرة رُؤى ذلك الجرم المنظور إليه: هش› صغيراً
والأجرام التى ترى بزوايا من ذلك الشعاع النورىّ مختلفة
ترى مختلفة الأقدار أيضاً
[٤-٤] والأجرام التّى ترى بزوايا من ذلك الشعاع النورىّ متساوية
ترى متساوية الأقدار أيضاً
[٤-٣] والأجرام التى ترى بزوايا من ذلك الشعاع النورىّ كثيرة
تظهَر كثيرة

From *Kitāb fī 'Ilal mā ya'riḍu fī al-marāyā min ikhtilāf al-manāẓir*
(Qusṭā ibn Lūqā al-Yūnānī)
Mashhad: Āstān-i Quds 392 (5593), 1, pp. 4–6:

كتاب فى علل ما يعرض فى المرايا من اختلاف المناظر (قسطى بن لوقا):
[١] البصر يكون بشعاع ينبثّ من العين
ويقع على المبصرات فتُبصَر بالشعاع الواقع عليها
فما وقع عليه الشعاع البصرىّ يبصره الإنسان
وما لم يقع عليه الشعاع البصرىّ لم يبصره الإنسان
وكذلك لا يبصر الأشيآء التّى بينها وبين البصر حواجز
تمنع وتعوق شعاع البصر أنْ يقع عليها . . .
[٢] الشعاع البصرىّ ينبثّ من العين فى صورة شكل مخروط
مستحدّه يلى العين الباصرة وقاعدته تلى المبصرات التّى يقع عليها
[٣] فما وقع عليه قاعدة المخروط الشعاعى أدركه البصر
وما لم يقع عليه الشعاع البصرىّ لم يدركه حاسّة البصر
وهذا المخروط البصرى ينفذ من العين الباصرة على خطوطٍ مستقيمةٍ
لا اعوجاج فيها
وله زاوية يحيط بها ضلعان من أضلاع المخروط
وتلك الزاوية تلى المبصرات . . .
المبصر الواحد يرى فى صورةٍ مختلفةٍ
لاختلاف زوايا الشعاع التى يرا بها
لأنّ زوايا المخروط مختلفة فى الصغر والكبر
[٤-١] فإذا رأى المبصر بزاويةٍ عظيمةٍ رأى عظيماً
[٤-٢] وإذا رأى بزاويةٍ صغيرةٍ لذلك يرى صغيراً . . .
[٤-٣؟] فالمبصر إذا وقع عليه شعاع واحد رأى واحداً
فإن وقع عليه شعاعين رأى إثنين
فإن وقع أكثر من شعاعين رأى أكثر من إثنين
فالمبصرات يرى بعدد الشعاعات الواقعة عليها
والشعاعات الواقعة على المبصرات تكون كثيرةً على جهات مختلفة . . .

From *Taḥrīr al-Manāẓir li-Ūqlīdis* (Naṣīr al-Dīn al-Ṭūsī)
Published in *Majmū‘ al-rasā’il*, Hyderabad, 1358 (=1939), p. 2:

تحرير المناظر لأقليدس (نصير الدّين طوسى):

[١] العين تُحدث باستمداد من الأجرام النيّرة فى الجسم الشفاف
المتوسّط بينها وبين المبصرات كالهواء وما شاكله شعاعاً
كما تُحدثه الأجرام النيّرة وحدها بعينه
ويكون ذلك الشعاع كأنّه منبعثٌ من العين وخارجٌ منها
ثمّ إنّه يصير آلةً لها فى الإبصار
فتختلف أحوال المناظر لاختلاف أوضاعه
فليُصدّق بذلكَ وليُتوهّم ذلك الشعاع
متّصلاً بالعين على خطوطٍ مستقيمةٍ
وليُحدِث سموتاً مستقيمةً لا نهايةَ لكثرتها

[٢] والشكل الشعاعى مخروطٌ رأسه يلى العين
وقاعدته تلى نهاية المبصرات

[٣] فالأشياء التى يقع عليها الشعاع تُبصَر
والتى لا يقع عليها لا تُبصَر

[٤-١، ٢] وما أبصِر من زاويةٍ عظيمةٍ ظهر عظيماً وبالعكس

[٤-٣] وما أبصِر من زوايا كثيرَةٍ ظهر كثيراً

[٤-٤] وما أبصِر من زوايا متساويةٍ ظهر متساويا

[٥، ٦] أقول وممّا ينبغى أنْ يُسلّم قولنا إذا اختلفت جهات الشعاعات
علواً وسفلاً ويميناً ويساراً رؤيَت المبصرات مختلفة الجهات بحسب ذلك
ما يقع عليه الشعاع أكثر فهو أصدق رؤيةً ممّا يقع عليه الشعاع أقلّ
ما يقع عليه سهم المخروط الشعاعى فهو أصدق رؤيةً ممّا حوله لكون
الشعاع الواقع عليه أكثر وأشدّ تراكماً
وما هو أقرب منه أصدق ممّا هو أبعد
ولذلك يقلّب الناظر سهم المخروط نحو ما يقصد رؤيته أوْ
يريد أنْ يحقّقه

[٧] إذا انعطف الشعاع من جسمٍ صقيلٍ كالمرآة حدَثت هناك زاويتان
متساويتان تُسمّى أحدهما زاوية الشعاع والأخرى زاوية الانعطاف

Sources in the History of Mathematics and Physical Sciences

Continued from page ii

Vol. 13: A.I. Dale
Pierre-Simon Laplace, Philosophical Essay on Probabilities, Translated from the fifth French edition of 1825, with Notes by the Translator

Vol. 14: K. von Meyenn (Ed.)
Wolfgang Pauli: Scientific Correspondence IV, Part I: 1950-1952

Vol. 15: K. von Meyenn (Ed.)
Wolfgang Pauli: Scientific Correspondence IV, Part II: 1953-1954

Vol. 16: E. Kheirandish
The Arabic Version of Euclid's Optics, Volumes I and II

MIX
Papier aus verantwortungsvollen Quellen
Paper from responsible sources
FSC® C105338

If you have any concerns about our products,
you can contact us on
ProductSafety@springernature.com

In case Publisher is established outside the EU,
the EU authorized representative is:
Springer Nature Customer Service Center GmbH
Europaplatz 3, 69115 Heidelberg, Germany

Printed by Libri Plureos GmbH
in Hamburg, Germany

Sources in the History of Mathematics and Physical Sciences

16

Springer
New York
Berlin
Heidelberg
Barcelona
Hong Kong
London
Milan
Paris
Singapore
Tokyo

Sources in the History of Mathematics and Physical Sciences

Continued after Index

Elaheh Kheirandish

The Arabic Version of Euclid's *Optics*

(*Kitāb Uqlīdis fī Ikhtilāf al-manāẓir*)

Edited and Translated with Historical Introduction and Commentary

Volume II

Springer

Elaheh Kheirandish
Harvard University
Cambridge, MA 02138
USA
kheirand@fas.harvard.edu

Sources Editor:
Gerald J. Toomer
2800 South Ocean Boulevard, 21F
Boca Raton, FL 33432
USA

With 1 figure.

Library of Congress Cataloging-in-Publication Data
Euclid.
[Optics. English & Arabic]
The Arabic version of Euclid's *optics* = Kitāb Uqlīdis fī ikhtilāf al-manāẓir/edited and translated with historical introduction and commentary Elaheh Kheirandish.
p. cm. — (Sources in the history of mathematics and physical sciences; 16)
Includes bibliographical references and index.
ISBN 0-387-98523-9 (alk. paper)
1. Optics—Early works to 1800. I. Kheirandish, Elaheh. II. Title. III. Title: Kitāb Uqlīdis fī ikhtilāf al-manāẓir. IV. Series.
QC353 .E8313 1998
535—dc21 98-17540

Printed on acid-free paper.

Production managed by Francine McNeill; manufacturing supervised by Jeffrey Taub.
Camera-ready copy prepared by the author using Microsoft Word.
Printed and bound by Braun-Brumfield, Inc., Ann Arbor, MI.
Printed in the United States of America.

9 8 7 6 5 4 3 2 1

ISBN 0-387-98523-9 Springer-Verlag New York Berlin Heidelberg SPIN 10677574

To my parents
Asadollah and Parvin-Dokht
(the "Lion of God" and the "Daughter of Pleiades")

آن قوم که راه بین فتادند وشدند

کس را بیقین خبر ندادند وشدند

آن عقده که هیچکس نتانست گشاد

هریك گرهی برآن نهادند وشدند

منسوب به خواجه نصیر الدّین طوسی

(نقل از احوال وآثار مدرس رضوی، ص ۶۱)

That group of people who, as guides, came and passed
and informed no one with certitude, as they passed
into that complex tangle, unraveled by anyone
each came to place a knot, and so they passed

Attributed to Naṣīr al-Dīn al-Ṭūsī (d. 672/1274)

(Persian poem quoted from Mudarris-i Raḍavī, *Aḥvāl va Āthār*, p. 61)

List of Illustrations

Manuscript folio of Ṭūsī's *Taḥrīr al-Manāẓir* on the following pages.

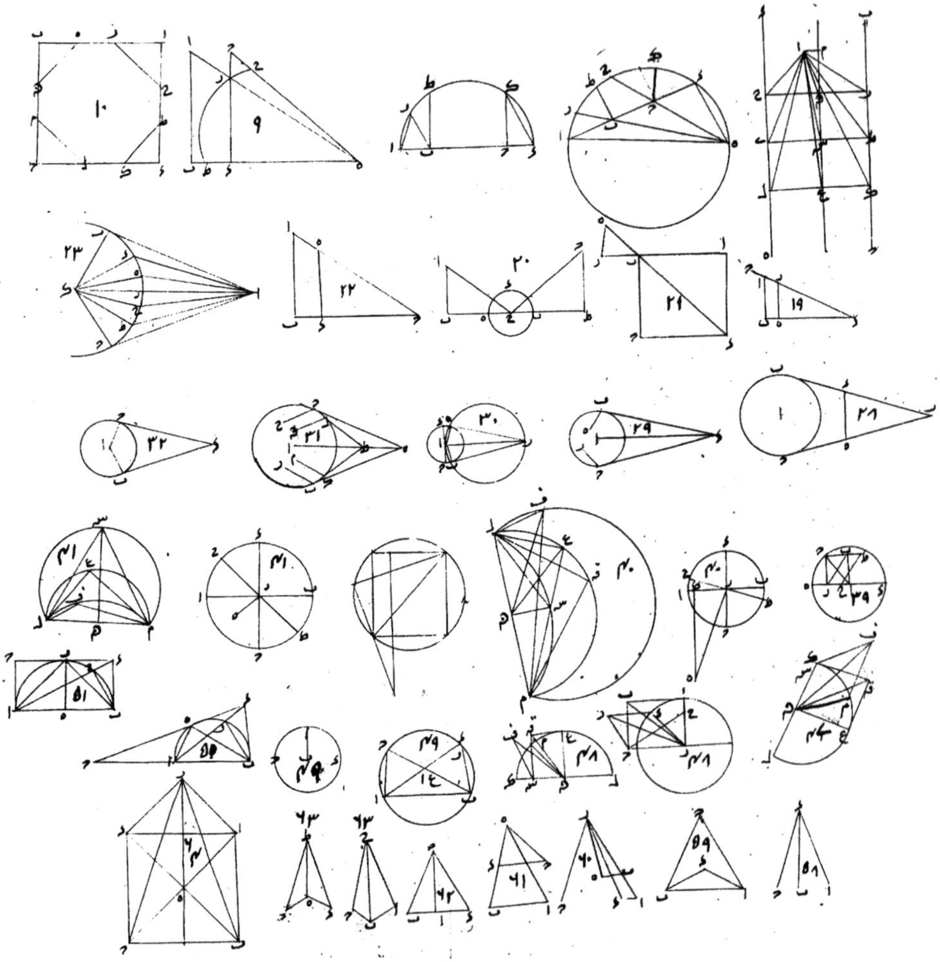

Manuscript Folio: By Permission of The British Library

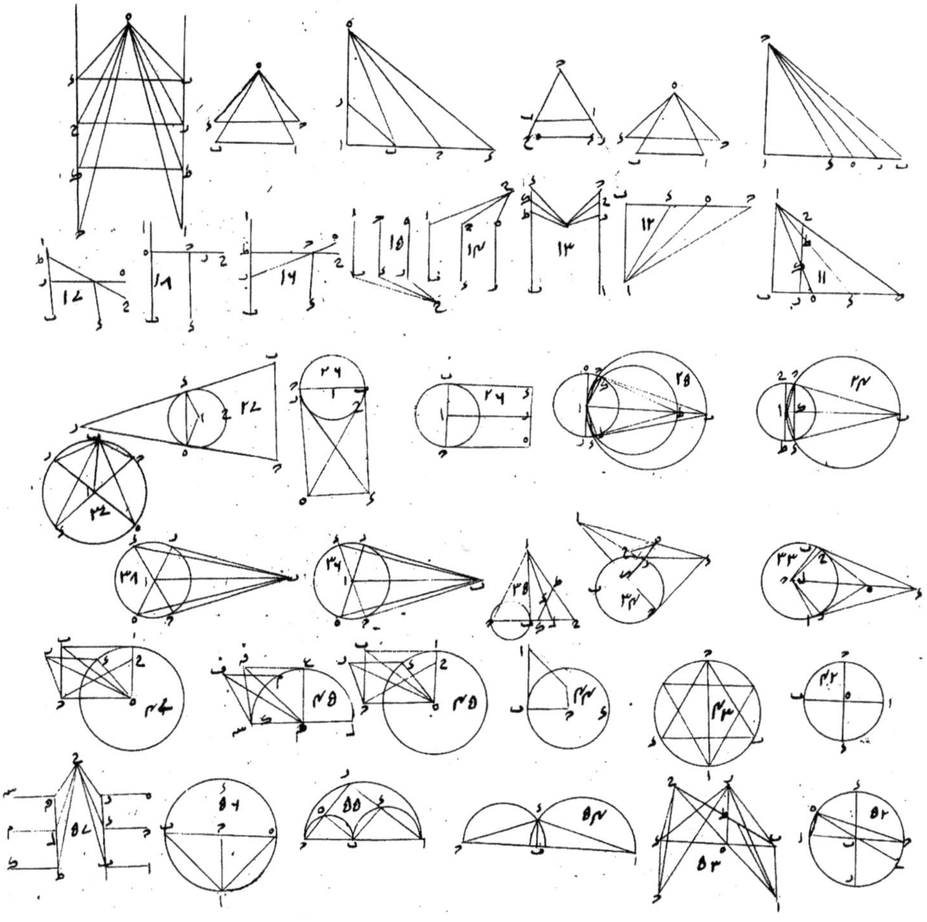

London: India Office 1249 (743), 2: 56, ca. 1100H t27/ت٢٧

Contents

Volume I

Contents

Volume II

Part III Outline of Key Points

Bibliography and Indexes

Volume II

Part I

Introduction to the Commentary

1 The Arabic "Version"

Kitāb Uqlīdis fi Ikhtilāf al-manāẓir[1] corresponds closely to Euclid's *Optika* – or rather Heiberg's *Euclidis Optica*[2] – as an Arabic version of the Greek. But not as a direct translation, as indicated by its title – *The Book of Euclid. . .* – or the association of the terms *tarjama* or *naql* with it,[3] as there are many places where the Arabic is not identical with the Greek edition, from the content to the arrangement of the definitions and propositions. All extant Arabic manuscripts contain 4 definitions and 64 propositions, in contrast to the 7 definitions and 58 propositions of the *Optics* passing under the name of Euclid or the 7 definitions and 57 propositions long known as its 4th-century recension by Theon of Alexandria. The apparently longer Arabic text does not contain all that there is in the first of these Greek versions, to which it is close in content, or all the divisions of the second, to which it is particularly close in structure;[4] it does not even contain all that is covered between the two; nor is it free of material not found in any of them.[5] Yet, for the Arabic text to which the name of Euclid is attached in the majority of its manuscript copies (m1, m2, m3, m5), Heiberg's *Optica* [version A] could have indeed supplied the original text and figures, while Heiberg's *Recensio* [version B], the version whose echoes are identifiable throughout the Arabic text, including critical places,[6] may have acted primarily as a source of revision. Accordingly, the Greek versions once called "genuine" and "vulgate" and more recently versions "A" and "B"[7] are here marked as "O" and "R", from the standpoint of the Arabic tradition.

---

[1] See Edition and Translation, v. 1, pp. 2–225; for variants of the title, see critical apparatus, p. 3, and Summary and Commentary: Title, p. 1, n. 37.

[2] Heiberg, *Euclidis Opera Omnia*, v. 7, pp. 1–121. For references to the recent re-examinations of Heiberg's position, see v. 1, p. xix, n. 1.

[3] The terms *tarjama* and *naql* appear, respectively, in the colophon of one of the manuscripts (m5) and in the title page of some copies of Ṭūsī's *Taḥrīr* (t3, t5, t10, t15, t50), as a reference to the earlier version; see List I, v. 1, pp. xxxvii–xliv.

[4] The Arabic lacks definitions from both Greek versions, as well as propositions and parts of propositions. On the different structure and order in each case; see Structural Correspondence and Correspondence Tables 1 and 2, pp. xxix–xxxiv.

[5] Some propositions only appear in the Arabic, with sections having various forms of correspondence with the Greek versions [see Structural Correspondence, p. xxxiv] and, in a few cases, a particularly close one to book VI of the *Collection* of Pappus; see Summary and Commentary: Propositions 36–42, p. 77 and p. 79.

[6] The first and most critical case is the opening definitions, particularly in m2 and m4, see Description, v. 1, pp. xxvi–xxvii.

[7] On these designations, see Heiberg, *Litterargeschichtliche Studien über Euklid*, pp. 129–30, and Heiberg, *Euclidis Optica*: *Prolegomena*, pp. xxx–xxxi; see also

Euclid's Book on Optics is not only an Arabic version of the Euclidean text, whichever Greek version may have been considered the original Euclid at the time; it turns out to be the Arabic "version" of the Euclidean visual theory as well. The so-called "translation" of Euclid's text represents a transformation much beyond a change in the number and order of definitions and propositions. The substantial *and* consequential transformation involves the very core of their founding assumptions, and one to be at least partly explained by incidental elements in transmission, including such linguistically imposed conditions as the presence of a chain of flexible terms and interchangeable scripts.[8] Thus it is difficult to think of this somewhat transformed version as a premeditated form of revision, such as the better known and much more widely available *Taḥrīr al-Manāẓir li-Uqlīdis* (*Recension of Euclid's Optics*) by the 7th/13th-century scholar Naṣīr al-Dīn al-Ṭūsī, with which it shares many features, including the exact number and order of propositions.[9]

Whether the supposed "translation" of the Euclidean text into Arabic transformed it by mediation of a late Greek source, an intermediate language, or in the course of an Arabic edition, and how exactly the shifts were transported into Arabic, or on into Latin, may never be determined conclusively; what the exact relationship of the Arabic text is to the Latin tradition, including the "Greco-Latin" version with which it shares some features,[10] will not be settled at present. But a consideration of certain features of the Arabic versions can finally remove, and perhaps even bury, some long-standing misconceptions.

Knorr, "On the Principle of Linear Perspective in Euclid's *Optics*," pp. 194–195; Knorr, "Pseudo-Euclidean Reflections in Ancient Optics," pp. 3–4, and Jones, "Peripatetic and Euclidean Theories," p. 49.

8 A good example is the occurrence of a term such as *kathīr*, with orthography close to *kabīr* (great) and a flexible enough meaning to correspond to Euclidean terms, πλείονα/*pleiona* (greater/more) and ἀκριβέστερον (more clearly), in the course of the few transmitted definitions. For other examples and their consequences, see Kheirandish, "The Arabic 'Version' of Euclidean Optics: Transformations as Linguistic Problems in Transmission," *Tradition, Transmission, Transformation*, Ragep and Ragep (eds.), pp. 227–247, and Summary and Commentary: Definitions 1–4 and Propositions 1, 2, 9, and 20, pp. 7–59.

9 This text is available in three more or less problematic published versions: Tehran, 1304(=1886-1887); *Majmū' al-rasā'il*, edn. Hyderabad, v. 1: 1358 (=1939): 2–24, and *Revue de l'Institut des Manuscrits Arabes* (*RIMA*), ed. Dimirdash, 1383 =1963, 9: 251–290; see *The Arabic Versions*, v. 1, p. xxii, and List I, v. 1, pp. xxxvii–xliv.

10 Theisen, *The Mediaeval Tradition*, contains editions of "Arabo-Latin" versions: *Liber de aspectibus* and *Liber de radiis visualibus*, pp. 336–384; 403–422, reference to a *De aspectuum diversitate*, p. 324, n. 10, and edition, translation, and commentary on *De visu*, pp. 66–319, the only one published so far; see Theisen, "*Liber de visu;*" on common features with the Arabic, see p. xxix, n. 32.

Most immediately, the equal length of all identified manuscripts, including those of the later recension of Ṭūsī, raises questions about the validity of all reports of "short" and "long" versions. The standard length of Euclid's *Kitāb al-Manāẓir* is 64 propositions, a number mentioned in historical sources as early as Ya'qūbī's *Ta'rīkh* (ca. 267/880) and as late as Ḥājjī Khalīfa's (d. 1068/1657) *Kashf al-ẓunūn.*[11] Yet a variety of lengths have been assigned not only to the early version [12, 23, 24, 34 or 36 propositions] but also to Ṭūsī's recension [73, 95 propositions], whose length is often reported as part of its title in the colophon of numerous of its manuscript copies as "*Taḥrīr al-Manāẓir, wa huwa arba' wa sittūn shakhlan.*"[12] Still, neither the presumed unequal length nor the markedly distinct opening section of Ṭūsī's *Recension* have prevented the curious confusion of the two Arabic versions separated by almost four centuries and the widespread documentation of one as the other.[13] And whereas the question of length can be linked to a possible confusion between the title of *al-Manāẓir* and those like *al-Masākin* or *al-Munāsabāt* with slight orthographic similarity, or even between it and other minor Euclidean works like *al-Ẓāhirāt* (*Phaenomena*),[14] the case of the repeated association of one of the well-known translators with it in widely consulted sources turns out to be rather arbitrary.

---

11 See al-Ya'qūbī, *Ta'rīkh*, ed. Houtsma, v. 1, p. 139; Ḥājjī Khalīfa, *Kashf al-ẓunūn*, ed. Flügel, v. 5, p. 159.

12 *Recension of Optics in Sixty-four Propositions*, the number sometimes appearing in *abjad* [i.e., alphabetical] form. For reports of versions with various lengths, some duplicated, see [in chronological order] Assemani, *Bibliotheca Medicea*, 1742, p. 392: 23 propositions; Wenrich, *De Auctorum Graecorum*, 1842, p. 183: 23 propositions [referring to Assemani]; Dorn, "Chanykov," 1865, p. 259: 12 propositions; Klamroth, "Schriftstellern bei al-Ja'qūbī," 1888, p. 9, n. 7: 24 propositions; Steinschneider, "Die arabischen Uebersetzungen," 1896, p. 171, n. 1, refers to versions by Thābit with 34 or 36 propositions for the early Arabic version and 95 propositions for Ṭūsī's *Taḥrīr*. Winter, "British Isles," 1956, p. 76, cites 73 "figures" [possibly for *ashkāl*: propositions]. Theisen, *The Mediaeval Tradition*, 1972, p. 322, adopts the hypothesis of short and long versions [24 and 64 propositions] "with some certainty," perhaps due to the existence of two extant Arabo-Latin versions with 38 and 54 propositions. The only version so far idenfied with a number of propositions other than 64 is a very late Persian version in 45 propositions, see D in List II, v. 1, p. lii, though the existence of other nonstandard lengths, whether extant or lost, cannot be ruled out.

13 Assemani, *Bibliotheca Medicea*, 1742, p. 392; Uri, *Bibliotheca Bodleiana*, 1787, p. 189; Winter, "British Isles," 1956, p. 76, all report what is in fact a copy of Ṭūsī's *Taḥrīr* as the early translation.

14 An examination of Chanykof 144 confirmed that Dorn, "Chanykov," p. 259, has reported as the 13th part of this codex *Kitāb al-Manāẓir* [with 12 propositions!] for *Kitāb al-Masākin*. The confusion between the length of *al-Manāẓir* and *al-Munāsabāt* has been pointed out by Steinschneider, "Euklid bei der Arabern," p. 100, who refers to Assemani and Casiri's citation of a short version; confusion with Euclid's *al-Ẓāhirāt* (*Phaenomena*) is mentioned by

There is no mention in any of the historical sources, including manuscript inscriptions, of the translator of the *Optics* being Ḥunayn ibn Isḥāq (d. 264/877),[15] his son Isḥāq ibn Ḥunayn (d. 298/910),[16] Thābit ibn Qurra (d. 289/901)[17], or Qusṭā ibn Lūqā (d. 300/912),[18] as a long list of names closer to our own times have posited in one form or another. The attribution of the "translation" (*tarjama*) to Hiliyā ibn Sarjūn [to be read for Hlīl ibn Sarḥūn; see transcription in Manuscript Plate V, v. 1, p. 240][19] in the colophon of one

---

Steinschneider in "Die mittleren Bücher," p. 469, and repeated by Heiberg in *Litterargeschichtliche Studien über Euklid,* p. 21, both of whom refer to Wenrich's report in *De Auctorum Graecorum* of a short version.

15 Uri, *Bibliotheca Bodleiana,* 1787, p. 189; Assemani, *Bibliotheca Medicea,* 1742, p. 392; Gartz, *De Interpretibus et explanatoribus Euclidis arabicis,* 1823, p. 1; Steinschneider, "Die mittleren Bücher," 1865, pp. 468-470 [discusses the source of the association of *al-Manāẓir* with Ḥunayn]; Steinschneider, "Die arabischen Uebersetzungen," 1896, p. 171; Brockelmann, *GALS I,* 1937, p. 368; Voorhoeve, *Handlist,* 1957, p. 184.

16 Wenrich, *De Auctorum Graecorum,* 1842, p. 182; Suter, "Die Mathematiker," 1900, p. 39-40; Meyerhof, "Die Optik der Araber," 1919, p. 19; Winter, "British Isles," p. 76; Steinschneider, "Die mittleren Bücher," 1865, p. 470, who mentions both the father and son, also in "Euklid bei den Arabern," 1886, p. 100, and "Die arabischen Uebersetzungen," 1896, p. 171; Theisen, *The Mediaeval Tradition,* p. 323, n. 6, includes a list of references to the father and son.

17 Uri, *Bibliotheca Bodleiana,* p. 1787, p. 189; Gartz, *De Interpretibus et explanatoribus Euclidis arabicis,* 1823, p. 1; Wenrich, *De Auctorum Graecorum,* 182; Steinschneider, "Die arabischen Uebersetzungen," 1896, p. 171; Suter, "Die Mathematiker," pp. 39–40; Brockelmann, *GALS I,* p. 368, Winter, "British Isles," p. 76; Theisen, *The Mediaeval Tradition,* p. 323. Thābit's name appears in some Hebrew manuscripts of the Euclidean text according to Steinschneider, "Euklid bei den Arabern," p. 101 [kindly reconfirmed by Tzvi Langermann].

18 Dorn, "Chanykov," p. 259. The source for the association of Qusṭā with the translation may be guessed, as he is the translator of *Kitāb al-Masākin* mistakenly reported as *Kitāb al-Manāẓir* as part of Codex 144 on the same page. Another source of confusion is the term *Mutawassiṭāt (Intermediate Books),* read as *maqūsṭāt* [derived from Qusṭā!, according to a letter from H. P. Kraus Private collection in New York, holding the codex containing m5, at the time of communication]; on m5, see Description, v. 1, p. xxvii.

19 The single appearance of the name of the translator in the manuscript colophon reads as Hlīl ibn Sarḥūn (هلىل ىں سرحون), which is to be understood as Hiliyā ibn Sarjūn (هليا بن سرجون) due to the following considerations. The Arabic letter *ḥā'* (حـ) in the second name, Sarḥūn, is close enough to the dotted form (جـ) for the name to be read as Sarjūn, the Arabized form of the common name Sergius. In the more complicated case of the first name, the reading of the initial letter as H

variant (m5) disqualifies many long-held convictions about the identity of the Arabic translator. This must be the same name documented, alongside al-Ḥajjāj ibn Yūsuf, as one of the earliest Arabic translators of Ptolemy's *Almagest* (ca. 212/827–828) and, with further minute variations of the name, also of Cassianus Bassus's *Geoponika*, there noted specifically for the high quality of his Arabic translation as well as a Persian translation of the same.[20] The correlation between the names of the translator of the *Optics* and, the other two works[21] – not to mention the interrelation of the subjects – suggests if not a

---

(ه), rather than Ḥ (ح), is due the cursive style of the script; that of the final letter as *alif* (ا), rather than *lām* (ل), is suggested by a common interchange between the two in Arabic manuscripts [kindly noted by professor Dimitri Gutas]. Such a reading is also suggested by the respective associations of the two possible forms. Both names [Hlīl (هليل) and Hiliyā (هليا)] are associated with specific translations of the period [see notes 20–23], the form Hiliyā [Hlyā in Kunitzsch, *Der Almagest,* p. 65; Hilīyā, in Rosenthal, "al-Kindî and Ptolemy," p. 48, and Ullmann, "Die Natur-und Geheimwissenschaften im Islam," p. 435] would particularly make sense as an Arabized form of the Greek name Elias (هليا), rather than Hlīl (هليل), the form comparable to Arabic names like Khalīl (خليل), or the rare form حليل [vowelled as Ḥulayl in al-Ya'qūbī's *Ta'rīkh*, ed. Houtsma, v. 1, p. 276].

20 On these translations and the manuscript variants of the translators' names, see Kunitzsch, *Der Almagest,* pp. 64–67. Note that the compositon date for the cotranslation of the *Almagest* [Leiden MS. Cod. Or. 680] is given as ca. 212/827–828 by Kunitzsch [*Der Almagest,* p. 65], but as 214/829 by Voorhoeve [*Handlist,* p. 212] and Sezgin [*GAS* VI, 1978, p. 89]. The details of the cotranslation and the other Arabic translations of the *Almagest* are given in a work by Ibn al-Ṣalāḥ (d. 549/1154) on Ptolemy's star catalogue, see Kunitzsch, "Ibn aṣ-Ṣalāḥ," p. 155 [ed. Kunitzsch, p. 6]. On the Arabic translation of the Greek *Geoponika* (agriculture) [Leiden MS. 1277 = Warner 414], see Kunitzsch, *Der Almagest,* p. 66 and n. 173, who identifies the original author as Cassianus Bassus and the translator as Sergius son of Elias the Greek; see also Ullmann, "Die Natur-und Geheimwissenschaften im Islam," p. 435; Voorhooeve, *Handlist,* p. 84, cites the title *(al-Filāḥa ar-Rūmiyya)* and composition date (ca. 212/827); de Jong and de Goeje, *Bibliotheca Academiae Lugduno Batavae*, pp. 211–213, include the opening text of the Leiden manuscript containing the remark about the translator that "his translation is more complete *(akmal)* and correct *(aṣlaḥ)*" than others [referring to Qusṭā ibn Lūqā and Yaḥyā ibn 'Adī], and that "he also translated this book into Persian."

21 Depending on manuscript variants, the transcriptions of the names appear in some cases almost exactly, in other cases only slightly different, or else in reverse order. The Leiden manuscript of the cotranslated Arabic *Almagest* has the name Sarjūn ibn Hiliyā ar-Rūmī (سرجون بن هليا الرومى) [the page from Leiden, Cod. Or.

single identity, at least a common family line,[22] which the dictates of current evidence would hardly allow one to place more than a generation up or down.[23]

680, is included in Kunitzsch, *Der Almagest*]. According to Kunitzsch [*Der Almagest*, p. 65, n. 168], the Latin version by Gerardus has Sergius filius Elbe [i.e., Sarjūn ibn Hiliyā]. The Leiden manuscript of the Arabic *Geoponika* [1277= Warner 414] has the name of the translator as Sarkhis ibn Hlyā al-Rūmī (سرخس بن هليا الرومى) [i.e., Sirjīs ibn Hiliyā al-Rūmī; see Kunitzsch, *Der Almagest*, p. 65, Voorhooeve, *Handlist*, p. 84, and de Jong and de Goeje, *Bibliotheca Academiae Lugduno Batavae*, p. 211]. All these forms have a reverse order from that of the Arabic *Optics*. But the same order and transcription appears in Ibn al-Ṣalāḥ's version of the name of the cotranslator, in at least two manuscripts of his text: "Hlyl ibn Sarjūn" (هليل بن سرجون) [Istanbul MS Seray 3455, 16, Kunitzsch, *Der Almagest*, n. 170] and هليل بن سرحون, the first letter being in cursive script, this time with no dots in the colophon in the manner of the *Optics* [Oxford MS Thurston 3; in Kunitzsch, *Der Almagest*, it is transliterated as Hlyl ibn Sr.wn]. The apparatus in Kunitzsch, "Ibn aṣ-Ṣalāḥ," includes only the variations of the first name [Hlyl in all MSs], and not the second, where the Oxford colophon exactly matches that of the *Optics*.

22 Sources for a close study of the family line are scant and scattered. There is no entry under Sarjūn, the arabized form of Sergius, in the *Encyclopedia of Islam*. And none of the references to this rather common name fit the profile and time line of our translator. For references to Sarjūn ibn Manṣūr and Manṣūr ibn Sarjūn as members of the Greek Christian family of Sergius translating government documents in Syria at the outset of Islamic history , see Ibn al-Nadīm, *Kitāb al-Fihrist*, ed. Flügel, v. 1, p. 242; tr. Dodge, *The Fihrist of al-Nadīm*, v. 2, p. 583; a slightly different translation of the same passage is included in Rosenthal, *The Classical Heritage of Islam*, p. 48; see also "Family of Sergius" in De Lacy O'Leary, *How Greek Science Passed to the Arabs*, pp. 138–142, especially, p. 139.

23 The date of the translation of the *Optics* cannot be established closely enough to determine the exact relationship of its translator to the other two names, both associated with composition dates of 212/827–828, see note 21. All that can be said about the case of the *Optics* is that, besides some apparently early vocabulary [see Summary and Commentary: Propositions 9 and 20, p. 41 and pp. 57–58], it has been cited in very early sources, from Ibn 'Īsā's *Kitāb al-Manāẓir wa al-marāyā al-muḥriqa* [before 250/864?] and the related works of al-Kindī (d. ca. 257/870) to al-Ya'qūbī's *Ta'rīkh* [composed 267/880]; on some of these dates, see Krause, "Stambuler Handschriften," p. 441, n. 4, and p. 513; see also The Arabic Versions, v. 1, p. xx. It is quite plausible for the translator of Euclid's *Optics* to have been the same as the translator of Ptolemy's *Almagest*. The former, being one of the "Middle Books" (*al-Mutawassitāt*), could conceivably have been translated as a preparatory text for the latter by one and the same author. Al-Kindī's consultation of Sarjūn ibn Hiliyā's cotranslation of Ptolemy's *Almagest*

In any case, the composition of the Arabic *Optics* would have to belong with the very earliest translations of the period and, on the face of it, also with those made directly from Greek. The explicit statements to that effect with reference to translation of both the *Almagest* and *Geoponika*[24] would suggest – at least on the assumption of a common translator – a similar conclusion for the *Optics*, a text for which the intervention of the most natural intermediate language, namely Syriac, does not seem likely.[25] Still, the existence of a revised Arabic version, as in the case of the Isḥāq-Thābit reproduction of other Euclidean works, or else a translation of the prefaced Greek version (R) suggested by an Arabo-Latin version[26] remains an open question.

[see Rosenthal, "al-Kindî and Ptolemy," p. 439, n. 2] is no indication that he consulted a translation by the same author for Euclid's *Optics*. If the two translators are not the same person, then the reversed order of the name suggests a father–son relationship. Ibn al-Ṣalāḥ's order of the name [the same order in the *Optics*; see note 22] has been spoken of as that of the father of the cotranslator of the *Almagest* [Kunitzsch, *Der Almagest*, p. 65, though he has expressed reservations about the assumption of a family of translators in a communication in which he has reconfirmed the published information].

24 In the Leiden manuscripts of the Arabic *Almagest* and Ibn al-Ṣalāḥ's reference to it [Kunitzsch, *Der Almagest*, p. 65, n. 171; "Ibn aṣ-Ṣalāḥ," ed. Kunitzsch, p. 6], it is explicitly stated that the text is a translation from Greek into Arabic (*min al-lisān al-rūmī ilā al-lisān al-'arabī* and *min al-yūnāniyya ilā al-'arabiyya,* respectively, both referring to the name of Caliph al-Ma'mūn (r. 198–218/813–833). In the case of *Geoponika*, the manuscript specifies that the translation is from the Greek language into Arabic (*min al-lugha al-rūmiyya ilā al-lugha al-'arabiyya*); see the full Arabic passage in de Jong and de Goeje, *Bibliotheca Academiae Lugduno Batavae*, p. 212.

25 The identfication of the name of Sergius of the *Almagest* with "Sergius of Ras 'Ain (6th century), the translator into Syriac" [see Steinschneider, "Die arabischen Uebersetzungen," p. 193] seems far-fetched, given the disagreement of the latter's dates, as observed by Kunitzsch [*Der Almagest*], pp. 65–66, and in the light of evidence about the former's engagement in direct translations from Greek into Arabic and in one occasion into Persian; see note 20. The mediation of Syriac in particular is questionable on the basis of textual analysis; I am grateful to Professor Dana Miller for his patient examination of the critical parts of the relevant texts.

26 There are points of overlap with some Isḥāq-Thābit terms or expressions, for example, the concluding premise – "*dhālika mā aradnā an nubayyina*" ("this is what we wished to demonstrate") [also in m3, m4; m2 has "*mā aradnāhu*", m5, "*mā aradnā bayānuhu*"] – and the verb introducing the geometrical construction if read in the first person plural as "*nukhriju. . .* " (we draw . . .) as in the Arabic translation of several books of Euclid's *Elements*, as edited in the [unpublished] works of Gregg De Young [Books VII-IX] and J. W. Engroff [Book V]. On the Arabo-Latin versions, see *The Arabic Versions*, v. 1, p. xxi and n. 6.

But even so, it is possible to make a case for other hands in the composition. The alternative terminology for the Euclidean visual rays as *al-manāẓir min al-'ayn,* literally "rays from the eye" [m2 and m4], alongside the more common *al-shu'ā'* or *al-shu'ā'āt* [also in m1, m3, m5], is somewhat suggestive, as a comparable form was employed by none other than Ḥunayn ibn Isḥāq in more than one of his compositions.[27] There are, however, problems with a blanket assignment of the translation to any one of the usual candidates, whose engagement with it is not just unreported in historical sources, but also unverifiable on a philological basis. There are cases, for example, where the vocabulary of the Arabic version is other than those employed by Ḥunayn or Isḥāq.[28] There are also cases where the nonstandard terminology of the Arabic text may be contrasted to the rather technical vocabulary of the more established translators: the term *in'iṭāf* is used instead of the more common form *in'ikās* (reflection) used by Ḥunayn, among many others, and the expression *'alā aqdār* for the more technical form *'alā nisbat al-aqdār* (ratio) by Isḥāq-Thābit. A striking feature of the Arabic text is in fact the nonstandard and at times inconsistent nature of its terminological apparatus.

This is a text in which the translator [if it is in fact a single translator, according to the colophon of m5] displays a rather noncommittal style of translation [if it is indeed a single translation and not a compilation]. The cases of different Arabic terms being used for the same concept, and of the same Arabic terms being used for different concepts, have already been mentioned with reference to the sensitivity involved in transmitting subtle distinctions through English translation [see Editorial Procedures and Apparatus, v. 1, p. xxxi]. For the case of different terms representing the same concept, we have, in addition to *mubṣar*, *manẓūr ilayh*, and *shay'* [all used for the object of vision],

---

27 Ḥunayn ibn Isḥāq employs the terminology of *al-manāẓir* for visual rays in two independent works: *Kitāb fī Manāfi' al-a'ḍā'*, the Arabic translation of Galen's *De usu partium* [Paris MS BN 2853, fol. 183a], and his own composition, *Kitāb al-'Ashar maqālāt fī al-'ayn*; see Meyerhof, *The Book of the Ten Treatises on the Eye Ascribed to Hunain ibn Is-hâq*, pp. 109–110, and Summary and Commentary: Title, p. 3, n. 47.

28 One example is the representation of the visual cone as *makhrūṭ* in all manuscript copies and as *ṣanawbar* in Ḥunayn's translation of Galen's *De usu partium*; another is the employment of *in'iṭāf*, as opposed to Ḥunayn's *in'ikās*, for the bending of rays; see Summary and Commentary: Proposition 20, p. 57, n. 205. The form *ṣanawbar*, together with the nonstandard vocabulary for the cone's apex [*mustaḥadd* and *zujj* used as synonyms], appear in the Arabic version of the pseudo-Euclidean *De speculis*: *Kitāb al-Mir'āh li-Uqlīdis*; see Sabra, "A Note on Codex Biblioteca Medicea Laurenziana Or. 152," where it is distinguished from the pseudo-Euclidean *Catoptrics* extant in Greek; for the Latin text, see *Tractatus [pseudo-] Euclidis De speculis*, Björnbo and Vogl, *Alkindi, Tideus und Pseudo-Euklid*, pp. 97–173. All three terms appear in the optical section of Ibn 'Īsā's text, the first two also in Ibn Lūqā's; see Supplementary Passages, v. 1, pp. 243–245, Summary and Commentary: Definitions, p. 7, n. 61, and Definition 2, pp. 15–17.

and *munkhafiḍ*, *taḥt*, *dūn*, and *suflā* [all referring to the object's low position], many other instances: *a'lā*, *'āl*, *arfa'* and *fawq* for high positions, *mawḍi'* and *makān* for location, *al-'ayn* and *al-baṣar* for the eye, *akthar* and *aṣdaq* for clearer vision, *ẓahara* and *yurā* for something appearing as, *yuẓannu* [*fī al-ru'ya*] for something seen other than it is, *yutawahham* for something imagined as, *ubṣira* and *nuẓira* for something viewed as, *na'lim* and *na'rif* for something known, *nakhuṭṭu*, *narsumu*, *nudīru* for something to be drawn [in this case, a circle], *mutaqaddim* and *sābiq* for something overtaking, *muta'akhkhir* and *lāḥiq* for something falling behind, and *miqdār* and *'iẓam* for magnitude. In the case of the same term representing more than one concept, there is *baṣar* for both eye and sight, *al-shu'ā'* for both a single ray and a conic collection of rays, *kathīr* in at least three senses – great in number, size, and clarity – *yakhruj* for something issuing from, drawn, or extended to, *mukhtalif* for both different and unequal, *samt* for path/direction and side, and *mawḍi'* for both position and point [see Index of Arabic Terms, pp. 127–152].

A few of the instances are particularly notable for the context in which the specific term is employed: *al-shu'ā'* stands out as being not just a single ray and a conic collection of rays at once, but a term with a verbal root commonly used for all forms of radiation, not merely visual ones. The case of *samt*, in the less common sense of path or direction, is no less notable. Then there is *kathīr*, which has, in addition to distinct usages in the text as many, more, much, and clear, an orthography interchangeable with *kabīr* (great) [on these cases, see Summary and Commentary: Definitions 1 and 4, pp. 8–14; Proposition 2, p. 32, and Variation Samples 1, 2, pp. xxxvii–li]. And there are at least two unexpected cases: the use of the expression *in'iṭāf* (deflection) in place of what is in fact its conceptual and graphical twin *in'ikās* (reflection), and the expression *'alā aqdār* (according to the magnitudes of) for what is meant as *'alā nisbat al-aqdār* (in the ratio of the magnitudes. . .) [see Summary and Commentary: Propositions 9 and 20, p. 42 and 56, and Variation Samples 3 and 4, pp. xlii–xlv].

That the terminological occurrences were not caused by the same kinds of problems is clear from these very instances. The cases of the interchangeability of one transcription with another [*in'iṭāf* for *in'ikās*, *kathīr* for *kabīr*] or a shorthand (*'alā aqdār*) with a full expression (*'alā nisbat al-aqdār*) of certain concepts may be considered internal, but the other cases seem closely tied to the problematics of the Greek. From the standpoint of the translator/editor, what would be a better term than *al-shu'ā'* to render the Greek visual ray? This is an Arabic term whose verbal root captures enough of the anology between visual and luminous radiation to stand for the simultaneous occurrence of the Greek ὄψεις/*opseis* and ἀκτῖνες/*aktines* and whose singular case transmits the similarly double sense of the Greek ὄψις/*opsis*. The same can be said of the many senses of the term *manāẓir* to render those of the equivalent forms ὀπτικά/*optika* [the Greek title], ὀπτική/*optikê* [the Greek discipline], and ὄψεις/*opseis* [the Greek visual rays]. There is also the case of *samt*, an attempt at rendering the different functions of the διάστημα/*diastêma* in the two alternative Greek versions which this time becomes, in its Arabic application, neither distance [as in O] nor interval [as in R]. And in the most complex case of the verbal root $K^aTH^aR^a$, applied in the form *kathra* to the Greek μεγεθῶν/*megethôn*, and in the form *kathīr* to ἀκριβέστερον/*akribesteron* and πλειόνων/*pleionôn*, its multiple use can be at least partly explained in terms of the flexible sense of the Greek *pleionôn* as both greater *and* more.

But the many complexities of the Arabic version cannot always be explained in terms of specific problems in transmission or of terminological and orthographic considerations. Neither is there a way to determine whether the seemingly trivial – and often critical – shifts involved were guided primarily by conceptual or linguistic elements in the first place. In an Arabic text in which many of the inherited problems of the Greek were given new twists and turns, it is hardly possible to think of one to the exclusion of the other. If the translator of Euclid's *Optics* was indeed the cotranslator of Ptolemy's *Almagest*, he would have been too early (ca. 212/827–828) to have been exposed to the Arabic terminological apparatus elsewhere, say in an Arabic *Optics* by Ptolemy or Theon, even if such texts were circulating in the early part of the 3rd/9th century. Especially if the supposed translator took part in a direct translation from the Greek as in the cotranslation of the *Almagest* [see pp. xxii–xxv, n. 19–23], he would have been conversant enough in Greek for the conceptual *and* linguistic elements of the relevant literature to have underlain his particular selection of Arabic terms and expressions. Whatever the roots *and* routes of the terminological occurrences, it is remarkable that through them the alleged "translation" of a few assumptions and geometrical demonstrations on vision would come to produce its own "version" of visual theory,[29] perhaps more remarkable, that it would come to transmit its components in such decisive directions.[30]

The Arabic textual tradition has been a wide channel for the transmission of the transformed visual model, the dependent works of al-Kindī and Ibn al-Haytham being only two of the better known and more widely circulated sources of external transmission [see Textual Transmission, v. 1, pp. lxiv–lxv].

---

29 See Kheirandish, "The Arabic 'Version' of Euclidean Optics: Transformations as Linguistic Problems in Transmission," *Tradition, Transmission, Transformation*, Ragep and Ragep (eds.), pp. 227–247.

30 On the routes of transmission, see Textual Transmission, v. 1, pp. lxii-lxv.
A good measure of the spread of the Euclidean tradition is the nature of the surviving manuscript copies. In the case of Arabic, the manuscripts are not only considerable in number but also the only ones still available in contrast to other traditions; as for Latin, the number of manuscript copies of the Euclidean text is close to those of the Latin texts of Ptolemy, al-Kindī, and Ibn al-Haytham put together; see Textual Transmission, v. 1, p. lxiii, n. 55.

The direct Arabo-Latin transmission is similarly decisive and diverse: three Arabo-Latin Euclidean versions have so far been identified, and a "Greco-Latin" version, with a widespread *and* high-ranking path of circulation,[31] seems not entirely independent of the Arabic textual tradition.[32] Of special interest to the pressing problems of textual transmission is that the well-diffused Arabic "version," which in all but one (m4) of its manuscript copies is explicitly associated with Euclid's name,[33] is itself the by-product of what is at least a two-version transmission.

2 Structural Correspondence

Correspondence between the Arabic and the two Greek versions is not direct in content, order, or structure. The Greek versions O and R are not identical in these respects,[34] and the Arabic version M, which has a one-to-one structural correspondence among its five manuscript copies [m1, m2, m3, m4, m5], as well as between it and Ṭūsī's later recension T, does not correspond exactly to either O or R in order *or* structure.

For the definitions, neither the heading[35] nor all the definitions introduced under it are to be found in the Arabic: the fifth and sixth Euclidean definitions

---

[31] Theisen, *The Mediaeval Tradition*, pp. 15–16, refers to the wide circulation of the Euclidean *De visu* and to Roger Bacon, John Pecham, and Witelo and their respective *Perspectiva*, *Perspectiva communis*, and *Perspectiva* in that connection.

[32] Some common characteristics of M and *De visu* (V) are the singular verb of the third Euclidean definition, the alternative proof of the first Euclidean proposition, the two-part division of the sixth Euclidean proposition, the absence of the phrase "as is said in *Catoptrics*" in the proposition involving a mirror; see Correspondence Tables 1–2 and Summary and Commentary: Definitions 3; Propositions 1, 6, and 20, pp. 18–59. On coincidences between V and the Arabo-Latin versions in the enunciations, as well as between the latter and M, both structural [four rather than seven Euclidean definitions] and conceptual [eye movements across objects]; see *Summary and Commentary*: Propositions 1, 16, and 17, p. 22, and pp. 56–57.

[33] That is to say, considered as Euclid's own from the standpoint of members of *both* the odd- and even-numbered set of manuscripts [namely m1, m3, m5, and m2].

[34] On specific definitions and propositions, see Ver Eecke, *Euclide, l'Optique et la Catoptrique*, 1959, Introduction, pp. XXVI–VIII; see also Jones, "Peripatetic and Euclidean Theories," pp. 28–43 and Knorr, "Pseudo-Euclidean Reflections in Ancient Optics," pp. 49–56.

[35] Greek ὅροι/*horoi* introduces the seven definitions of both O and R in the manner of these and other Euclidean texts. The term ordinarily used to render Euclidean definitions in Arabic is *ḥudūd*, as in some Arabic versions of Euclid's *Elements*. But it is not uncommon for this label to be left out; see De Young, *The Arithmetic Books of Euclid's Elements in the Arabic Tradition*, v. 1, part 1, p. 2, and Engroff, *The Arabic Tradition of Euclid's Elements Book V*, p. 60 [both unpublished

[identical in both O and R] are missing from M altogether. These definitions, which concern appearance when rays are higher and lower [def. 5] and to the right and the left [def. 6], respectively, may have been transmitted in some form, for something close appears in combined form in Ṭūsī's later *Taḥrīr* [T] under what could be counted as *its* fifth definition.[36] For the remainder of the "corresponding" definitions, there is still not a full correspondence. Apart from terminological and phraseological disparities, which are most extreme in the case of the opening definition introducing the Euclidean visual ray hypothesis, there is also a structural breakdown in the merging of the seventh Euclidean definition on clarity of appearance with the fourth on size of appearance [see Summary and Commentary: Definitions 1 and 4, p. 7 and p. 20, respectively]. Ṭūsī's recension has additional definitions: a two-part definition on clarity of vision and one on reflection at equal angles, such that its much more modified version of Euclidean definitions [see Supplementary Passages, v. 1, p. 246] can still be divided into seven, the number of definitions in both O and R.

In the case of the Arabic propositions, their arrangement is such that they are close to or far from having a direct correspondence with the Greek versions, depending on the particular part of the text. In the first half, what starts as a close structural correspondence breaks down, with one or two cases where a Euclidean proposition is presented as two independent treatments [O6/R6 as M6, M7 and O28/R28 as M29, M30]. By the second half, it is not just the proposition numbers that move further and further apart every time a Greek proposition is presented as two or more independent treatments in Arabic [e. g. O 36/R37corresponding to M42]; there is also a change in their order [e. g. M48 being a continuation of propositions corresponding to M43 (O37/R41)]. Still, a great many of the Euclidean propositions are identifiable in the Arabic version, despite occasional shifts of emphasis and departures from the supposed original [see Summary and Commentary: Propositions 1, 3, and 10, v. 2, pp. 21–48].

Two tables have been prepared to represent the structural correspondence of the definitions and propositions of the Arabic version M to those in the two Greek versions O and R: *Optica* and *Recensio* [ed. Heiberg], the later Arabic recension T: *Taḥrīr* [ed. Hyderabad], as well as the "Greco-Latin" version V, *De visu* [ed. Theisen] that shares some features with it [see Correspondence Tables 1–2, pp. xxxiii–xxxiv]. The table cells contain the order and number of definitions or propositions in M as the reference point [middle] to be compared to the other four texts, O, R [above it], and T and V [below it].

Harvard Ph.D. dissertations]. For the few Latin variants that label the assumptions, see Theisen, *The Mediaeval Tradition,* p. 272.

36 Whether or not always part of the original Greek, the two definitions find expression in the definitions section of the mid-7th/13th-century recension but not in any of the manuscript copies of the earlier version, mostly copied around the same date. The related portions of the text of Aḥmad ibn 'Īsā [see The Arabic Tradition, v. 1, p. liv] suggest exposure to these same definitions early on.

The entries within each cell are abbreviated so that the letter representing a text [e.g., M] is followed by the number of the definition or proposition in that text [Mdef. X and MX, respectively]. When there is more than one part to a definition or proposition or the latter is supplied with an alternative treatment or lemma in the Greek, the abbreviated form is MX-Y, where X and Y represent the specific number and part of the definition or proposition. For example, Mdef. 4-3 in Correspondence Table 1, representing the third part of M's fourth definition, corresponds to O7, R7, T7 and V7 in the same cell; M7 in Correspondence Table 2, representing M's seventh proposition, corresponds to O6-2 and R6-2, the second part of the sixth proposition in O and R, but still to T7, V7, which similarly treat this part as a separate proposition. The same applies to propositions involving many parts; for example, M41, corresponding to O35-4, R36-5, T41, and V40-5 in the same cell, means that the forty-first proposition in M corresponds to the fourth part of proposition 35 in O, the fifth part of proposition 36 in R, the fifth part of proposition 40 in V, and the forty-first proposition in T, which as always, has the same number as M [the proposition number in the sixth book of Pappus's *Collection* is only reflected in the commentary on the Arabic proposition corresponding to it, in this case, Summary and Commentary: Proposition 41]. Correspondences to enunciations or demonstrations alone are marked by E or D after the numbers.

It is immediately clear from these tables that there are definitions and propositions in O and R with no correspondence in M. In addition to the two Euclidean definitions already mentioned [O5, O6 and R5, R6], a few propositions [e.g., O48D, R45, R51] and several parts of propositions [e.g., O23-2, 3; O34-3; O37-2; O42-2; O44-2] are not to be found in M. There are, on the other hand, certain propositions in M with no correspondence in O [e.g., M46, M47] and others with no correspondence in R [e.g., M52, M53]. Again, there are propositions with correspondence only to the enunciation or demonstration of a particular Greek proposition [e.g., M56 corresponding to O48E and O49D, and M29, M30 corresponding to R28E, R28D, respectively.

Following are other points to bear in mind with regard to the Correspondence Tables immediately following this section.

1. None of the Arabic definitions are numbered in any manuscript copies of either M or T, in contrast to the propositions, which are supplied with side numbers in alphabetic (*abjad*) form [the numbers in m3 are at the end of each proposition] from one to 64. The definitions, setting the point of reference for Correspondence Table 1, have been arranged such that they can be added up to seven; they can be grouped differently with a different count. This number has been adopted for the number of cells in that table as a means of matching corresponding definitions, whether or not it represents the actual plan of the texts, as is the case for the originally numbered 64 propositions of Correspondence Table 2.

2. The four surviving manuscript copies of the definitions [m1, m2, m4, m5; m3 lacks the definitions page] contain the first three Euclidean definitions and all three parts of the fourth one; the first three of these (m1, m2, m4) include the seventh Euclidean definition as part of the fourth [m5 has no corresponding definition]. Of the seven cells of Correspondence Table 1, the first three present the direct correspondence of the first, second and third Arabic definitions [Mdef. 1–Mdef. 3] to the numbers in O, R, T, and V [in this case being the same]; the four remaining cells are treated, in accordance with the supposed plan of the majority of manuscript copies, as parts of the fourth Arabic definition [Mdef. 4-1 to Mdef. 4-4], where Mdef. 4-1, Mdef. 4-2 and Mdef. 4-4 correspond to the three parts of the fourth definition in O, R, and V; Mdef. 4-3 corresponds to def. 7 in all three cases].

3. All manuscript copies of T include a combined form of the fifth and sixth Euclidean definitions and two additional definitions [not the same as the two additional ones in V]. The long definition section of the later version T has also been arranged, for the purposes of Correspondence Table 1, according to seven divisions; there the sixth and seventh definitions (T6, T7) have no counterpart in either of the two Greek texts O and R, or in the Arabic and Latin versions, M and V. The fifth definition (T5), on the other hand, is matched as a combined representation of definitions five and six by O5,6, R5,6, and V5,6 with no entry for M, which lacks these particular definitions.

Correspondence Table 1: Definitions

1	**2**	**3**	**4-1**	**4-2**	**4-3**	**4-4**
Odef. 1 Rdef. 1 Mdef. 1 Tdef. 1 Vdef. 1	Odef. 2 Rdef. 2 Mdef. 2 Tdef. 2 Vdef. 2	Odef. 3 Rdef. 3 Mdef. 3 Tdef. 3 Vdef. 3	Odef. 4-1 Rdef. 4-1 Mdef. 4-1 Tdef. 4-1 Vdef. 4-1	Odef. 4-2 Rdef. 4-2 Mdef. 4-2 Tdef. 4-2 Vdef. 4-2	Odef. 7 Rdef. 7 Mdef. 4-3 Tdef.4-3 Vdef.7	Odef. 4-3 Rdef. 4-3 Mdef.4-4 Tdef. 4-4 Vdef. 4-3

5	**6**	**7**
Odef. 5,6 Rdef. 5,6 not in M Tdef. 5 Vdef. 5,6	Tdef. 6	Tdef. 7

O= *Euclidis Optica* [ed. Heiberg]
R= *Opticorum recensio Theonis* [ed. Heiberg]
M= *Kitāb Uqlīdis fī Ikhtilāf al-manāẓir*
T= Ṭūsī, *Taḥrīr al-Manāẓir* [ed. Hyderabad]
V= *Euclid, De visu* [ed. Theisen]

Correspondence Table 2: Propositions

1	2	3	4	5	6	7	8
O1 R1 M1 T1 V1	O2 R2 M2 T2 V2	O3 R3 M3 T3 V3	O4 R4 M4 T4 V4	O5 R5 M5 T5 V5	O6-1 R6-1 M6 T6 V6	O6-2 R6-2 M7 T7 V7	O7 R7 M8 T8 V8
O8 R8 M9 T9 V9	O9 R9 M10 T10 V10	O10 R10 M11 T11 V11	O11 R11 M12 T12 V12	O12 R12 M13 T13 V13	O13 R13 M14 T14 V14	O14 R14 M15 T15 V15	O15 R15 M16 T16 V16
O16 R16 M17 T17 V17	O17 R17 M18 T18 V18	O18 R18 M19 T19 V19	O19 R19 M20 T20 V20	O20 R20 M21 T21 V21	O21 R21 M22 T22 V22	O22-1 R22E M23 T23 V23-1	O23 R23 M24 T24 V26
O24 R24 M25 T25 V27	O25 R25 M26 T26 V28	O26 R26 M27 T27 V29	O27 R27 M28 T29 V30	O28-1 R28E M29 T29 V31	O28-2 R28D M30 T30	O29 R29 M31 T31 V33	O30 R30 M32 T32 V34
O31 R31 M33 T33 V35	O32 R32 M34 T34 V36	O33 R33 M35 T35 V37	O34-1 R34 M36 T36 V38	O34-2 R35-1 M37 T37 V39	O35-1 R36-1 M38 T38 V40-1	O35-2 R36-2 M39 T39 V40-2	O35-3 R36-4 M40 T40 V40-4
O35-4 R36-5 M41 T41 V40-5	O36 R37 M42 T42 V41	O37 R41 M43 T43 V42	O39 R38 M44 T44 V44	O40-1 R40-1 M45 T45 V45	 R40-2 M46 T46 V45	 R40-3 M47 T47 V45	O38 R42E M48 T48 V43
O41 R39 M49 T49 V46	O42-1 R43 M50 T50 V47-1	O43D R44D M51 T51 V47-3	O44-1 M52 T52 V48-1	O46 M53 T53 V50	O45 R46 M54 T54 V49	O47 R47 M55 T55 V51	O49D R/O48E M56 T56 V52
O50 R49 M57 T57 V53	O51 R50 M58 T58 V54	O53 R52 M59 T59 V56	O54-2 R53E M60 T60 V57	O55 R54 M61 T61 V58	O56 R55 M62 T62 V59	O57 R56 M63 T63 V60	O58 R57 M64 T64 V61

3 Textual Divergence

Differences between the Greek and Arabic versions are not limited to a lack of direct correspondence between their respective definitions and propositions. There is, besides the issue of noncorresponding parts, the case of the textual divergence of corresponding parts. The latter issue is of a more subtle form here, and also of a more critical nature in terms of dependent developments. The variance of the two Greek versions O and R, carried through to Arabic as part of the two-version transmission, is only one dimension of the problems involved. The variance of the Arabic versions among themselves, first between manuscript copies of the two main variant sets of the early version [m1, m3, m5 and m2, m4], then among members within each set [see Description and Analysis of Manuscripts, v. 1, pp. xxvi–xxvii], and finally among these and those of Ṭūsī's later recension T, bring further layers of complexity to the largely invisible problems of incremental transformations and multiple distributions.

The part that presents the most complex and indeed curious problem is that of the opening definitions. Not only is this the most critical part of the Euclidean text in terms of *the* statement of the visual ray hypothesis, it is the most problematic part in terms of textual transmission. It is in the same part of the text that the two main Greek versions are most divergent. It is also in the same part, though not in the same manner, that the variance of two slightly different Arabic versions, and in fact all Arabic variants and their marginal notes, is most striking. Again, the same part has been considered defective in the Greek, and it is the same part for which restoration efforts can be documented in Arabic [see Summary and Commentary: Definition 1, p. 6, n. 60; p. 8, n. 63].

It seems that the defects or complexities of the opening section, apparent in the case of *both* Greek and Arabic, have had something to do with the many manuscript variants and various early Arabic formulations of what is always the same part of the text [see Summary and Commentary: Definitions, p. 7, n. 61].

Part II, Summary and Commentary, is devoted to the subject of textual divergence, a subject that forms a crucial part of both the textual and historical commentaries on Arabic definitions and propositions, with emphasis on the layers of variance packed into the most critical parts. Each case is given individual treatment, which in the case of the propositions is supplied with a summary of its contents [under Synopsis] presented in the same order as the text's divisions: enunciation, illustration . . . conclusion. The commentary begins with a textual treatment [under Textual Correspondence], comparing M to O, R, and T [and C in few cases], with reference to corresponding numbers and pages in the respective editions. In a few select cases [all four transmitted definitions, Mdef. 1–Mdef. 4, and some propositions: M1–3, M9–10, M20, M23], the textual treatment is followed by a historical commentary [beginning with *Historical Context*] with a focus on the many linguistic and orthographic elements involved in textual transmission and the consequences in each case.

Of the cases treated, the most critical ones [two definitions, Mdef. 1 and Mdef. 4-3, and two propositions, M9 and M20] are supplemented with a side-by-side presentation of the original Greek and Arabic texts and their modern translations [see Variation Samples 1–4, pp. xxxvii–xlv].

The first of these [Variation Sample 1, see p. xxxvii] is the striking and critical case of the opening Euclidean definition [Odef. 1, Rdef. 1, Mdef. 1, Tdef. 1]. It is striking as the only part that is not identical either in the two Greek editions or in any of the Arabic versions, including Ṭūsī's later recension, and it is a critical case because it contains the main assumptions of their respective visual theories. To lay bare some of the sources and effects of the textual variations of this crucially important definition, Variance Sample 1 presents, in addition to the original and interpreted versions of the opening definition [namely the two Greek versions O and R, the two Arabic versions M and T, and their English translations], a variety of "versions" in Arabic based on textual variants and marginal additions.

Variation Sample 2 [see p. xl] presents the case of the closing Euclidean definition on clarity of appearance [Odef. 7, Rdef. 7, Mdef. 4-3, Tdef. 5,6], with a comparably intriguing history, largely linked to specific problems in textual transmission, including that of the second Euclidean proposition for which the relevant premises are also included.

The next two Variation Samples [Variation Sample 3 and Variation Sample 4] are concerned with the transmission of specific propositions. Variation Sample 3 [see p. xlii] presents the case of a Euclidean proposition commonly associated with linear perspective [O8, R8, M9, T9], and Variation Sample 4 [see p. xliv] the case of one of the so-called "surveying" Euclidean propositions involving a plane mirror [O19, R19, M20, T20]. In both cases, slight textual differences between Greek and Arabic versions represent a decisive shift in the statement of a critical part of that proposition.

Variation Sample 1: Definition 1

O: *Euclidis Optica, Euclidis Opera Omnia*, eds. Heiberg and Menge, v. 7, p. 2

Ὑποκείσθω τὰς ἀπὸ τοῦ ὄμματος ἐξαγομένας εὐθείας γραμμὰς φέρεσθαι διάστημα μεγεθῶν μεγάλων.

Ver Eecke, *Euclide, l'Optique et la Catoptrique*, p. 1
Supposons que les lignes droites qui émanent de l'œil se propagent à divergence des grandes grandeurs.

Burton, "The Optics of Euclid," p. 357
Let it be assumed that lines drawn directly from the eye pass through a space of great extent.

Cohen and Drabkin, *A Source Book in Greek Science*, 1958, p. 257
Let it be assumed that the rectilinear rays proceeding from the eye diverge indefinitely.

Jones, "Peripatetic and Euclidean Theories," p. 52
Let it be postulated that the straight lines drawn out from the eye travel an interval of great magnitude.

Knorr, "Pseudo-Euclidean Reflections in Ancient Optics," p. 29, n. 48
Let it be postulated that the straight lines leading out from the eye are carried out a distance of great magnitude.

R: *Opticorum Recensio Theonis*, ed. Heiberg and Menge, v. 7, p. 154

Ὑποκείσθω τὰς ἀπὸ τοῦ ὄμματος ὄψεις κατ' εὐθείας γραμμὰς φέρεσθαι διάστημά τι ποιούσας ἀπ' ἀλλήλων.

Ver Eecke, *Euclide, l'Optique et la Catoptrique*, p. 57
Supposons que les rayons visuels émanés de l'œil se propagent suivant des lignes droites faisant quelque divergence entre elles.

Jones, "Peripatetic and Euclidean Theories," p. 52
Let it be postulated that the visual rays from the eye travel along straight lines, while making some interval from one another.

Knorr, "Pseudo-Euclidean Reflections in Ancient Optics," p. 29, n. 48
Let it be postulated that the visual rays from the eye are carried in straight lines that make a certain distance from each other.

M: *Kitāb Uqlīdis fī ikhtilāf al-manāẓir*, v. 1, pp. 2–3 [Mdef. 1(3–4)]

م١ : (زعم اقليدس ان : هش) الشعاع (المؤدى؟ : فس) يخرج من العين
على خطوط مستقيمة ومعه يحدث (بعد : فس)
سموت مستقيمة لا نهاية لكثرتها

m1: [margin: Euclid maintained that] A [collection of] ray[s] [above line: *al-mu'addī?*] issues from the eye in straight lines while producing straight [above line: a distance (*bu'd*), or then (*ba'd*)] paths/directions of infinite multitude.

م٢ : المناظر من العين هى أنّ
الشعاع يخرج من العين على خطوط مستقيمة مختلفة الأبعاد والمقادير
وتحدث به سموت مستقيمة لا نهاية لكثرتها

m2: Visual rays from the eye is [to say] that a [collection of] ray[s] issues from the eye in straight lines of unequal distances and magnitudes, producing by it straight paths/directions of infinite multitude.
m3: lacks the first page containing the definitions.

م٤ : إنّ الشعاع يخرج من العين على خطوط مستقيمة مختلفه الابعاد
والمقادير وتحدث به سموت مستقيمة لا نهاية لكثرتها

m4: [Title:Visual rays from the eye] A [collection of] ray[s] issues from the eye in straight lines of unequal distances and magnitudes, producing by it straight paths/directions of infinite multitude.

م٥ : الشعاع يخرج من العين على خطوط مستقيمة
ويحدث بعده سموت مستقيمة لا نهاية لكثرتها

m5: A [collection of] ray[s] issues from the eye in straight lines, and its distance produces straight paths/directions [or then straight paths/directions are produced] of infinite multitude.

al-Ya'qūbī, *Ta'rīkh*, ed. Houtsma, v. 1, p. 139

إنّ الشعاع يخرج من العين على خطوط مستقيمة
وتحدث بعد سموت لا نهاية لكثرتها

A [collection of] ray[s] issue from the eye in straight lines, and then paths/directions of infinite multitude are produced.

T: Ṭūsī. *Taḥrīr al-Manāẓir li-Uqlīdis*, ed. Hyderabad, 1358=1939, p. 2

العين تُحدِث باستمداد من الأجرام النيّرة فى الجسمِ الشفاف المتوسّط
بينها وبين المبصرات كالهواء وما شاكله شعاعاً كما تُحدِثه الأجرام النيّرة
وحدها بعينه ويكون ذلك الشعاع كأنّه منبعثٌ من العين وخارجٌ منها
ثمّ إنّه يصير آلةً لها فى الإبصار فتختلف أحوال المناظر لاختلاف أوضاعه
فَلْيُصدّق بذلكَ وليُتوهّم ذلك الشعاع متّصلاً بالعين على خطوطٍ مستقيمةٍ
ولْيُحدِث سموتاً مستقيمةً لا نهاية لكثرتها

The eye produces (*tuḥdith*), through the aid of self-luminous bodies (*al-ajrām al-nayyira*), a ray (*shu'ā'*) in transparent bodies (*al-jism al-shaffāf*) such as air and the like, which are between it [the eye] and the visible objects (*al-mubṣarāt*), just as luminous bodies themselves produce it [the ray]. And it is as if that ray is sent out of the eye (*ka'annahu munba'ithun min al-'ayn*) and coming out from it (*khārijun minhā*), then becomes an instrument (*āla)* for it in vision, so that the states of aspects (*aḥwāl al-manāẓir*) change according to its [the ray's] different positions. So let this be granted (*yuṣaddaqu bi dhālika*) and let that ray be imagined (*wa l-yutawahham dhalik al-shu'ā'*) to be connected to the eye (*muttaṣilan bi al-'ayn*) by straight lines (*'alā khuṭūṭ mustaqīma*), and let it produce straight paths/directions (*sumūt mustaqīma*) of infinite multitude (*lā nihāyata li-kathratihā*).

J: *Tajrīd Uqlīdis fī al-Manāẓir* by Ibn Abī Jarāda [Dār al-Kutub, DR 638]

الشعاع يخرج من العين على خطوط مستقيمة
وتحدث سموت لا نهاية لكثرتها

A [collection of] ray[s] issue from the eye in straight lines, and paths/directions of infinite multitude are produced.

Variation Sample 2: Definition 4

O: *Euclidis Optica, Euclidis Opera Omnia*, ed. Heiberg and Menge, v. 7, p. 2

R: *Opticorum Recensio Theonis*, ed. Heiberg and Menge, v. 7, p. 57

τὰ δὲ ὑπὸ πλειόνων γωνιῶν ὁρώμενα ἀκριβέστερον φαίνεσθαι.

Ver Eecke, *Euclide, l'Optique et la Catoptrique*, 1959, p. 2
Enfin, que les grandeurs vues sous des angles plus nombreux apparaissent plus distinctement.

Burton, "The Optics of Euclid," 1945, p. 357
But that things seen within several angles appear to be more clear.

M: *Kitāb Uqlīdis fī ikhtilāf al-manāẓir*, v. 1, pp. 2–3 [Mdef. 4-3 (11)]

م ١ : ما أبصِر من زاوية كثيرة/كبيرة ظهر كثيراً/كبيراً
ما أبصِر من زوايا كثيرةٍ ظهر كثيراً (هش)

m1: That which is seen from a great/large angle, appears more [clearly] [margin: That which is seen from many/great (large) angles, appears more [clearly].

م ٢ : ما أبصِر من زوايا كثيرة/كبيرة ظهر كثيراً

m2 : That which is seen from many/great (large) angles appears more [clearly].

m3: Lacks the first page containing definitions.

م ٤ : ما أبصِر من زوايا كثيرةٍ ظهر كثيراً

m4 : That which is seen from many/great angles appears more [clearly].

m5: Lacks this particular definition.

T: Ṭūsī, *Taḥrīr al-Manāẓir li-Uqlīdis*, ed. Hyderabad, 1939, p. 1

ما أبصِر من زوايا كثيرةٍ ظهر كثيراً

That which is seen from many/great angles appears more [clearly].

From Proposition 2:

O: *Euclidis Optica, Euclidis Opera Omnia*, ed. Heiberg and Menge, v. 7, p. 4

οὐκοῦν τὸ ΓΔ ὑπὸ πλειόνων ὄψεων ὁρᾶται ἤπερ τὸ ΚΛ.
ἀκριβέστερον ἄρα φανήσεται τὸ ΓΔ τοῦ ΚΛ.
τὰ γὰρ ὑπὸ πλειόνων γωνιῶν ὁρώμενα ἀκριβέστερον φαίνεσθαι.

Ver Eecke, *Euclide, l'Optique et la Catoptrique*, 1959, p. 2
. . . donc, le grandeur GD est vue sous plus de rayons visuels que la grandeur KL. En conséquence, la grandeur GD apparaîtra plus distinctement que la grandeur KL; car les grandeurs vues sous des angles plus nombreux apparaissent plus distinctement.

Burton, "The Optics of Euclid," 1945, p. 357
So GD is seen by more rays of the eye than KL.
So GD will appear more clear than KL;
for objects seen within more angles appear more clear.

R: *Opticorum Recensio Theonis,* ed. Heiberg and Menge, v. 7, p. 156

οὐκοῦν τὸ ΓΔ ὑπὸ πλειόνων ὄψεων ὁρᾶται ἤπερ τὸ ΚΛ.
ἀκριβέστερον ἄρα φανήσεται τὸ ΓΔ τοῦ ΚΛ.
Ver Eecke, *Euclide, l'Optique et la Catoptrique*, 1959, p. 58
Par conséquent, la grandeur GD est vue sous des rayons plus nombreux que la grandeur KL;
donc, la grandeur GD apparaîtra plus distincte que la grandeur KL.

M: *Kitāb Uqlīdis fī ikhtilāf al-manāẓir*, v. 1, pp. 6–7 [M2(8–10)]

فالواقع على جد من الشعاع أكثر من الواقع على اب
وما وقع عليه الشعاع أكثر فرؤيته أصدق
فرؤية ج أصدق من رؤية اب

Now more of the ray falls on GD than on AB;
And that upon which more of the ray falls is seen more accurately.
So GD is seen more accurately than AB.

T: Ṭūsī, *Taḥrīr al-Manāẓir li-Uqlīdis*, ed. Hyderabad, 1358=1939, p. 3

فيكون الواقع على جد من الشعاع أكثر من الواقع على اب
فلذلك يكون رؤيته أصدق من رؤية اب

Now more of the ray falls on GD than on AB;
And because of that, it is seen more accurately than AB.

Variation Sample 3: Proposition 9

O: *Euclidis Optica, Euclidis Opera Omnia*, ed. Heiberg and Menge, v. 7, p. 14

τὰ ἴσα μεγέθη καὶ παράλληλα ἄνισον διεστηκότα ἀπὸ τοῦ ὄμματος οὐκ ἀναλόγως τοῖς διαστήμασιν ὁρᾶται.

Burton, "The Optics of Euclid," 1945, p. 358
Lines of equal length and parallel, if placed at unequal distances from the eye, are not seen in proportion to the distances.

Ver Eecke, *Euclide, l'Optique et la Catoptrique*, 1959, p. 6
Les grandeurs égales et parallèles, inégalement distants de l'œil, ne sont pas vues proportionellement aux distances.

Tobin, "Ancient Perspective and Euclid's Optics," 1990, p. 19
Equal and parallel magnitudes set out unequally from the eye are not seen proportionately with the distances.

R: *Opticorum Recensio Theonis,* ed. Heiberg and Menge, v. 7, p. 164

τὰ ἴσα μεγέθη ἄνισον διεστηκότα οὐκ ἀναλόγως τοῖς ἀποστήμασιν ὁρᾶται.

Ver Eecke, *Euclide, l'Optique et la Catoptrique*, 1959, p. 61
Les grandeurs égales, inégalement distantes, ne sont pas vues en proportion des distances.

Tobin, "Ancient Perspective and Euclid's Optics," 1990, p. 19
Equal magnitudes uequally set out are not seen according to a proportion with the distances.

M: *Kitāb Uqlīdis fī ikhtilāf al-manāẓir*, v. 1, pp. 26–27 [M9(2–4)]

الأقدار المتساوية المتوازية التى أماكنها مختلفة البعد
لا يرى اختلاف مناظرها على أقدار (قدر) أبعادها

[For] equal and parallel magnitudes, unequally distant [from the eye], the difference of their aspects is not seen according to the magnitude(s) of their distances.

T: Ṭūsī, *Taḥrīr al-Manāẓir li-Uqlīdis*, ed. Hyderabad, 1358=1939, p. 6

المقادير المتساوية المتوازية المختلفة الأبعاد لا يكون اختلافها فى الرؤية على نسبة اختلافها فى الأبعاد

[For] equal and parallel magnitudes unequally distant [from the eye], their difference in appearance (*ikhtilāf fī al-ru'ya*) is not in the same ratio as the difference of their distances.

Variation Sample 4: Proposition 20

O: *Euclidis Optica, Euclidis Opera Omnia*, ed. Heiberg and Menge, v. 7, pp. 29–30

καὶ προσπιπτέτω ἀκτὶς ἀπὸ τοῦ ὄμματος τοῦ Γ ἡ ΓΗ, καὶ ἀντανακεκλάσθω, ἄχρις οὗ συμβαλεῖ τῷ πέρατι τοῦ ΑΒ μεγέθους τῷ Α . . . ἐπεὶ οὖν προσπέπτωκεν ἀκτὶς ἡ ΓΗ καὶ ἀντανακέκλασται ἡ ΗΑ, πρὸς ἴσας γωνίας ἀνακεκλασμέναι εἰσίν, ὡς ἐν τοῖς Κατοπτρικοῖς λέγεται.

Burton, "The Optics of Euclid," 1945, p. 360
And let a ray fall, GL, from the eye G, and let it be reflected to the point where it touches A, at the end of AB . . . Now, since the ray GL has fallen and the ray LA has been reflected, they have been reflected at equal angles, as it is said in the *Catoptrica*.

Ver Eecke, *Euclide, l'Optique et la Catoptrique*, 1959, p. 13
Faisons tomber le rayon GH de l'oeil G; réfléchissons-le jusqu'où il recontre l'extrémité A de la grandeur AB . . . Dès lors, puisque le rayon GH est incident et rayon HA réfléchi, ils sont infléchis à angles égaux, comme il est dit dans les *Catoptriques*.

R: *Opticorum Recensio Theonis*, ed. Heiberg and Menge, v. 7, p. 176
καὶ ἀπ' αὐτοῦ προσπιπτέτω ἀκτὶς ἡ ΔΘ καὶ ἀνακεκλάσθω ὡς ἡ ΘΒ ἐπὶ τὸ Β πέρας, καὶ ἀπο τοῦ Δ ὄμματος κάθετος ἡ ΔΖ. οὐκοῦν ἴσαι εἰσὶν αἱ πρὸς τῷ Θ γωνίαι ἀλλήλαις· τοῦτο γὰρ δείκνυται ἐν τοῖς Κατοπτρικοῖς.

Ver Eecke, *Euclide, l'Optique et la Catoptrique*, 1959, p. 67
Soit D l'œil, que le rayon DY tombe de celui-ci et se réfléchisse, tel que YB, sur l'extrémité B . . . Dès lors, les angles situés au point Y sont égaux entre eux, car cela est démontré dans les *Catoptriques*.

M: *Kitāb Uqlīdis fī ikhtilāf al-manāẓir*, v. 1, pp. 58–59 [M20(6–11)]

ونخرج من نقطه ج شعاع البصر مثل جح ولينعطف الشعاع من موضع
نقطه ح. فلانّ شعاع جح وقع على نقطة ح وانعطف الى نقطه ا
يكون انعطافه على زوايا متساوية

And we draw from point G a visual ray such as GH and let the ray be deflected from the position of point H.
Now, because ray GH falls upon point H and is deflected to point A,
its deflection is at equal angles.

T: Ṭūsī, *Taḥrīr al-Manāẓir li-Uqlīdis*, ed. Hyderabad, 1358=1939, p. 9

وليكن الشعاع جح والمنعطف منه ح ا . . .
وزاويتا ح الشعاعية والانعطافية متساويتين

And let there be the [visual] ray GH, and the deflected [ray] HA . . . then the two angles [at] H, the visual [angle] and the deflecting [angle], are equal.
T: Ṭūsī, *Taḥrīr al-Manāẓir li-Uqlīdis*, ed. Hyderabad, 1358=1939, p. 3

إذا انعطف الشعاع من جسمٍ صقيلٍ كالمرآة حدَثت هناك زاويتان
متساويتان تُسمّى أحدهما زاوية الشعاع والأخرى زاوية الانعطاف

When the [visual] ray is deflected from a polished body such as a mirror, there are produced two equal angles, one called the [visual] angle and the other the angle of deflection.

Part II

Summary and Commentary

4 Title

There are at least two main variations to the title of the extant copies; one variation has the expression *Ikhtilāf al-manāẓir* corresponding to Greek *Optika* (m1, m3, m5), the other has the short form *al-Manāẓir* (m2, m4) [m6 is reported with the title *Kitāb al-Manāẓir li-Uqlīdis*]. The full title and the place where it appears in the manuscripts are not the same within each set, either.[37] The title *Kitāb Uqlīdis fī Ikhtilāf al-manāẓir* is adopted in the critical edition [see Edition and Translation, v. 1, p. 3] as a form closest to the original title due to considerations beyond the dominance of this form in the available manuscripts.

Kitāb Uqlīdis: The first part of the title is a reference to the "Book of Euclid." The author's name, which appears in four out of the five available manuscript copies (m1, m2, m3, m5), is transliterated in two forms: *Uqlīdis* [in m1], and *Ūqlīdis* [in m2, m3, and m5[38]]. The first form, adopted here from m1, also appears in some early references to Euclid's text[39] and with increasing frequency in later sources, including many copies of Ṭūsī's *Taḥrīr al-Manāẓir* [see List I, v. 1, pp. xxxvii–xlv].

---

[37] In the odd-numbered set, m1 (625H) has *Kitāb Uqlīdis fī Ikhtilāf al-manāẓir wa al-shu'ā'āt* in the incipit, title page, and colophon; m3 (600H) has *Manāẓir* as title, and *Kitāb Ūqlīdis fī Ikhtilāf al-manāẓir* at the colophon; m5 (ca. 680) has *Kitāb Ūqlīdis fī Ikhtilāf al-manāẓir* as title, *Kitāb Uqlīdis fī Ikhtilāf al-manāẓir* in the title page and *Kitāb al-Manāẓir li Ūqlīdis* at the colophon. In the even-numbered set, m2 (692H) has *Kitāb al-Manāẓir li Ūqlīdis al-Ṣūrī* only at the colophon; m4 has *al-Manāẓir min al-'ayn* as title and in title page and *Kitāb al-Manāẓir* at the colophon: see Manuscript Plates, v. 1, pp. 232–241; for details of the Arabic manuscripts, see Description and List I, v. 1, p. xxvi and p. xxxiv, respectively.

[38] In the case of m5, this form is only in the title and the colophon; see Manuscript Plate V, v. 1, pp. 241–242. The title page has it without *wāw* after the initial *alif.*

[39] For example, Aḥmād ibn 'Īsā's *Kitāb al-Manāẓir wa al-marāyā al-muḥriqa*, ḥ1 [copied ca. 600H?, see Ḥ, List II, v. 1, p. xlvi], fol. 89a, or al-Ya'qūbī's *Ta'rīkh*, ed. Houtsma, v. 1, p. 139. The manner of reference to Euclid's book in Houtsma's edition is "*li Uqlīdis hadhā al-kitāb* . . . " ("by Euclid, [there is] *this* book . . . "), as if he had a copy of the work at hand. The full statement is: "*li Uqlīdis hādhā al-Kitāb fī al-Manāẓir wa ikhtilāfuhā min makhārij al-'uyūn wa al-shu'ā'*;" see Supplementary Passages, v. 1, p. 243. Common features of al-Ya'qūbī's *Ta'rīkh* with m1 other than the form of Euclid's name are the terms *ikhtilāf* and *wa al-shu'ā'* [the latter appearing as *wa al-shu'ā'āt* in the title of m1]; the expression *makhārij al-'uyūn* is comparable to the similarly uncommon construction *makhraj al-shu'ā'* in m1's text of one of the propositions; see Summary and Commentary: Proposition 4, p. 37.

Ikhtilāf al-manāẓir: The second part of the title, appearing as such in three out of five manuscript copies (m1, m3, m5), is a common expression for both *Optika* the text and *optikê* the discipline, just as it is with the short form *al-Manāẓir* of the other two (m2, m4) [or three, to count m6]. In the case of this part of the title, there is also some evidence that the longer expression represents a form closer to the title of the original translation. Early references to Euclid's text appear as *Kitāb Ikhtilāf al-manāẓir* in Aḥmad ibn 'Īsā's *Kitāb al-Manāẓir wa al-marāyā al-muḥriqa* (before 250/864?) and Ibn al-Nadīm's *Kitāb al-Fihrist*[40] (ca. 377/987) and in the form *Kitāb fi al-Manāẓir wa ikhtilāfuhā . . .* in Ya'qūbī's *Ta'rīkh* (ca. 267/880).[41] *Ikhtilāf al-manāzir* is also the form in which the title of Euclid's *Optics* appears in the title pages of the "Middle Books" collections,[42] as well as in the title pages of some manuscript copies of its 13th-century recension *Taḥrīr al-Manāẓir,* where reference is made to the "old translation" (*al-naql al-qadīm*) in those very terms.[43]

The predominantly longer and apparently older form of the title, *Ikhtilāf al-manāẓir*, introduces many puzzles. That it was used to represent the discipline of optics in addition to the title of Euclid's *Optics* is verifiable.[44]

---

40 The reference to Euclid's book in the form "*Kitāb Ikhtilāf al-manāẓir*" appears in fol. 89a of ḥ1 [see Ḥ, List II, p. xlv]. The same form appears in Ibn al-Nadīm's list of Euclid's works [see *Kitāb al-Fihrist*, ed. Flügel, p. 266; tr. Dodge, v. 2, p. 636], where *Ikhtilāf al-manāẓir* is translated as "Inversion of Visions"!

41 The passage in al-Ya'qūbī offers too many possible readings to illuminate the problem; see Supplementary Passages, v. 1, p. 243. If the part *wa ikhtilāfuhā* is read together with *Kitāb fī al-Manāẓir*, then the resulting combination would be another way of referring to *Kitāb fī Ikhtilāf al-manāẓir*; but the same part can be read as part of the phrase *wa ikhtilāfuhā min makhārij al-'uyūn . . .* (and their difference from visual extensions/emissions . . .), a reading that still leaves the problematic interpretation of the last part, *wa al-shu'ā'*, unresolved.

42 This is the case for codices containing m1 and m5 as well as manuscripts of the Arabic translation of Euclid's *Elements* that include the *Mutawassiṭāt* titles. On the availibility of the Arabic text of *Optics* as early as those of Euclid's *Elements* and Ptolemy's *Almagest*, the two texts to be studied before and after it, see The Arabic Versions, v. 1, p. xix.

43 The statement that "in its old translation this was (*kāna fī al-naql al-qadīm*) *Kitāb Uqlīdis fī Ikhtilāf al-manāẓir*" is contained in the following copies: t3, t5, t10, t14, t4; see T in List I, v. 1, pp. xxxvii–xlvi. The expression *al-naql al-qadīm* is often associated with the earliest phase of the translation movement.

44 Some examples are the heading of the optical treatments of Aḥmad ibn 'Īsā and Qusṭā ibn Lūqā [see Ḥ and L in List II, v. 1, pp. xlv–xlvi] respectively as *Kalām fī ikhtilāf al-manāẓir* (ḥ1, fol. 62b) and *Fī 'Ilal mā ya'riḍu fi al-marāyā min ikhtilāf al-manāẓir* (L, p. 1).

Yet it is not immediately clear why the term *ikhtilāf* (difference, variation) was sometimes included in and other times excluded from the Arabic title.[45] Establishing the intended meaning of the combined expression is no less problematic, as there can be at least as many senses to it as that of its second component, *al-manāẓir*. On the one hand, *manāẓir* is the plural form for two slightly different terms, *manẓar* and *manẓara* [from the verbal root $N^aẒ^aR^a$ (to look at)], each with at least two senses, one specifically in connection with the eye and one more general.[46] On the other hand, the meaning of the plural term, *al-manāẓir*, ranges from a strict sense of visual rays[47] to the more general sense of aspects or appearances.[48] It is in the more restricted sense that the term *manāẓir* may be contrasted to *shu'ā'āt* [part of the title in m1], as corresponding pairs to Greek ὄψεις/*opseis* and ἀκτῖνες/*aktines* or Latin *visus* and *radii,*

---

45 A few cases are the titles of m2, m4, and Ibn al-Haytham's reported *Summary of the Optics of Euclid and Ptolemy* (*Talkhīṣ Kitābay Uqlīdis wa Baṭlamyūs fī al-Manāẓir*); see Sabra, *The Optics of Ibn al-Haytham*, v. 2, p. xxxii, no. II. 18 and Ibn al-Qifṭī's (d. 646/1248) citation of Euclid's *Optics* in *Ta'rīkh al-Ḥukamā'*, ed. Lippert, p. 63.

46 For a discussion of the general and specific senses of the singular *manẓar* and *manẓara*, together with examples from relevant literature; see Sabra, "Manāẓir, or 'Ilm al-manāẓir," *EI*[2], 6, p. 376 and Sabra, "Ibn al-Haytham," *DSB*, 4, p. 203, n. 9.

47 In the Arabic translation of Galen's *De usu partium*, the plural, *al-manāẓir*, and its singular form, *al-manẓar*, render the Greek terms ὄψεις/*opseis* (visual rays) and ὄψις/*opsis* (visual ray); see *Kitāb fī Manāfi' al-a'ḍā'* (*On the Usefulness of the Parts of the Body*), Paris MS *Bibliothèque Nationale*, Arabe 2853, fol. 182b. The same strict meaning is suggested by the constructions *in'ikās al-manāẓir* (reflection of visual rays) in Ḥunayn ibn Isḥāq's own discussion of vision, see *Kitāb al-'Ashar maqālāt fī al-'ayn*, ed. and tr. Meyerhof, pp. 109–110. [Note that the term *manāẓir* is translated into English as "images" on pp. 36–37 of Meyerhof's translation, as noted in Sabra, "Manāẓir, or 'Ilm al-manāẓir," p. 376, even though its Greek equivalent ὄψεις is given in the Glossary, pp 202–203.] The same can be said of the expression *manāẓir min al-'ayn* (rays from the eye) in some variants of Euclid's *Optics* [m2, m4, see Variation Sample 1, p. xxxviii], as well as *manẓar al-ibṣār* appearing in sources as late as Tahānawī, *Kashshāf iṣṭilāḥāt al-funūn* (*Dictionary of Scientific Terms*), edn. Calcutta, p. 486.

48 A more general sense of the term *al-manāẓir* would have to be understood by the constructions *Manẓar al-kawākib*, the alternative title for Ibn al-Haytham's *Ru'yat al-kawākib*; see Sabra, "On seeing the Stars," and Sabra, "Ibn al-Haytham," *DSB*, 6, p. 203, n. 9, where the terms "view or appearance" are suggested for *manẓar*, and where it is stated that "in the context of Arabic optics *manāẓir* is the plural, not for *manẓar* (view, appearance) but of *manẓara*, that by which vision is effected, an instrument of vision."

respectively.[49] It is also in the same sense that *manāẓir* can be distinguished from the more general meaning of aspects or appearances. The more general meaning seems to be the sense in which *'ilm al-manāẓir* and *'ilm al-shu'ā'āt* come to correspond to *optikê*, the Greek science,[50] perhaps even one in which *Kitāb al-Manāẓir* becomes at once the standard title for *Optika* the Greek text and the prototype for *De aspectibus* or *Perspectiva* of the Latin titles.[51]

Manāẓir may have been a happy choice for translating the Greek terms ὄψεις/*opseis* and ὀπτικά/*optika* at the same time, while maintaining the double sense of both.[52] But its different shades of meaning in Arabic make the interpretation, and so the translation of the title *Ikhtilāf al-manāẓir*, a particuarly sensitive undertaking. The single appearance of the same construction within the Euclidean text[53] does not reveal enough of its meaning to explain its adoption for the title. Nor does the similarly phrased astronomical principle *ikhtilāf al-manẓar* (parallax) help determine the exact sense of the optical expression[54] or the terms of the relationship between the two.[55]. Other comparable

---

[49] The Greek and Latin corresponding terms are provided in Lejeune, *Euclide et Ptolémée*, pp. 19–21, and Theisen, *The Mediaeval Tradition*, p. 272, n. 1.

[50] For early references to *'ilm al-manāẓir* and *'ilm a-shu'ā'āt*, see, respectively, Fārābī's *Iḥṣā' al-'ulūm*, ed. Amīn, p. 98, and Qusṭā ibn Lūqā, *Fī 'Ilal . . .* [see L in List II, v. 1, p.xlvi], p. 1. The addition of *al-shu'ā'āt* at the end of m1's title may also be taken as a reflection of the currency of such an expression for a "science of rays" in the early period.

[51] Many texts including the Latin versions of Euclid and al-Kindī's *Optics* become Latinized in the form *De aspectibus*, whereas in the case of Ibn al-*Haytham's Kitāb al-Manāẓir*, the form *Perspectiva* is also used as the Latin title.

[52] ὄψις: sight, vision or aspect, appearance; ὀπτικός: of or for sight; ὀπτικά the theory of the laws of sight. See Liddell and Scott, *Greek-English Lexicon*, new edition, Oxford, 1968, p. 1282 and p. 1242, respectively.

[53] See Summary and Commentary: Proposition 9 and Variation Sample 3, p. 41 and p. xlii, respectively. There, the principle *ikhtilāf al-manāẓir* is presented as the ratio of the apparent magnitudes of two equal and parallel objects determined through their respective visual angles.

[54] Its meaning as the apparent displacement of a heavenly body makes it closer to the first sense. But in the more common singular form *ikhtilāf al-manẓar*, and description in terms of angles formed by visual lines, it is closer to the second.

[55] It is not immediately clear which discipline had borrowed the expression from which and in what form. Since *Ikhtilāf al-manāzir* was among the *al-Mutawassiṭāt* works, principles in optics were to be prior to and studied in preparation for astronomy. That optics was long perceived to be relevant to astronomy can be further concluded from the statement of Ibn Khaldūn in his *Muqaddimah* that '*ilm al-manāẓir* (optics) determines, among other things, *ikhtilāf al-manẓar* [i.e., the astronomical principle]; see *The Muqaddimah*, ed. Wāfī, v. 3, p. 1100, tr.

constructions, from *inḥirāf al-manāẓir* of Fārābī's *Iḥṣā' al-'ulūm* to *aḥwāl al-manāẓir* of Ṭūsī's *Taḥrīr*, introduce more ambiguities than insights.[56]

Indeed, the exact sense of *ikhtilāf al-manāẓir* cannot be easily determined at any given period or with any certainty. What can be gathered is that there were at least two senses to the expression. The more general sense, which is encountered in some related studies, is "variations in appearance,"[57] which, as such, would be an appropriate designation for a text primarily concerned with the appearance of variously situated objects. A more restricted meaning of *ikhtilāf al-manāẓir* as "the angle between visual rays" would be particularly consistent with the Euclidean representation of the appearance of objects through visual angles.[58]

Rosenthal, v. 3, pp. 132–133. But optics also borrowed from astronomy, though in a different form and pace; on the notion of *i'tibār* (experiment), see Sabra, "The Astronomical Origin of Ibn al-Haytham's Concept of Experiment" and "The Physical and the Mathematical in Ibn al-Haytham's Theory of Light and Vision," pp. 458–459.

56 In Fārābī's *Iḥṣā' al-'ulūm*, it is not clear whether "*inḥirāf al-manāẓir*" refers to the deviation of visual rays or of aspects; see ed. Amīn, p. 99. Sabra's English translation of the full passage includes the Latin text and Arabic variants of this problematic part; see *The Optics of Ibn al-Haytham*, v. 2, pp. lvi–lviii. On reformulations of dependent Latin texts, see Grant (ed.), *A Source Book in Medieval Science*, p. 17. The puzzle of the opening passage of Ṭūsī's *Taḥrīr* is whether it is *aḥwāl al-manaẓir* as the "modes of aspects" or "orientation of visual rays," that change according to the positions of ray[s]; see Variation Sample 1, p. xxxix.

57 The following examples include their modern interpretations in this particular sense: a passage in Thābit ibn Qurrā's translation of *Ptolemy's Planetary Hypotheses*, ed. Goldstein, pp. 34–35, English translation, Sabra, "Psychology vs. Mathematics: Ptolemy and Alhazen on the Moon Illusion," pp. 221–222 [transliteration, ibid. p. 244]: "as for the reason why what appears to sight and is imagined in it regarding the size of the planets is not in proportion to their distances, we must recognize that this is an error that appears to sight on account of *difference of perspective*[*s*]". On the title of Qusṭā ibn Lūqā's *Fī 'ilal mā ya'riḍu min ikhtilāf al-manāẓir*, translated as "On the reasons for *the variations in appearance* which occur in mirrors," see Toomer, *Diocles on Burning Mirrors*, p. 27; on the definition of *ikhtilāf al-manẓar* in Bīrūnī's *Tafhīm* (composed 420/1029), see Wright, pp. 158–159: "the observation of an object at the same time from different places involving *different points of view*." [all italics added]

58 This seems to be the case in Definition 4 and Proposition 9 in particular, whether relative appearance is meant for different objects or the same object in different locations, and especially, in the context where the unique expressions *fī al-manẓar* and *fī al-zāwiya* occur in the same variant (m1); see M9(7, 27), v. 1, p. 27 and p. 29, and Summary and Commentary: Proposition 9, p. 41 and n. 162–163.

The translation adopted for *Ikhtilāf al-manāẓir* of the Arabic title of Euclid's *Optics* is *Difference of Aspects*, in accordance with all the above considerations, as in the Latin titles *Diversitatum aspectus* and *De aspectuum diversitate*.[59]

5 Definitions

The definitions section is indeed the most critical and complex part of the whole text. It is the most critical for containing the founding assumptions of Euclidean optics; it is the most complex for representing various problems of textual transmission and interpretation. The more straightforward of problems are those of textual correspondence and variance: of the seven Euclidean definitions in Heiberg's edition, only five are fully represented in the Arabic tradition; the transmitted definitions represent, just as in the case of the title, at least two main variations [the pair m1, m5 (m3 lacking the definitions page), and the pair m2, m4]; and for the opening and closing Euclidean definitions, with the most substantial *and* consequential, transformations there is also variance within each set [see Edition and Translation, v. 1, p. 3, and Variation Samples 1 and 2, p. xxxviii and p. xl]. Far more complex are problems of the manner of individual occurrences and their respective consequences.

The originals of the transmitted definitions, the assumptions that visual rays extend along straight lines to a distance of great magnitude (def. 1), that this extension is in the form of a cone with its apex in the eye and its base reaching objects (def. 2), and that the visual angle determines the apparent size (def. 4) and the clarity of objects (def. 7), were not devoid of problems of transmission and of interpretation in the main two Greek Euclidean versions.[60] When these definitions were passed on into Arabic in their respective variations, not only were the Arabic "versions" not exactly what was in either of the two Greek

---

59 These Latin forms indicate links to the Arabic Euclidean tradition. The first form appears in the full Latin title of al-Kindī's *De aspectibus* as *De causis diversitatum aspectus* [*et dandis demonstrationibus geometricis super eas*; see Björnbo and Vogl, *Alkindi, Tideus und Pseudo-Euklid*]; the second form is the title of the Arabo-Latin MS Vatican Urbinas 1329, fols. 1r–19r, 1458; see Lindberg, *A Catalogue of Medieval and Renaissance Optical Manuscripts*, p. 21 and p. 47.

60 The complexity of such problems is reflected in the case of the first and last Euclidean definitions in Variation Samples 1–2, pp. xxxvii–xli, which include modern translations for this very reason. Jones, "Peripatetic and Euclidean Theories," contains a discussion of the variance of the opening definition in Greek with reference to problems of textual defectiveness; see, p. 52. Some cases [the pairs ὄψεις κατ' εὐθείας γραμμὰς (visual rays in straight lines), ἐξαγομένας εὐθείας γραμμὰς (straight lines leading out) and τι ποιούσας ἀπ' ἀλλήλων (making from each other), μεγεθῶν μεγάλων (of great magnitudes)] are referred to as possible restoration attempts and further discussed by Knorr in the same terms; see "Pseudo-Euclidean Reflections in Ancient Optics," p. 29, n. 48–49. Other problems linked to the same definition are included in Brownson, "Euclid's *Optics* and Its Compatibility with Linear Perspective."

versions associated with Euclid's name, there emerged a new set of problems and possibilities. The problems, some carried over from the Greek, others unique to the Arabic text, and none guided in either language by accompanying figures as in the case of the propositions, are here directed to the many components of the definitions and to their individual *and* combined effects; the possibilities, partly occasioned by comparable formulations or transcriptions, are addressed with reference to the problematics in both Greek and Arabic.

The treatment of the opening Euclidean definition is most extensive, as the bearer of *the* statement of the visual ray hypothesis and the seat of a range of complex of problems in textual transmission [see Textual Divergence, p. xxxv]. This part provides the fullest range of divergent formulations of the most central assumption of Euclidean optics and the occasion for the assumption of a problematic text anywhere from unintelligible, to illegible, corrupt, or simply missing, at least in the case of the opening lines.[61] The first Euclidean definition also captures not just the first, but the most critical instances of the rather simultaneous nature of transmission and transformation. It is, moreover, where the occurrence of the slightest shifts in the Greco-Arabic transmission of few Euclideam concepts, from Greek ὄψεις/*opseis* (visual rays) to Arabic *al-shu'ā'* (ray), from διάστημα/*diastêma* (distance/interval) to *samt* (path/direction), and from μεγεθῶν/*megethôn* (magnitude) to *kathra* (multitude), seem to have had consequences of some magnitude.

Definition 1: The Visual Ray

There is a ray (*al-shu'ā'*) [or collection of rays] that issues from the eye in straight lines (*'alā khuṭūṭ mustaqīma*), producing straight paths/directions (*sumūt mustaqīma*), [literally] with no limit to their multitude (*lā nihāyata li-kathratihā*). The closing portion, appearing in all copies and in most subsequent formulations of the opening Euclidean definition [see Variation Sample 1, pp. xxxvii–xxxix], is particularly problematic due to the multiple sense of *kathra* [on *kathīr* and *akthar,* see below; for historical occurrences of each, see Ullmann, *Wörterbuch der Klassischen Arabischen,* v. 1, pp. 64–69]. Here, *lā nihāyata li-kathratihā* is translated as "of infinite multitude" according to the dominant sense of the expression [see Summary and Commentary: Definition 1, p. 14].

---

[61] Variation Sample 1 includes, in addition to the Greek and Arabic variants and modern translations of the first definition, the many marginal notes in the Arabic manuscripts; see pp. xxxviii-xxxix; see also Supplementary Passages, v. 1, pp. 243–245, where the early alternative formulations of the Arabic definitions are close enough to each other to suggest that the overlapping parts of a *Kitāb al-Mir'āh* passing under the name of Euclid may have acted as a supplement to a defective opening in the manuscripts of Euclid's *Kitāb al-Manāẓir*. On the evidence supporting the latter case, see p. 8, n. 63; on possible defects in the Greek, see references to the related works of Jones and Knorr on p. 6, n. 60.

Textual Correspondence:[62] In this definition, where the two Greek versions O and R differ in the most striking manner, the Arabic represents elements from both, in no apparent order. The subject (*al-shu'ā'*) [or its alternative form, *al-manāẓir min al-'ayn* in m2, m4], the first adverb (*'alā khuṭūṭ mustaqīma*) [immediately followed by "unequal in distance and magnitude" (*mukhtalifat al-ab'ād wa al-maqādīr*) in the same two variants], and the second verb (*yaḥduth*) [the first letter, undotted in m1, m5; *taḥduth* in m2, m4] are [in either case] closer to R, while the first verb (*yakhruj*) and the second adverb (*lā nihāyata li-kathratihā*) are closer to O [the latter still quite different from O, even in terms of something extending many times]; in the case of the object (*sumūt*), while its presence in the statement parallels R, the sense it takes on here is closer to O. The same terms are to be found in many subsequent formulations, including Ṭūsī's much longer version of the opening definition. By contrast, the alternative terminology of variants m2, m4 – *al-manāẓir min al-'ayn* as the opening words [straight lines in O, visual rays in R] and *mukhtalifat al-ab'ād wa al-maqādīr*, as the description of the straight lines [absent in O, different in R] – does not occur elsewhere, not even in the margins of any of the numerous copies of T, thereby indicating the limited circulation of the even-numbered variants. Yet, the very existence of such variants, especially in the opening definition, is by itself of some significance. Hand in hand with the distinct senses of the alternative formulations of the text's most central assumption is the pressing issue of the defective nature of this particular part of the text, in Greek, Arabic, or both.[63] No less significant is that some components of the Arabic opening definition, together with a few others, provide the first and most definitive case against a strict one-to-one assignment of the two main Arabic variants to the textual traditions of the two Greek versions O and R.

al-Shu'ā': The subject of the first definition [*al-shu'ā'*, preceded by *al-manāẓir min al-'ayn* in m2, m4], corresponding to the Euclidean visual ray, is closer to R: O postulates (Ὑποκείσθω/*hupokeisthô*) straight lines (εὐθείας γραμμάς/ *eutheias grammas*). The expression "visual rays" (ὄψεις/*opseis*) appears only in the next two definitions of O. By contrast, R has visual rays (ὄψεις/*opseis*) as

62 Mdef. 1 compared to Odef. 1, Rdef. 1, and Tdef. 1: Mdef. 1: v. 1, pp. 2–3; Odef. 1, ed. Heiberg, p. 2, tr. Ver Eecke, p. 1; Rdef. 1: ed. Heiberg, p. 154, tr. Ver Eecke, p. 57; Tdef. 1, edn. Hyderabad, p. 2.

63 The colophon statement of m2 (692H) on the "poor" (*saqīm*), "obscure" (*mughlaq*), and "cut-off" (*batr*) nature of the "original" (*aṣl*)[see alternative readings in Description, v. 1, p. xxvi; the reading of the second term as "*mu'allaq*" (suspended) in de Jong and de Goeje, *Bibliotheca Academiae Lugduno Batavae*, p. 43, makes much less sense than *mughlaq* (obscure) or even *muqlaq* (mixed-up)], the missing opening page of m3 (600H), and early reformulations of the opening portion, all point to textual defects; but they do not reveal the exact nature or source of problems. For signs of restoration efforts involved in Greek and Arabic, see pp. 6–7, n. 60–61, and The Arabic Versions, v. 1, p. xxi, n. 10.

part of its first definition, a form that reappears predominately as *al-shu'ā'* (ray), in both the early and late Arabic versions of the opening Euclidean definition.[64]

The Arabic term *al-shu'ā'* [translated here as "the ray", but also meaning a collection or cone of rays elsewhere; see Index of Arabic Terms, pp. 139–140], differs from the Greek corresponding term in both case and root. Not only is the term *al-shu'ā'* in the singular form,[65] whereas ὄψεις/*opseis* is plural in both O and R;[66] the verbal root of *al-shu'ā'*, *SH*$^{a''a}$ (to radiate), no longer exclusive to visual radiation, now also differs from that of Greek *opseis* and may be considered closer to the alternative form ἀκτῖνες/*aktines*, which runs throughout the Greek texts alongside it.[67] In the much more elaborate version of the definition in Ṭūsī's *Taḥrīr*, the occurrence of *al-shu'ā'* is actually preceded by a long preface on the analogy between rays produced by the eye and luminous bodies [see Variation Sample 1, p. xxxix].

The adoption of the term *shu'ā'* to render the Greek visual ray of the first definition is one of few cases to be considered quite relevant to the subsequent development of the Euclidean visual theory. The root of the term, *SH*$^{a''a}$, being linked, according to such native Arabic writers as al-Kindī and Ibn al-Haytham, to the ancient analogy of visual and luminous radiation,[68] the adopted term carries with it a long history of connotations and associations. The predominance

---

64 The variant m2 contains an alternative terminology as well: "visual rays from the eye is [to say] that the ray . . . " ("*manāẓir min al-'ayn hiya anna al-shu'ā'. . .*"); m4 has "visual rays" (*manāẓir min al-'ayn*) as the title and only *al-shu'ā'* as the subject of its first definition. All copies of T contain the latter (*shu'ā'*) to the exclusion of the former (*manāẓir*); on this form, see Summary and Commentary: Title, pp. 3–4, n. 46–49.

65 Unless it is read as *al-shi'ā'* (rays); see Hava, *Al-Farā'id: Arabic-English Dictionary*, p. 366 [reading and reference kindly suggested by Professor Len Berggren]. But in the light of accompanying singular verbs in the first three definitions and in consistence with the rather common Greek singular term ὄψις/*opsis* with the same collective connotations [see Lejeune, *Euclide et Ptolémée*, pp. 19–21], a singular form *al-shu'ā'* is more likely to represent the term's intended application.

66 The plural form *al-shu'ā'āt* does appear in at least one variant (m1), both within the text, where it is adjacent to the singular in the construct "*nukhriju al-shu'ā' shu'ā'at . . .* " ("we draw the [collection of] ray[s] as rays . . ."), and in the title, where it could stand for '*ilm al-shu'ā'āt* (science of rays).

67 On the different constructions and the historical context of the two Greek terms, as well as the ancient analogy of visual and luminous radiation, see Lejeune, *Euclide et Ptolémée*, pp. 19–21.

68 Al-Kindī writes that "the ray" (*radius*), which he describes as "the impression of luminous bodies on opaque bodies,", takes its name from the name lumination (*SH*$^{a''a}$?); see *De aspectibus*, sec. 11a, ed. Björnbo, p. 13; in the case of Ibn al-Haytham, his exact words are available in the original Arabic: "Mathematicians have applied (*sammū*) ray (*shu'ā'*) to the ray of sight (*shu'ā' al-baṣar*) only by analogy (*tashbīh*) with the ray of the

of the form *al-shu'ā'* over its alternative *al-manāẓir* is suggested not just by the more frequent appearance of *al-shu'ā'* as an expression for visual radiation, but also by its simultaneous and at times interchangeable application to other forms of radiation from early on.[69] The eventual switch in the direction of radiation from the surface of the eye to that of the luminous source may not be reducible to the choice or prevalence of one form of technical vocabulary (*al-shu'ā'*) over another (*al-manāẓir*) for one and the same Greek concept. But the development of visual theories further and further removed from the "hypothesis of visual rays"[70] must be viewed in the context of this and other critical elements of the Arabic opening defintion.

yakhruju min al-'ayn: The verb describing the behavior of the rays in the opening part of the definition is another notable case in the development of visual theory. Read as *yakhruju*[71] *min al-'ayn* [literally "exits from the eye," here translated as "issues . . . ," elsewhere as "drawn . . .," as in geometrical constructions], it is closer to O insofar as it corresponds to the verb ἐξαγομένας/*exagomenas* (led out) more closely than the form φέρεσθαι/ *pheresthai* (carried) in R. Yet even as plural (*takhruju*) or passive (*yukhraj*), the root $KH^{a}R^{a}J^{a}$ (to come out) still differs from O in terms of its physical connotations. A good measure of the effect of the Arabic root is the subtle yet distinct senses of the early and late formulations of the same part of the opening definition: in the early version (M), the ray issues from the eye [or "maintained" (*za'ama*) to do so (m1, margin)], with no more physical overtones than is to be expected from the statement of a definition in one of "the more physical of the mathematical sciences" [Aristotle, *Physics* 194a7]; but Ṭūsī's later recension (T)

---

sun or of fire;" Sabra, *The Optics of Ibn al-Haytham*, v. 2, p. 9, includes English translation of the full passage; see *Qawl fī al-Ḍaw' (Discourse on light)*, *Rasāi'l*, p. 9; French translation, Rashed, "Le 'Discours de la Lumière' d'Ibn al-Haytham (Alhazen)."

69 Qusṭā Ibn Lūqā's *Fī 'Ilal* [see L in List I, v. 1, p. xlvii] has a separate chapter on "kinds of ray" (*aqsām al-shu'ā'*), which discusses *al-shu'ā' al-baṣarī* (visual ray), *al-shu'ā' al-shamsī (solar rays)* and *al-shu'ā' al-nārī* (fiery rays). In Aḥmad Ibn Īsā's *Manāẓir wa al-marāyā al-muḥriqa* [see Ḥ in List I, v. 1, p. xlv], the alternative forms *al-nūr* (light) and *quwwa nūriyya* (luminous power) are used in the part corresponding to the first Euclidean definition, and the latter form appears in the Arabic translation of the Pseudo-Euclidean *De speculis*; see Supplementary Passages, v. 1, pp. 243–244.

70 See Sabra, "Ibn al-Haytham and the Visual-Ray Hypothesis," p. 204; for the passage containing Ibn al-Haytham's decisive position on the subject, see Ibn al-Haytham's *Kitāb al-Manāẓir*, bk. I, chap. 6, sec. 62, ed. Sabra, pp. 158–159, tr. Sabra, *The Optics of Ibn al-Haytham*, v. 1, p. 82.

71 It is possible to read this active masculine singular form as passive (*yukhraju*) or as plural (*takhruju*), depending on the position of points and voweling of letters. Of the five copies, only one (m4) is dotted [two below the first letter] to provide a definitive reading.

of the same *Arabic* text reads: "it is [only] *as if* (*ka- annahu*) the ray is sent out of the eye (*munba'ith min al-'ayn*) and coming out of it (*khārij minhā*)," so such a ray (*dhālika al-shu'a'*) is [just] to be imagined (*wa l-yutawahham*)."

The verbal root $KH^{a}R^{a}J^{a}$ is applied to the extension of geometrical entities such as lines repeatedly throughout the text. Still, there are physical connotations in the literal sense of its very first occurrence as something "coming out" of the eye, just as it is in the case of its "spreading out" (*yanbathth*) in some early reformulations of the first Euclidean definition.[72] And it is around the verbal form *yakhruj* [literally exits] that many discussions and debates on the founding assumptions of visual ray theory take place. This is the case in the extensive literature on the visual ray hypothesis, itself coming to be called "*al-shu'ā'*," whether it was defended, refuted, or replaced.[73] That the physical connotations of the Arabic verb circulating as part of a prominant Euclidean definition were disturbing, even to visual ray theorists of a reformed kind, can be deduced from Ṭūsī's insistence on the metaphorical nature of its language.[74] Responses to the same language are at the base of the key argument with which this most central and problematic component of visual ray theory is disposed of altogether. Ibn al-Haytham's eventual abandonment of the "hypothesis of the visual ray," begins with the common argument "if vision occurs only through something that goes out (*yakhruj*) from the eye . . . ," and it concludes with the decisive formulation "therefore nothing issues (*yakhruj*) from the eye to perceive visible objects."[75]

---

72 The verbal form *yanbathth* is used in the Arabic version of the Pseudo-Euclidean *De speculis*, and the corresponding texts of Aḥmad ibn 'Īsā and Qusṭā ibn Lūqā [see Supplementary Passages, v. 1, pp. 243–245]. The transcription of *yanbathth* is close to *yanba'ith* [from the same root as Ṭūsī's *munba'ith*], both appearing in the Arabic edition of the lost treatise of Alexander of Aphrodisias with the revealing title *Fī al-Radd 'alā man yaqūlu inna al-ibṣār yakūnu bi al-shu'ā'āt al-khārija 'inda khurūjihā min al-baṣar*: *Commentaires sur Aristote perdus en Grec et autres Épîtres*, ed. Badawī, p. 28; another close form *yanbat* in Sbath's edition of 'Ubaydullāh Bakhtīshū''s *al-Rawḍat al-ṭibbiyya*, p. 69, otherwise following the wording of Ibn Lūqā's *Fī 'Ilal*, is a possible misreading.

73 For standard arguments, see Rosenthal, "On the Knowledge of Plato's Philosophy in the Islamic world," pp. 412–416, which includes in addition to the arguments of the theory of visual ray (*shu'ā'*) and its immediate rival, the theory of forms (*inṭibā'*), those of the illuminationist theory of immediate perception (*mushāhada*) with a translation of Suhrawardī's related discussions; on the debate, see also Lindberg, "The Intromission-Extramission Controversy in Islamic Visual Theory: Alkindi versus Avicenna."

74 The argument in Ṭūsī's *Talkhīṣ al-Muḥaṣṣal* is that "early philosophers support its exit from the eye (*khurūjihi 'an al-'ayn*) only by metaphor (*majāz*);" see ed. Nūrānī, p. 173.

75 *Kitāb al-Manāẓir*, bk. I, chap. 6, sec. 56, Sabra, ed. pp. 158–159; tr. *The Optics of Ibn al-Haytham*, v. 1, p. 80 [facing Latin and English translation of the full passage, Sabra, "Ibn al-Haytham and the Visual-Ray Hypothesis," sec. 5, pp. 198–199]. On various versions of the visual ray hypothesis as described in a colorful passage recasting some coporeal descriptions of it as consisting of "fine bodies" (*ajsām diqāq*) or as a "luminous

'alā khuṭūṭin mustaqīmatin: This part of the first definition [identical in all copies] corresponds more closely to R: κατ' εὐθείας γραμμάς/*kat eutheias grammas* carried (φέρεσθαι/*pheresthai*) on straight lines [in O, the straight lines are being led out (ἐξαγομένας/*exagomenas*)]. The accompanying description "unequal in distance and magnitude" (*mukhtalifat al-ab'ād wa al-maqādīr*) in m2 and m4 is not matched by any of the two Greek versions, though it is comparable to the last part of R's definition. The Arabic differs from the Greek in containing both "straight lines" (*khuṭūṭ mustaqīma*) [corresponding to εὐθείας γραμμάς/*eutheias grammas*) in O and R] *and* "straight paths/directions" (*sumūt mustaqīma*) produced by them [the part *sumūt*, apparently corresponding to διάστημα/*diastêma* as constrcuted in R, though as meant in O]. The distinction between the straight "lines" (*khuṭūṭ*) and "paths/directions" (*sumūt*) leads to developments of some consequence. The "lines" (*khuṭūṭ*) upon which rays extend describe the rectilinear form of radiation,[76] while the "paths/directions" (*sumūt*) "produced" by the rays have more diverse associations, ones that in this case assume a rather critical function.

wa ma'ahu yaḥduthu sumūtun mustaqīmatun: Whether it is the straight lines [reading plural *taḥduth*] or a conic form [reading singular *yaḥduth*] that produce[s] straight paths or directions (*sumūt*), the verbal root *Ḥ*ᵃ*D*ᵃ*TH*ᵃ is close enough to the one used in the Greek [ποιούσας/*poiousas* (produce) in R]. There is more of a difference between the Arabic and the Greek in *what* is produced: in R, it is διαστήματα . . . ἀπ' ἀλλήλων/*diastêmata ap'allelôn* (intervals from one another) [the possible source for the description just noted]; but in the Arabic, it is the "straight paths/directions" (*sumūt mustaqīma*), a sense that even as the object of the verb in R (produced) is more in line with the διάστημα/*diastêma* in O, if understood in the sense of a geometrical interval.

For the corresponding Arabic term, *sumūt*, occurring without the description *mustaqima* in al-Ya'qūbī's *Ta'rīkh* and Ibn Abī Jarāda's *Tajrīd* [see Variation Sample 1, v. 1, p. xxxvii and p. xxxix, respectively], there is besides an unclear meaning and function in the corresponding Arabic definition, a range of

---

power" (*quwwa nūriyya*) issuing forth [as in the Arabic translation of Galen's *De usu partium*, MS *Bibliothèque Nationale* ar. 2583, fol. 183a, and in the Pseudo-Euclidean *De speculis*, MS Biblioteca Medicea Laurenziana or. 152]; see *Kitāb al-Manāẓir*, bk. I, sec. 4, ed. Sabra, p. 61, tr. Sabra, *The Optics of Ibn al-Haytham*, v. 1, p. 4, and v. 2, p. 8.

76 These seem comparable to the expressions *shu'ā'āt khaṭṭiyya* (linear rays) and *khaṭṭ shu'ā'ī* or *khuṭūṭ al-shu'ā'*" (radial lines), appearing respectively in Ibn Sīnā's *al-Shifā'*, bk. 3, ed. Anawati and Zayed, p. 102, Ibn 'Īsā's Ḥ [see List II, p. xlv, ḥ1, fol. 7b], and Ibn al-Haytham's *Kitāb al-Manāẓir*, see Sabra, *The Optics of Ibn al-Haytham*, v. 2, p. 9; Qusṭā's *Fī 'Ilal* has the qualification "without deviation" (*lā i'wijāj fīhā*); see Supplementary Passages, v. 1, p. 245.

manuscript variations and marginal additions [ibid., p. xxxviii],[77] each prompting further formulations and discussions. Qusṭā ibn Lūqā's formulation of "rays falling on objects in different directions (*jihāt*)" [see Supplementary Passages, v. 1, p. 245] seems to be a reading of *sumūt* of the first definition as directions; al-Kindī's reaction against "intervals" (*intervalla*) between rays, could be interpreted as a response to a particular reading of the term *bu'd* (distance, interval),[78] introduced as a synonym to *samt* [singular for *sumūt*] in some variants of the same part of the definition.[79]

A critical feature of this segment of the opening Arabic definition is not only that it has produced two different terms [*bu'd* and *samt*] with different connotations for one and the same Greek original (διάστημα/*diastêma*), but that it has produced two distinct classes of "straight" (*mustaqīm*) entities with different functions. A clear distinction between the "straight lines" (*khuṭūṭ mustaqīma*) upon which rays extend and the "straight paths/directions" (*sumūt mustaqīma*) produced by them reappears together with the same terminology in Ibn al-Haytham's radical solution to the extramission-intromission puzzle: *khuṭūṭ* become imaginary lines constituting the geometry of the Euclidean visual cone, *sumūt*, the set of lines perpendicular to the eye's surface, charting, as it were, the preferred "direction" for the reception of the Aristotelian forms.[80]

---

[77] One variant (m1) has the term *bu'd* (interval, distance) [or *ba'd* (then)] as a marginal addition above *yaḥduth*; in m5 the term appears in the definition itself as *bu'duhu* [or *ba'dahu* (and after that)], immediately following the verb. Al-Ya'qūbī's text has *bu'd*, the verb *taḥduth*, and only *sumūt*. Ibn Abī Jarāda's *Tajrīd al-manāẓir* [J in List I, v. 1, p. xliv] also has *sumūt* by itself. See Supplementary Passages, v. 1, p. 243 and p. 246.

[78] In this passage [*De aspectibus*, sec. 11a–11c, ed. Björnbo, pp. 13–14], which is to be read in the light of the Galenic and Ptolemaic continuity of the visual cone, the original Arabic term for the concept of interval (*intervalla*) under attack may be the term *bu'd*.

[79] So could the curious language of "*ittiṣāl*" (continuity) in Ṭūsī's version of the same definition: "Let that ray be imagined (*yutawahham*) as connected to the eye (*muttaṣilan bi al-'ayn*) along straight lines (*'alā khuṭūṭ mustaqīma*);" see Variation Sample 1, p. xxxix. In Ṭūsī's *Shu'ā'* [SH in List II, v. 1, p. xlviii], "continuity" is part of the definition of a cone. In 'Āmirī's (d. 382/992) *al-Qawl fī al-Ibṣār wa al-mubṣar* [Q in List II, p. xlvi, ed. Khalīfāt, p. 361], *ittiṣāl* is associated with the theories of Euclid, Ptolemy, and Galen, in a passage that is of much historical interest if only for an explicit reference to Ptolemaic optics as early as the mid–4th/10th century. There, as in Ṭūsī's *Taḥrīr*, *ittiṣāl* is expressed as a contact between the sense (*ḥāss*) and object of sense (*maḥsūs*), as opposed to the continuous body of the cone as a whole, as in Ptolemy, al-Kindī, and Ibn al-Haytham.

[80] Note the wording of Ibn al-Haytham's combined treatment (*tarkīb*) of the methods of the mathematicians (*aṣḥāb al-ta'ālīm*) and of natural philosophers (*aṣḥāb al-'ilm al-ṭabī'iyya*): "These (*khuṭūṭ al-shu'ā'*) are only imaginary lines and through them alone the eye perceives the forms of visible objects . . . the eye is disposed to be affected by these forms, but in a certain perceptible direction (*sumūt*) which is that of the perpendiculars to its surface . . . [so] "radial lines (*khuṭūṭ al-shu'ā'*) are imaginary lines that determine the

lā nihāyata li-kathratihā: This concluding part of the definition [identical in all copies of M and T, and running through a great number of texts in between and beyond the early and late versions], is closer to O where the corresponding expression μεγεθῶν μεγάλων/*megethôn megalôn* (great magnitudes) occurs, with no comparable expression in R. The degree to which the Arabic expression differs from the Greek can itself be determined according to the sense in which the term *kathra,* replacing μεγεθῶν/*megethôn* (magnitude), is to be interpreted in the Arabic phrase. This is by far the most problematic and critical part of the Arabic definitions, if not of the entire text. *Kathra,* having more than one sense in Arabic, the phrase *lā nihāyata li kathratihā,* could mean an indefinitely great *magnitude* [i. e. something extending many times, as in the Greek], or *multitude,* in which case the full statement of the first definition can be read as "straight lines or paths" *multiply* rather than extend, indefinitely. The evidence provided by dependent texts in both Arabic and Latin reveals that it is in the latter sense that it has been often read. The multiplying effect is preserved in at least two verbatim Latin forms: *multitudini non est finis* in Euclid's *Liber de aspectibus* and *multitudine infinitae* in al-Kindī's *De aspectibus.*[81] Internally it is expressed explicitly and variously, now as "straight lines extending from the visual cone's apex in infinite multitude (*lā nihāyata li kathratihā?*)" [al-Kindī], then as "many rays (*shuʻāʻāt kathīratan*) falling on objects in different directions" [Ibn Lūqā], and again as "light issuing from the object to the surface of the eye on an infinity of (*bi lā nihāya*) variously situated straight lines" [Ibn al-Haytham].

An "infinite multitude," applied to visual and luminous radiation models alike, marks a decisive transition from the single cone of Euclidean optics to multiple-point radiation models of the later figures, models that at least in the case of al-Kindī and Ibn al-Haytham, are associated with an important optical principle known as "punctiform analysis of light radiation."[82]

---

direction (*sumūt*) in which the eye is affected by the form;" see *Kitāb al-Manāẓir*, bk. I, chap. 6, sec. 62, ed. Sabra, p. 160; tr. Sabra, *The Optics of Ibn al-Haytham,* v. 1, p. 82; see also Sabra, "Ibn al-Haytham and the Visual-Ray Hypothesis," p. 204.

81 *Liber de aspectibus,* ed. Theisen [*The Mediaeval Tradition*], p. 336: "Radius egreditur ab oculo super lineas equales rectas et accidit post ipsum rectitudo recta cuius *multitudini non est finis*;" *De aspectibus,* prop. 11f, ed. Björnbo, p. 15: "Si enim a puncto sinapis lineae rectae protraherentur, egrederentur ab eo lineae in multitudine infinitae." The part *multitudine infinitae* is enough, even in the absence of the original Arabic, to confirm the much debated access of al-Kindī to at least this part of the early Arabic version of Euclid's *Optics.*

82 See al-Kindī: "If straight lines were extended from the point of origin, the lines would go forth from there in infinite multitude," *De aspectibus,* prop. 11f; on Qusṭā's formulation, see Supplementary Passages, v. 1, p. 245. Ibn al-Haytham's statement is taken from *Maqāla fi al-Ḍaw'*, *Majmūʻ al-rasā'il,* p. 10, English translation of passage, Sabra, *The Optics of Ibn al-Haytham,* v. 2, p. 10. The comparable statement that "lines of

Definition 2: The Visual Cone

The shape (*shakl*) of the ray is a cone (*makhrūṭ*), whose apex is next to the eye (*yalī al-'ayn*) [on the eye (*'alā al-ayn*), according to m5], and whose base is next to the extremity of the visible object (*talī nihāyat al-mubṣar*).

Textual Correspondence:[83] In this definition, where O and R differ only with regard to the position of the visual cone's apex [O has ἐν τῷ ὄμματι/*en tôi ommati* (in the eye), R has πρὸς τῷ ὄμματι/*pros tôi ommati* (at/near the eye)], M is closer to R in having the cone's apex "next to the eye" (*yalī al-'ayn*) [*'alā al-'ayn* (on the eye) in one of the variants (m5) is closest to R]. The Arabic, identical in the early and late versions, still differs from the Greek corresponding definition in minor, yet notable instances.

al-shakl alladhi yuḥīṭu bihi al-shu'ā' makhrūṭun: The first half of the definition, stating that "the shape enclosed by the ray is a cone," contains one such notable instance: the term *makhrūṭ* describing the conic shape of the ray. The occurrence of the singular *al-shu'ā'* here, in place of the plural ὄψεων/*opseôn* in both O and R, already creates some ambiguity as to whether the shape of a cone is being granted to a single ray [a literal reading of the full statement], or to a number of rays bunched together [a more collective understanding of *al-shu'ā'* in that statement].[84] But a crucial difference between the Arabic and Greek

---

[light] infinite in number issue in every point in the visual field" is quoted from Kepler, *Ad Vitellionem paralipomena*, p. 157, in Lindberg, *Theories of Vision*, p. 193, n. 75. On the principle and the visual models, see Lindberg, *Theories of Vision*, p. 30 and p. 226; Lindberg, "Alkindi's Critique of Euclid's Theory of Vision;" pp. 484–485, and Sabra, *The Optics of Ibn al-Haytham*, v. 2, p. 23, where the discussion is extended to include reflection and refraction. See also Summary and Commentary: Definition 3, p. 19.

[83] Mdef. 2 compared to Odef. 2, Rdef. 2, and Tdef. 2: Mdef. 2: v. 1, pp. 2–3; Odef. 2, ed. Heiberg, p. 2, tr. Ver Eecke, p. 1; Rdef. 2: ed. Heiberg, p. 154, tr. Ver Eecke, p. 57; Tdef. 2, edn. Hyderabad, p. 2.

[84] On the "*sens vague*" of the Greek, see Lejeune, *Euclide et Ptolémée*, p. 42. Some examples are the conceptions underlying the expressions *al-makhrūṭ al-shu'ā'ī* (the cone of ray[s]) appearing in Qusṭā Ibn Lūqā's *Fī 'Ilal* and Ṭūsī's *Taḥrīr* [L and T in Lists II and I, v. 1, p. xvli and p. xxxvii]. Related discussions include chapters on "what is the shape (*shakl*) of *al-shu'ā'*," [*Fī 'Ilal*, L, p. 5], as well as discussions of configuration as in the historical accounts of the cone "as a solid (*muṣmat*) body (*jism*), continuous (*muttaṣil*), [and] compact (*multa'im*) consist[ing] of straight lines which are fine bodies (*ajsām diqāq*)," some extending to the Greek tradition; see *Kitāb al-Manāẓir*, bk. I, ed. Sabra, pp. 60–61, tr. Sabra, *The Optics of Ibn al-Haytham*, v. 1, pp. 4–5. Ibn al-Haytham qualifies *makhrūṭ al-shu'ā'* (radial cone) with the statement that "each ray (*shu'ā'*) through which a visible object is perceived has, as a whole (*jumla*), the shape of a cone."

counterparts is the hardly discussed shape of the cone's base: when the Arabic *makhrūṭ* replaces the circular-based Greek κῶνον/*kônon* [in both O and R], it becomes a geometrical figure that, just as in its Latin corresponding term *pyramis*, represents a cone without any specification for the shape of its base.

The distinction between the Greek *kônos*, Arabic *makhrūṭ*, Latin *pyramis* and the alternative forms, Greek *puramis*, Arabic *ṣanawbar*, and Latin *conus*, as well as between these and the modern term "cone" by which all are commomly translated, is a subtle, but significant one.[85] That the Greek κῶνος/*kônos* is a circular-based cone is explicit in Greek works, both before and after Euclid,[86] a feature that makes it different from the Greek alternative *puramis*, a cone with an angled (triangular, quadrilateral, polygonal) base [Euclid, *Elements* bk. XI, def. 12, 18, Heath, *The Thirteen Books of Euclid's Elements*, v. 3, pp. 268–270].

By contrast, the term *makhrūṭ* represents a more flexible shape in terms of its base, both within and outside the present text. Internally, this is suggested by the the propositions involving a cone as an object of vision that include the specification "the cone whose base is a circle" (*al-makhrūṭ alladhī qā'idatuhu dā'iratin*).[87] External to the text, in Bīrūnī's *Tafhīm* (mid-4th/10th century), where such definitions are likely to be found, *makhrūṭ* is defined as "a body (*jism*) whose base is a circle *or another figure*" (*ayya shaklin/shakl-ī dīgar)*. And in Quṭb al-Dīn Shīrāzī's *Nihāyat al-idrāk* (late 7th/13th century), there are at

[85] On the Greek and Latin forms, see Heath, *The Thirteen Books of Euclid's Elements*, v. 3, pp. 268–270, and Lejeune, *Euclide et Ptolémée*, p. 34, n. 3. The following passage from Damianus's work is on p. 38: "The emission from us which we habitually call *opseis* propagates on straight lines and under the form of a rectangular cone; Ptolemy in his treatise on *Optics* has demonstrated it by means of apparatus and one can additionally consider it by reasoning;" see also Theisen, *The Mediaeval Tradition*, p. 38, and Brownson, "Euclid's *Optics* and Its Compatibility with Linear Perspective," p. 168.

[86] Pseudo-Aristotelian *Problemata*: *Problems*, Book XV, Loeb edition, p. 83, 872a37: "Vision travels in a circle owing to the natural formation of the eye. For it is a cone (*kônos*), *the base of which is a circle*;" see also *Problems*, Loeb edition, p. 333, 911b3–7: "Why is it that when the sun passes through quadrilaterals, as for instance in wickerwork, it does not produce a figure rectangular in shape but circular? Is it because the sun's rays fall in the form of a cone and *the base of the cone is a circle*? [italics added]; Galen, *De usu partium*, tr. May, p. 492: "From the pupil to the line which bounds the circle and is also called its circumference imagine a series of very many other lines extending like thin cobwebs. Call the figure . . . a cone, and think of the pupil as its apex and *the circle as its base* . . . " [italics added].

[87] These are Arabic propositions 32–35; see v. 1, M32(2), pp. 104–105; M33(5), pp. 106–107; M34(2), pp. 108–109; M35(2), pp. 114–115, has *al-makhrūṭ al-mustawā* (straight cone) in variants m2, m4.

least four variations: *makhrūṭ al-mustadīr* (circular cone), *makhrūṭ al-ṣanawbarī* (pine-like cone), *makhrūṭ al-usṭuwāna al-mustadīra* (circular/leveled cone) [described as synonyms], and *makhrūṭ muḍala'* (sided cone).[88] The impression that a strictly circular cone was more closely associated with the terminology of *ṣanawbar*, as opposed to *makhrūṭ*, can be gathered from early sources as well.[89] Yet, from the standpoint of critical developments, it is as the term *makhrūṭ* that the visual cone of the second Euclidean definition was most widely transmitted, and it is to the unspecified shape of its base that the Latin term *pyramis*, so crucial to the development of linear perspective, was most closely tied.[90]

ra'suhu yalī al-'ayn wa qā'idatuhu talī al-mubṣarāt: The closing part of the definition concerns the orientation of the conic shape. The first portion, *ra'suhu yalī al-'ayn* defines the position of the cone's apex (*ra's*), the second, that of its base (*qā'ida*) with respect to the eye (*'ayn*) and the visible object (*mubṣar*). The terms for the eye and the object appear elsewhere in the text with the alternative terminology of *al-baṣar* [Proposition 9] and *shay'* (object) [Definition 3] or *al-manẓūr ilayh* (viewed object) [Propositions 48–51; see Index of Arabic Terms, pp. 127–152]. But apex, expressed as *mustaḥadd* or *zujj* in some other early texts,[91] is rendered as *ra's* throughout and in subsequent treatments.

Of particular interest here is that the Arabic definition does not place the apex (*ra's*) in the eye (*fī al-'āyn*) [as in ἐν τῷ ὄμματι/*en tôi ommati* in O], but next to the eye (*yalī al-'ayn*)[*'alā al-'ayn* (on/at the eye) in one variant (m5)[92] [as in πρὸς τῷ ὄμματι/*pros tôi ommati* (at/near the eye) in R]. Given that the exact place of the cone's apex is the only part in the definition where the two Greek versions differ, the appearance of a form closer to R in all manuscript copies [*yalī al-'ayn* in m1, m2, m4 (m3 lacks the definitions section), and *'alā al-'ayn* in m5] is another piece of concrete evidence against the simple assumption that

---

88 Bīrūnī, *Al-Tafhīm li awā'il al-ṣinā'at al-tanjīm*, ed. Wright, pp. 19–20; Shīrāzī, *Nihāyat al-idrāk*, MS Köprülü 957: fol. 4b, see Şeşen, *Köprülü*, p. 486; al-Kindī's refers to the circular base in *De aspectibus*. sec. 11h, 12, ed. Björnbo, pp. 16–17.

89 This term in used in the Arabic translations of the pseudo-Euclidean *De speculis*, Galen's *De usu partium*, and such early Arabic texts as Aḥmad ibn 'Isā's al-*Manāẓir wa al-marāyā al-muḥriqa* [Ḥ in List II, v. 1, p. xlvi]; see Supplementary Passages, v. 1, pp. 243–44, and 'Āmirī, *al-Qawl fī al-ibṣār wa al-mubṣar* [Q in List II, v. 1, p. xlvi].

90 Theisen, *The Mediaeval Traditon*, p. 38, n. 87 states that "the Latin word *pyramis* came to be employed for *conus* . . . especially in those texts translated from the Arabic;" on the unspecified shape of the base and its discussion with respect to linear perspective, see Brownson, "Euclid's *Optics* and Its Compatibility with Linear Perspective."

91 These terms appears in the overlapping section of the Arabic text of *De speculis* and Ibn 'Īsā's close formulation; see Supplementary Passages, v. 1, pp. 243–244.

92 Note that *fī al-'ayn* (في العين), *yalī al-'ayn* (يلي العين), *'alā al-'ayn* (على العين) can have a close transcription in manuscripts.

each set of Arabic variants belongs strictly to one of the two Greek traditions O and R [or A and B]. So far, the Arabic variants were divided slightly unevenly between the two Greek traditions; now *all* match one better than the other. The shift itself, that of the position of the cone's apex from inside the eye [a position commonly associated with Euclid, Galen, Ptolemy, Damianus] to its outer surface [long associated with Theon],[93] has effects beyond the single switch of a point. An important component of the visual theories of al-Kindī, Ibn al-Haytham, and Ṭūsī is that their respective multiple-point radiation models [tied to the transmission of the first definition; see Summary and Commentary: Definition 1, p. 14] all involve the *outer surface* of radiation [as in the second Arabic definition].

Definition 3: The Condition for Vision

Objects (*al-ashyā'*) upon which the ray/cone of rays (*al-shu'ā'*) falls (*yaqa'u*) are seen (*tubṣaru*), and those upon which the ray/cone of rays (*al-shu'ā'*) does not fall (*lā yaqa'u*) are not seen (*lā tubṣar*).

Textual Correspondence:[94] In this definition, where O and R are identical, the Arabic matches the Greek closely in both versions M and T, except that now both subject (*al-shu'ā'*) *and* verb (*yaqa'u*) are singular, in contrast to the corresponding Greek forms ὄψεις/*opseis* (visual rays) and προσπίπτωσι/*prospiptôsi* (fall).

al-ashyā' allatī yaqa'u 'alayhā al-shu'ā' tubṣaru: The transmission of this part of the third Euclidean definition is also of some interest. Whereas the verb describing the decisive condition of the "falling" of rays on object in the Greek (προσπίπτωσι/*prospiptôsi*) is in plural form, the Arabic corresponding verb is transmitted as singular *yaqa'u* (falls), even if meant as the plural form *taqa'u* [orthographically interchangeable in undotted transcriptions as in m1 and m2].[95]

---

[93] Lejeune, *Euclide et Ptolémée*, pp. 19–21.

[94] Mdef. 3 compared to Odef. 3, Rdef. 3, and Tdef. 3: Mdef.3: v. 1, pp. 2–3; Odef. 3: ed. Heiberg, p. 2, tr. Ver Eecke, p. 1; Rdef. 3: ed. Heiberg, p. 154, tr. Ver Eecke, p. 57; Tdef. 3: edn. Hyderabad, p. 2.

[95] Note that the transcriptiion of the singular *yaqa'u* (يقع) is different from the plural form *taqa'u* (تقع) only through the position of the dots, though this is not the case with the singular form *waqa'a* (وقع), sometimes employed alongside it, as in Aḥmad ibn 'Īsa's *al-Manāẓir wa al-marāyā al-muḥriqa* [see Ḥ in List II, p. xlv], ḥ1, fol. 6a: "And the eye sees (*rāha al-baṣar*) that upon which the luminous ray (*shu'ā' al-nūriyya*) falls (*waqa'a*) . . . but not when it does not (*lam yaqa'*);" Ibn Lūqā, *Fī 'Ilal* [L in List II, p. xlvi], p. 5: "So one sees that upon which the visual ray (*al-shu'ā' al-baṣarī*) falls (*waqa'a*), but not when it does not (*lam yaqa'*);" see Supplementary Passages, v. 1, pp. 244–245.

The transmission of this crucial condition for vision with a singular verb, is not without consequence. The shift in the verbal case, not only affects the terms of the condition, it is carried over in singular forms (*cadit*, *inciderit*) in all the Latin versions including the so-called "Greco-Latin version."[96] In Arabic, once the singular verb is matched with the singular subject *al-shuʿāʿ* [itself replacing with many of its own consequences, the plural *opseis* in the first three definitions], the statement of the third definition literally becomes: "that upon which *a* ray falls, is seen." It may be no accident that the critical requirement of a one-to-one correspondence between illuminating and illuminated points, a major component of the celebrated principle of "punctiform analysis of light radiation" [see Summary and Commentary: Definition 1, p. 14 and n. 82] is described by early Arabic authors like al-Kindī and Ibn ʿĪsā as a "straight-line" correspondence between these points.[97]

wa allati lā yaqaʿu ʿalayhā lā tubṣaru: The stated assumption here that "those things (*allatī*) upon which it [visual ray] does not fall (*lā yaqaʿu*) are not seen," runs parallel to the first part, including the singular form of the Arabic verb, which is transmitted and discussed according to the same terms.

The language of "obstruction" related to the negative part of the definition, is not however applied strictly to point-like objects like the "grain" or "millet" of the Galenic tradition. Mediums such as water can also become an "obstacle" (*ḥājiz*) to the direct "falling" of rays on objects.[98]

---

96 The first Latin form *cadit* appears in *Liber de aspectibus Euclidis*, ed. Theisen, *The Mediaeval Tradition*, p. 336; *Liber de radiis visualibus* lacks the third definition, Theisen, *The Mediaeval Tradition*, p. 403; the form *inciderit* is used in the widely circulated "Greco-Arabic" text, *De visu*, a form closer to the Arabic singular than to the Greek plural corresponding term; see *Liber de visu*, ed. Theisen, p. 62, also Theisen, *The Mediaeval Tradition*, p. 66, p. 185, and p. 272.

97 Aḥmad ibn ʿĪsā's *al-Manāẓir wa al-marāyā al-muḥriqa*: "So from the surface of every part of the short span of the pupil . . . the ray of light that we described spreads out to every part to which it would be possible to draw something like a straight line," ḥ1 [Ḥ in List II], fol. 7a; see also Supplementary Passages, v. 1, p. 244, where the singular forms *waqaʿa* and *yaqaʿu* (falls) are used. Al-Kindī's formulation in *De aspectibus* is as follows: "We have illustrated how each part of the luminous body illuminates that which is opposite to it, namely that to which a straight line can be drawn;" see ed. Björnbo, sec. 13, p. 57, English translation, Lindberg, *Theories of Vision*, p. 30.

98 Qusṭā ibn Lūqā, *Fī ʿIlal*, L, p. 5: "This way, one does not see those objects with obstacles (*ḥawājiz*) between them and the eye, preventing and hindering the visual ray[s] (*al-shuʿāʿ*) from falling (*yaqaʿu*) on them" [elsewhere, the visual rays penetrate (*yanfudh*) from the eye, rather than issue from it]; see also Galen, *De usu partium*, tr. May, p. 492.

Definition 4: The Size and Clarity of Appearance

Objects viewed under "large, small, or equal angles" (*zawāyā 'aẓīma, ṣaghīra, mutasāwiya*) appear "large, small, and equal" [Mdef. 4-1–Mdef. 4-4 respectively]; objects under many/great angles (*zawāya/zāwiya kathīra*) appear more [clearly] (*kathīran*) [Mdef. 4-3, absent in m5].

Textual Correspondence:[99] In this definition, where its corresponding parts in Greek are identical in O and R, the Arabic differs from them in the following manner: The Arabic definition combines the three parts of the fourth Euclidean definition on size of appearance with that of its seventh definition on clarity of appearance [also identical in O and R], though not in the same order as the Greek. Of the four distinct parts of the Arabic definition, the first, second and fourth [Mdef. 4-1, Mdef. 4-2, Mdef. 4-4 on the appearance of the size of objects viewed under large, small, or equal angles] follow their Greek prototypes, but the third part [Mdef. 4-3] on the clarity of appearance "under many/great angles" (*zāwiya/zawāyā kathīra*) [*zāwiya* only in m1; m5 lacks this part], corresponding to the last Greek definition [def. 7 in both O and R], involves far more than a simple change in the internal structure of the Greek text.

mā ubṣira min zawāyā kathīratin ẓahara kathīran: The third portion of the fourth Arabic definition applies the term *kathīr* related to the same problematic root $K^aTH^aR^a$ [see Summay and Commentary: Definition 1, p. 14], to replace two other Greek terms, this time in one and the same definition: the Euclidean forms πλειόνων/*pleionôn* (greater, more) and ἀκριβέστερον/*akribesteron* (more clearly), each with separate sets of problems.

The problem with the first *kathīr* begins with the question of whether it is the *magnitude* or *multitude* of visual angles that determines visual clarity. With the double sense of the Arabic term, which is already present in some form in its Greek counterpart *pleionôn* (greater/more) [i.e., in the form πλειόνων γωνιῶν/*pleionôn goniôn* (greater/more angles)], the statement of the definition can be read not only as "what is viewed under *great* angles is seen more clearly" but also as "what is viewed under *many* angles is seen more clearly." To the problem of the terminological ambiguity of the Greek *pleionôn* and the Arabic *kathīr* [both can mean *greater* and *more* at once] a new dimension is added by the orthographic similarity between the Arabic terms *kathīr* (more) [dotted as such in m4] and *kabīr* (greater) [the latter being another possible reading in the undotted forms of m1 and m2; m3 and m5 lack this part].[100]

---

[99] Mdef. 4 compared to Odef. 4, Rdef. 4, and Tdef. 4: Mdef. 4: v. 1, pp. 2–3; Odef. 4: ed. Heiberg, p. 2, tr. Ver Eecke, pp. 1–2; Rdef. 4: ed. Heiberg, p. 154, tr. Ver Eecke, p. 57; Tdef. 4: edn. Hyderabad, p. 2.

[100] *Kathīr* (كثير), *kabīr* (كبير). See also Variation Sample 2, p. xl.

The situation becomes further complicated with the appearance of the second *kathīr* [now dotted as such also in m2] to mean "more clear." The connotations of multitude in the second occurrence of the term *kathīr* make the statement of the definition particularly problematic, at times leading to otherwise incomprehensible reformulations of this part of the definition as a statement no longer in terms of the *clarity* of an object's appearance [as in both the Greek and Arabic definitios], but of the *multiplicity* of the objects appearance. For Qusṭā ibn Lūqā, whose section on the Euclidean definitions is particularly striking in reflecting further bends from the already twisted Arabic terminology, it is not only the case that "the rays (*al-shuʿāʿāt*) falling on objects (*wāqiʿatun ʿalā mubṣarāt*) are many (*kathīratan*), and in different directions (*ʿalā jihāt mukhtalifatan*)," a model clearly bound to his reading of the first three Arabic definitions [see Summary and Commentary: Definition 1, p. 14 and Definition 3, pp. 18–19]; it is in fact, the *number* of "falling rays" that determine *how many* object are seen: "If one ray falls (*yaqaʿu*) upon an object," states Ibn Lūqā in *his* version of the Euclidean definitions, "it is seen as one, if two rays fall, it [the object] is seen as two" (*raʾā athnayn*), and "if more than two rays (*aktharu min shuʿāʿayn*) fall, it is seen as more than two" (*raʾā akthara min athnayn*).[101] That the problematic root $K^{a}TH^{a}R^{a}$ of the corresponding definition spread into other parts of the text with remarkable developments in the subject of visual clarity is suggested by the transmission of a closely related Euclidean proposition [see Summary and Commentary: Proposition 2, p. 30 and Variation Sample 2, p. xli].

6 Propositions

The Arabic propositions follow the definitions in representing a nonlinear mode of transmission. In their case, a form of contact with the Greek version long known as Theon of Alexandria's *Recensio* (R) is further suggested by the particular arrangement of the Arabic propositions, which is close in all available copies (m1–m5). The contact is most pronounced in the nature of structural correspondence between specific propositions and the two Greek versions O and R [e. g. M6–7 corresponding to O6/R6–7; M36–37 to O34/R34–35; M45–47 to O40/R40-1, R40-2, R40-3] and details of certain figures [e.g., M10, M16, M17, and M20, corresponding to O9/R9, O15/R15, O16/R16 and O19/R19]. The content [enunciation, demonstration, figures and lettering] of the Arabic reflects close correspondence with the Euclidean version O. This holds for cases with most striking differences between the two Greek versions [proof of M4 closer to O4 than R4; proof of M23 corresponding only to O22; figure of M8 closer to O7 than R7, figure of M12, closer to O11 than R11] and for less obvious cases [concluding statements of M2 closer to O2 than R2; enunciation of M9 closer to O8 than R8; lettering of M13–15 closer to O12–14 than R12–14].

---

101 Qusṭā ibn Lūqā, *Fī ʿIlal*, L, p. 6; see Supplementary Passages, v. 1, p. 245. It is possible to read this statement as the same object being seen once, twice or more.

There are, on the other hand, exceptional cases, for example, the case of the movement of far and near objects with respect to a moving eye, a crucial component in M61 [the eye moving] that is the same as that in R54 and the opposite to that in O55 [the eye being at rest]. There are also other forms of anomaly, ranging from the absence of elements in the Arabic that are present in both Greek versions [e.g., the critical reference to the *Catoptrics* in O19/R19; alternative treatments in O22/R22; see Summary and Commentary: Propositions 20 and 23, p. 55 and p. 62, respectively, and Structural Correspondence, p. xxxiv], and even the presence of elements in the Arabic beyond the main Euclidean versions [e.g., the closer correspondence to parts of the sixth book of Pappus's *Collection* (C); see Summary and Commentary: Propositions M36–42, corresponding to O34–36/R34–37/C50, C42, 44, C51, p. 77 and p. 79; see also Correspondence Table 2, p. xxxiv, and entries under individual propositions].

The transmission of the 64 Arabic propositions, representing the 58 and 57 propositions of the Greek versions O and R, is further characterized by the variety of forms the differences between the Greek and Arabic versions take. A Euclidean proposition is in some cases presented as divided [e.g., O6/R6, R7 corresponding to M6, M7; O40/R40-1, R40-2, R40-3 to M45, M46, M47], downsized [e.g., O22-1, O22-2, O22-3/R22-3 corresponding to M23], or absent altogether [e.g., O48/R48]; in other cases it is as if it were physically displaced [e.g., O38/R42 corresponding to M48], its proof replaced [e.g., O1, O3, O9/R1, R3, R9 corresponding to M1, M3, M10], or partly mutilated [e.g., O54–55/R53–54 corresponding to M60–61]. In yet other cases, the proposition is linguistically transformed to a point where the original sense is altered [O2/R2 corresponding to M2, O19/R19 to M20], or occasionally reversed [O8/R8 to M9; see Variation Samples 2–4, pp. xl–xlv].

Transmitted as such, the Arabic propositions represent both intended *and* unintended forms of transformation. Editorial revisions, structural re-arrangements, or alternative treatments, partly dictated by the complications of at least two versions being transmitted, are more in the nature of consciously-guided shifts. But more notably from the standpoint of critical developments, there are also shifts that are half-conscious at best, being like so many instances in the transmission of the definitions, as passive as the product of a linguistic or orthographic "slip" in the very process of translation or transcription [see Summary and Commentary: Definitions 1, 4, Propositions 9, 20, p. 20, 40, 55].

A comparatively straightforward and rather uniform feature of the Arabic propositions is their individual format. The equivalents to the formal division of a Euclidean proposition are *khabar* (enunciation), *mithāl* (illustration), *khalaf* (specification) [illustrated enunciation], *tartīb* [or *naẓar*] (set-up) [geometrical construction], *burhān* (demonstration), and *tamām* (conclusion).[102] These

[102] Sometimes the Arabic expressions include *faṣl*, to mean "distinguishing between a possible and impossible enunciation;" see al-Ya'qūbī, *Ta'rīkh*, ed. Houtsma, v. 1, p. 136. Also note that the third division (*khalaf*) could literally mean what comes after [i.e., after the illustration], to present an illustrated enunciation; on the Greek terms for such

divisions are represented, more or less uniformly in all copies and identifiable as such by specific expressions running throughout the Arabic propositions. The setting out of illustrations often begins with *fa-l-yakun* (let there be) [*mithāluhu* (its illustration) actually precedes this in variants m2 and m4], the specification with *fa aqūlu inna* (I say that), the proof with *burhānuhu* (its demonstration), geometrical constructions with *nukhriju* (we draw), and the statement of the conclusion with *dhālika mā aradnā an nubayyina* (this is what we wished to demonstrate) [in m1, m3, m4; *mā aradnāhu* in m2, *mā aradnā bayānuhu* in m5].

Propositions 1–3: Special Problems in Appearances

Proposition 1: Relative Time of Appearance – Parts of the Same Object

Synopsis: The enunciation is that nothing among visible objects is seen all at once. The illustration is that of a straight line (AB) [representing the object]. The demonstration (*burhān*) begins with the geometrical construction of the [cone of] ray[s] drawn as rays GA, GD, GE, GW, GB, from the eye (G) to the points on the object (AB). Then follows, one of these rays (GE) falls [on AB] "before" (*qabla*) others (GD, GA) and (GW, GB) [adjacent to the left and right, respectively], so that some portions of the object (ED, EW) are seen before (*qabla*) others (DA, BW). The conclusion is that [because parts of the object upon which rays closer to the perpendicular ray fall are seen before others], nothing among visible objects is seen all at once.[103] This is followed by the statement that the object [i.e., all its parts] is still imagined (*yutawahham*) to be seen simultaneously, due to the speed of sight's glance (*sur'at lamḥ al-baṣar*).

Textual Correspondence:[104] This proposition represents substantial differences from the Greek [close in O and R, Ver Eecke translates the identical final premises slightly differently; see ed. Heiberg, p. 4 and 156, and Ver Eecke *Euclide, l'Optique et la Catoptrique*, p. 2 and p. 58, respectively]. In the Arabic demonstration, there is no mention of "intervals" (διάστημα/*diastêma*) between the visual rays. The unequal rays [presented as five rather than four divergent

---

divisions see Heath, *The Thirteen Books of Euclid's Elements*: *The Formal Divisions of a Proposition,* v. 1, pp. 129–131, where the fourth division is translated as "construction."

103 The implicit assumptions are that, (a) every point on an object is seen by the falling of a visual ray on it [internal assumption: def. 3]; (b) some rays fall on objects before others [i.e., the perpendicular ray before adjacent ones, external assumption]; and (c) the *time* of the falling of rays on any point of the object directly affects the *time* that the same point is seen [external assumption]. The part M1(10–11), see v. 1, p. 5 [only in m1], is corrupt.

104 M1 compared to O1, R1, and T1: M1, v. 1, pp. 4–5; O1, ed. Heiberg, pp. 2–4, tr. Ver Eecke, p. 2; R1, ed. Heiberg, p. 156, tr. Ver Eecke, p. 57; T1, edn. Hyderabad, p. 3.

straight lines; see v. 1, p. 4] fall on the object at different *times*, and that affects the *time* the corresponding parts of the object are seen, such that the shortest ray [i.e., the perpendicular] reaches the object *before* others, and its corresponding point on the object is seen first, and so on.[105] None of this is in the Greek. The proposition ends by explaining the act of imagining (*wahm*) [thinking (δοκεῖ/*dokei*) in Greek] the visual process to be simultaneous (*jamī'an ma'an*) in terms of the rapidity of sight's glance (*lamḥ al-baṣar*), not the movement of visual rays (ὄψεων/*opseôn*), as in the Greek. The abridged version of Ṭūsī's *Taḥrīr* (T) differs from M in the following cases: in the enunciation, the subject is *al-mubṣarāt al-kathīra* (many visible objects), rather than *shay'un min al-mubṣarāt* (any visible object or parts thereof), and there is the added description, *bi qaṣdin wāḥidin* (by a single intention) at the end. The reason for AD being seen before DE is given as its "being closer in position" [i.e., to the perpendicular GA],[106] and the reason for simultaneous vision is given as "the speed of the glance *and* its movement (*intiqāl*)."

Historical Context: A remarkable feature of the opening Arabic proposition is the radical difference between its demonstration and that of the original Greek in the edition of Heiberg. Of particular interest is that the alternative proof bypasses the notions of "intervals," "spaces," or "divergences" associated with Euclidean visual rays in favor of their relative lengths, covered distances, and temporal aspects. As a result, the problem of the Arabic proposition cannot be reduced to that of "checkered vision," as in the case of the Greek.[107] The first Arabic proposition is *not* about filling in the blanks, so to speak, of those spots

---

105 The Arabic variants share the same figure and lettering, though oriented differently in the manuscripts. Of these, m1 is the only one to refer to GE [the shortest ray in the diagram] falling on the object first [a marginal note adds that this is because it is perpendicular]; m2–m5 [with same letters and similar diagrams as m1] refer to ray GA [the longer peripheral ray in their corresponding diagrams] reaching the object first, then GD, and so the corresponding portion (AD).

106 The diagram in T is that of a right-angled triangle, where GA [the perpendicular side of the right angle] is mentioned first [as in m2–m5, where GA is the longest rather than shortest ray] and termed "the radial cone's axis" (*sahm al-makhrūṭ al-shu'ā'ī*).

107 The expression together with the discussion appears in Brownson, "Euclid's Optics and Its Compatibility with Linear Perspective," pp. 174–175, where reference is made to modern interpretations of this Euclidean proposition according to the notion of divergence of rays with specific reference to Burton, "The Optics of Euclid," p. 353, Lindberg, *Theories of Vision*, p. 13, Cohen and Drabkin, *Source Book*, p. 258, and Ver Eecke, *Euclide, l'Optique et la Catoptrique*, p. 1, note 2, among others. Brownson's own position [see p. 174] is that "if we try to understand this interpretation of the proof [i.e., assuming "a single and momentary glance which is made up of diverging rays"] in relation to visual experience, it seems to suggest that we should see in a checkered fashion (this effect being alleviated only by the movement of our gaze)."

on an object *not met by visual rays* [def. 3] due to spaces between them [def. 1 in R]; it is rather about "imagining [i.e., judging] a thing to be seen all at once" (*qad yutawaham al-shay' mubṣaran jamī'an ma'an*)[108] when it [i.e., its different parts] is not reached by visual rays *at the same time* [external assumption]. Still, the alternative model may be not so irrelevant to the notions of spaces or intervals so fully at work in "checkered vision." If it is a conscious deviation, it may even be explained in terms of a step back, not necessarily from the philosophical implications of such notions, but from the standpoint of a translator/editor, from the linguistic complexities arising from the flexible meanings and diverse connotations of such terms as διάστημα/*diastêma* and φέρονται/*pherontai* already avoided in the case of the first Euclidean definition.[109]

The identification of the sources of the divergent features of the alternative Arabic proof is just as complex as the explanation of their occurrence. Such non-Euclidean assumptions as the different magnitudes of visual rays and distances covered by them can be considered directly related to the internal problems just discussed.[110] The sources of other assumptions, such as the speed of the eye's glance or the direct relation between that and the rays' respective lengths, cannot be determined beyond resonances between them and certain statements associated with some historical texts or figures.[111] For the more striking notions of time, no such association may even be easily made.

---

108 For the discussion of the term "*wahm*" in Ibn al-Haytham, as well as "*quwwa wahmiyya*" (faculty of judgement) in historical context, see Sabra, *The Optics of Ibn al-Haytham*, v. 2, pp. 64–66. For a more general discussion, see Sabra, "Sensation and Inference in Alhazen's Theory of Visual Perception," and Wolfson, "The Internal Senses in Latin, Arabic and Hebrew Philosophical Texts."

109 The former is used as both radius or distance in Euclidean texts, including the opening definition of O, and as intervals or spaces in the corresponding part in R [the term *sumūt* (paths, directions) of the first Arabic definition seems to correspond more closely to the first sense]. The latter [the standard Aristotelian vocabulary for locomotion] is less problematic. See also Summary and Commentary: Proposition 63, p. 98.

110 Besides such a representation in all the figures of the first Euclidean proposition, there is the qualification in two variants of the first Arabic definition (m2, m4) regarding the "unequal distances and magnitudes" (*mukhtalifat al-ab'ād wa al-maqādīr*) of rays in their extension along straight paths; see Variation Sample 1, p. xxxviii.

111 Hero of Alexandria (1st century A.D.) and Damianus of Larissa (4th century A.D.) are, respectively attributed with the statements, that "visual rays manifesting themselves in extreme speed must necessarily follow a short path" and that "this [rectilinearity] is useful for seeing something rapidly." On Hero's *Catoptrics* and Damianus' *Optics*, see Lejeune, *Euclide et Ptolémée*, p. 38, n. 2–3. References to other relevant elements including the related treatments of the Stoic school and Ptolemy are included in Theisen, *The Mediaeval Tradition*, p. 274. The prominent place of the central region of the visual field is also to be found in Ptolemy's assertion that what lies on the visual cone's axis must be seen more clearly [*Optica*, bk. II, sec. 20, ed. Lejeune, pp. 20–21] and

Whatever the sources or occasions for the resulting divergences, the first Arabic proposition becomes a channel for parallel developments, some of no apparent relevance. Some imported features reappear as alternative proofs or additional definitions in the Latin Euclidean versions, including some variants of the Greco-Latin version *De visu* (V),[112] while comparable arguments underly many discussions through which the terms of the problem change once more.[113] Characteristic of the Arabic discussions is that the discourse of simultaneous vision from the first proposition is often combined with, and at times overshadowed by, that of clarity of vision – the focus of the next Euclidean proposition. This often occurs by way of the recurring example of "letters in a book," the same example used in the prefaced version of the Euclidean text [i.e., R], where nonsimultaneous appearance is discussed, according to the text's own formulation of the opening definition, in terms of the separation of the visual rays.[114] By contrast, Arabic discussions of the same phenomenon and of the same example, wed the first proposition not to the definition preceding it but to the proposition immediately following it, in the manner of the Pseudo-Euclidean *De speculis*.[115] This is not merely the physical patching of a discourse on the

---

Damianus's slightly different formulation that "axial rays, being shorter than the peripheral, perceive more clearly" [Theisen, *The Mediaeval Tradition*, p. 274, n. 6].

112 Theisen, *The Mediaeval Tradition*, p. 275, reports two Florentine variants with "a considerably different proof about [visual rays] sweeping across an object from one end to another." The Latin version *De visu* has an additional definition on the equal speed of all visual rays; see *Liber de visu*, ed. Theisen, p. 62, English translation, Theisen, *The Mediaeval Tradition*, p. 185.

113 References to statements about clear apprehension being achieved by shifting the visual axis over the surface of the object and to the related works of John Pecham and Roger Bacon are made in Theisen, *The Mediaeval Tradition*, p. 275, n. 8, and to those of Witelo in Ver Eecke, *Euclide: l'Optique et la Catoptrique*, p. xvii.

114 On why objects seen are not seen entirely at the same time, the unknown author of the preface reports that he [the author], basing this on those who look at books attentively, said that they do not see all the letters in the same page because visual rays are not directed on all the letters but rather separated so that a great number of aligned letters are not seen. Later in the passage, he adds that "it seems that one sees these [i.e., the letters] entirely because visual rays move with too much speed without omitting anything, which means in a continuous manner and without jumps. See ed. Heiberg, pp. 144–154, tr. Ver Eecke, pp. 53–56.

115 In the Pseudo-Euclidean *De speculis*, partially extant in Arabic [see The Arabic "Version," p. xxvi, n. 28], the respective subjects appear together, and with reference to another example: "Often sight, when it looks for a small object that has fallen on the ground, does not find it, even though the eye sees a larger portion of the earth. The reason for this is that the eye does not see it until the perpendicular ray of the triangle, which is stronger than the other rays, falls upon it. As it was shown, equal quantities that are closer are seen more clearly than more distant ones. Nor is every part of the visual object seen at

totality of appearance [Proposition 1] onto that of *clarity* of appearance [Proposition 2], but a logical extension of the problem of "simultaneous vision" with the prominent place of the perpendicular ray of the first Arabic proposition to the problem of "accuracy of vision" (*ṣidq al-ru'ya*), to use the language of the second Arabic proposition itself.

The privileged status of the central region of the visual field entered discussions about the first Euclidean proposition early on, together with its companion example.[116] The treatement of Aḥmad ibn 'Īsā, which includes explicit references to "the letter in the script" (*al-ḥarf min al-kitāb*), though without naming the "group among the ancients" (*qawm min al-awā'il*) against which the argument is directed, is a good reflection of early developments. In seeking to provide the [correct] reason "why sight does not perceive its objects all at once," his explanation is offered in terms of the distance from the central area of the pineshape (*ṣanawbara*), more precisely its perpendicular (*'amūd*).[117] In al-Kindī's *De aspectibus*, where the discourse is now at the same time against the "checkered" version of Euclidean theory[118] *and* the main rival to any

---

the same time as the rest;" see Proposition 15 in *Tractatus [pseudo-] Euclidis De speculis*, Björnbo and Vogl, *Alkindi, Tideus und Pseudo-Euklid*, p. 106, English translation, Theisen, *The Mediaeval Tradition*, p. 276, where reference is made to Vogl's description of the text as a "compilation of selections from Hero's *Catoptrics*, Euclid's *De visu*, Theon's *Catoptrics*, and perhaps a truly Euclidean *Catoptrics*." The extant text of *Kitāb al-mir'āh li-Uqlīdis*, MS *Medicea Laurenziana* or. 152 does not contain such a passage.

116 The appearance of the same example of "letters in a book" in these early texts suggests the early availability of the preface, or its prototype, in some form.

117 The full statement is as follows: "A group among the ancients were wrong in saying that sight does not perceive its objects all at once but rather one after another, and their mistake in this matter is due to a reasoning (*'illa*), and that is, at times they would search to see a letter in a script, and the eye would not fall upon that letter at first, while the whole of the script is under the eye, and it only falls upon it a bit later; then they thought that the eye had struck other objects before falling upon that letter; also when a small object is being searched by the eye, either in the place where it is situated, or where it is descending from, such as the moon's crescent when it is first seen, or when a small gold or silver object is on the ground, the same thing happens. But the reasoning (*'illa*) is not what they thought, but rather the reasoning is that the desired object may be far from proximity to the central section of the pineshape, and so when the eye moves so that the desired object touches or nearly touches the pineshape's perpendicular, it perceives the opaque object at that place, because of the extreme manifestation of the light, and so that upon which more (*akthar*) rays fall, its vision is more clear (*abyan*) and accurate (*aṣdaq*) than what is farther from the perpendicular or whose distance from the pupil is larger, as we shall now show;" ḥ1 [see Ḥ in List II], fols. 63a–64a.

118 The argument against intervals with reference to the letters in a script appears in *De aspectibus*, sec. 11e, ed. Björnbo, pp. 14–15 .

version of it, namely intromission of forms,[119] the subjects of the first two propositions are further lined up with the third, once again in line with the *De speculis* passing under the name of Euclid in Arabic. He states not only that we comprehend objects only one by one [prop. 1] and that things falling directly under sight appear more clearly to it [prop. 2] but that as an object is moved farther away from the eye, to the extent that it is moved away from the central line of sight, it is concealed until it is no longer comprehended [prop. 3].[120]

The earlier approaches and their distinctly geometrical demonstrations[121] can be contrasted to the treatment of Ibn al-Haytham, who subjects the same two problems, united through the same example, to an experimental verification.[122] In the first three books of Ibn al-Haytham's *Optics*, where echoes of the earlier arguments are identifiable in various passages,[123] the problem of the manner in which sight perceives its objects "simultaneously" (*ma'an*) and most "clearly" (*abyan, aṣdaq*) is further extended to discussions of vision by contemplation.[124]

---

119 The reasoning is that if the reception of the objects of sense were simply the result of the rapid passage of the forms going to the eye, it would follow that, provided there were no obstacles for sight, the forms of objects at a closer distance would make their passage more clearly than those at very far distances, which is not the case. For al-Kindī's position, together with his related arguments, see Lindberg, *Theories of Vision*: "Al-kindi's Defense of the Extramission Theory," pp. 19–24.

120 The apprehension of the letters in a script "after some time" appears in sec. 9, ed. Björnbo, pp. 11–12, where al-Kindī further speaks of clarity of appearance in terms of what is on a straight line to the center of our sight, and of disappearance as a function of the farther one moves away from it. The privileged status of the central region of the visual field for clarity of vision is also discussed in sec. 11e, ed. Björnbo, pp. 14–15, still with reference to letters in a book.

121 Ibn 'Īsā's demonstration, or rather graphic representation, that vision is more accurate under the illuminated perpendicular, is presented in ḥ1, fols. 63b–64b. Al-Kindī's treatment is expressed in terms of the central point being seen more clearly than points in the periphery, because it is illuminated from three sides causing more light to fall upon it; see *De aspectibus*, sec. 14, ed. Björnbo, pp. 23–25.

122 "It is evident from this experiment (*i'tibār*) [with equal scripts on equal paper strips] that of the visible objects facing the eyes and binocularly perceived, the clearest is that which lies where the two axes meet, and the closer the object is from this point, the clearer (*abyan*) it will be"; see *Kitāb al-Manāẓir*, bk. III, 2, sec. 61, ed. Sabra, p. 366; tr. Sabra, v. 1, p. 245.

123 On the relationship of Ibn al-Haytham's *Optics* with al-Kindī's *De aspectibus* and Aḥmad ibn 'Īsā's *Optics*, see Sabra, *The Optics of Ibn al-Haytham*, v. 2, p. 26.

124 "If a survey is made of the conditions of visible objects, and if the manner in which sight simultaneously (*ma'an*) perceives its objects and that in which it perceives the parts of one object are discerned, these conditions will be found to agree, uniformly and with no variation or contradiction, with the state of affairs we have defined. For if the observer (*nāẓir*) keeps his eyes stationary while facing a multitude of objects (*mubṣarāt kathīra*) at the same time (*fī al-waqt al-wāḥid*), he will find those objects opposite to the middle of

A probable offshoot of the temporal elements introduced into the first Arabic proposition [as hinted by the particularly close language], Ibn al-Haytham's psychological treatment goes beyond mere observations that in vision there is time involved between intention and apprehension.[125] When he states that perception by contemplation (*idrāk bi al-ta'ammul*), recognition (*ma'rifa*), and discernment (*tamyīz*) takes place only "in time" (*fī zamān*), it is not just the "one by one" inspection of the letters of a word that strikes a familar note with discourses of the first Arabic proposition.[126] The correlation between contemplation and "the motion of the eye" (*ḥarakat al-baṣar*), and that between contemplation by prior knowledge and occupation of "a shorter time" (*zamān aqṣar*) now carry inescapable overtones of the few imported elements of the first Arabic proposition.[127] Ṭūsī's later and more limited treatment[128] is a good measure of how by his time the troubling problem of "checkered vision" had long disappeared in favor of the more lasting problem of "accurate vision."[129]

---

his eye clearer than those lying aside from that middle object, and those closer to the middle object will be clearer than those more remote from it. This state of affairs becomes even clearer if the line on which these objects lie is long, there being a fairly large interval between the extreme objects and the middle one;" see *Kitāb al-Manāẓir*, bk. II, 2, sec. 27, ed. Sabra, pp. 214–215; tr. Sabra, v. 1, p. 124.

125 Al-Kindī, *De aspectibus*, sec. 15, Björnbo pp. 25–27. This is also where the argument for the proportional nature of the time involved and the object's distance is developed.

126 "Perception by contemplation (*idrāk bi al-ta'ammul*) can take place only in time (*fī zamān*), for contemplation is achieved only by discernment and the motion of the eye (*ḥarakat al-baṣar*), and these can take place only in time . . . [also] perception by recognition (*ma'rifa*) and perception by discernment (*tamyīz*) can take place only in time; see *Kitāb al-manāẓir*, bk. II, 4, sec. 20, ed. Sabra, p. 328, tr. Sabra, v.1, p. 216.

127 Other passages include discussions of words recognized from total forms before one-by-one (*ḥarfan ḥarfan*) inspection of letters, and of species perceived in a shorter time than their individual parts; see *Kitāb al-manāẓir*, bk. II, 4, sec. 21–23, ed. Sabra, pp. 329–331, tr. Sabra, v. 1, p. 217–218.

128 The presence of some common elements here is rather curious. Ibn al-Haytham talks of facing a multitude of visible objects (*mubṣarāt kathīra*) at the same time (*fī al-waqt al-wāḥid*), Ṭūsī, of *mubṣarāt kathīra* not being seen at once with a single intention (fī *qaṣdin wāḥidin*)." The additional definition in T on clarity of vision contains the term "*yuḥaqqiq*," also used by Ibn al-Haytham: "In this way the observer (*nāẓir*) turns the axis of the cone to what it intends to see or wishes to verify (*yuḥaqqiq*)." Both treatments also base the formulation of the problem of visual accuracy on the visual cone's central ray, now raised to the status of an axis (*sahm*). In T, the visual cone's axis is noted as being the first ray falling on the object, and the word *intiqāl* (movement) is specifically used in discussing the speed of the eye's glance. Some manuscripts of T contain marginal comments on the enunciation.

129 The proposition as treated in the late Persian version *Dar 'Ilm-i manāẓir* (on the science of optics) [composed ca. 1250/1835; see D in List II, v. 1, p. lii] reflects other

Proposition 2: Relative Clarity of Appearance – Equal objects, Unequal Distances

Synposis: The enunciation is that the closer (*aqrab*) of two equal magnitudes is seen more accurately. The illustration is that of two equal magnitudes (AB, GD) [drawn parallel], GD closer to the eye (E). The demonstration (*burhān*) begins with the geometrical construction of the [cone of] ray[s] drawn as rays EG, EG, EB, EA. Then follows, more (*akthar*) of the ray falls upon GD than AB. The conclusion is that because that upon which more of the ray falls (*mā waqaʿa ʿalayhi al-shuʿāʿ akthara*) is seen more accurately, GD is seen more accurately than AB. The implicit assumptions are that (a) the closer of two magnitudes (GD) subtends a larger angle [external assumption]; (b) the larger of two visual angles contains a greater number of rays [external assumption]; (c) what is seen under larger/more angles is seen more clearly [internal assumption: def. 4-3].

Textual Correspondence:[130] In this proposition, where O and R are close, M is not identical to either. Among minor differences, M lacks the term "parallel" [in both O and R] in the illustration part of all copies; it also contains a different lettering for the figures [identical in O and R].[131] Much more significant is the case of causal permise [only in O], which is the only place where the Greek versions differ. For the reasoning behind the clear appearance of the closer magnitude, M is closer to O in containing such a premise, though not in its manner of application. The Arabic demonstration that visual clarity is dependent on the extent of visual rays falling on objects [as in both O and R] contains a causal premise that is now expressed not only as the middle term rather than the final statement [as in O, R lacks this], but also in a language that is no longer a restatement of the last Euclidean definition [as in O]. The relationship of the content of the Arabic premise and of the concluding statements to the Greek versions depends on how their respective parts in Arabic are read.

---

shifts. The enunciation states that "when a visual object (*mubṣar*) is seen by both eyes and a single intention (*bi qaṣd-i vāhid*) [the same expression in Ṭūsī's Proposition 1], it is seen as clear (*jalī*);" the treatment provides the reasoning (*ʿilla*) for simultaneous vision (*ru'yat-i maʿan*) in terms of the instantaneous (*dafʿatan*) union (*ittiḥād*) of the axes (*sahm*) of the two cones (*makhrūṭ-i shuʿāʿ*). The union of the two cones (*ittiḥād-i makhrūṭān*) is also the subject of discussion in an earlier anonymous Persian text [see R in List II, p. 1], which challenges the discourse (*kalām*) of the "most noble of recent scholars" (*afḍal-i muta'akhkharīn*) [i.e., ʿAlī Qūshjī" (d. 879/1474); see Q, List II, p. 1].

130 M2 compared to O2, R2, and T2: M2, v. 1, pp. 6–7; O2, ed. Heiberg, p. 4, tr. Ver Eecke, p. 2; R2, ed. Heiberg, p. 156, tr. Ver Eecke, p. 58; T2, edn. Hyderabad, p. 3.

131 In the figure of one variant (m1) the closer magnitude (GD) is drawn greater than the farther one (AB), and the intersection of the visual rays EA, EB with GD is marked with Z and H, respectively.

The causal permise is phrased as *mā waqaʿa ʿalayhi al-shuʿāʿ akthara* (that upon which more of the ray falls) is seen more accurately *(fa ruʾyatuhu aṣdaq)*," a curious construction that seems closely tied to the problematics of the Greek. The critical part of that premise for providing the cause of visual clarity, namely *mā waqaʿa ʿalayhi al-shuʿāʿ akthara* [slightly less problematic in the form *mā waqaʿa ʿalayhi al-shuʿāʿ al-akthar* (that upon which more rays fall)[132]] is verbalized such that the determining factor [the part *akthar*] in that phrase can refer to the greater *number* of visual rays falling on objects *or* the greater *size* of the angle enclosed by the visual rays. The first would be a construction corresponding to πλειόνων ὄψεων/*pleionôn opseôn* [in both O and R, see Variation Sample 2, p. xli, containing Ver Eecke's alternative translation of the same Greek expression as "plus de rayons visuels" and "des rayons plus nombreux"; *Euclide, l'Optique et la Catoptrique,* p. 2, pp. 57–58]. The second would correspond to the similarly flexible Greek construction πλειόνων γωνιῶν/*pleionôn goniôn* (greater/more angles) [of the causal premise in O, and def. 7 in both O and R; see Variation Sample 2, p. xli – translated alternatively as "des angles plus nombreux" (Ver Eecke) and "several angles" (Burton)].

The reading of the Arabic expression in the sense of a greater *number* of rays [see M2(9), v. 1, pp. 6–7] for *pleionôn opseôn* [in O2 and R2], rather than for *pleionôn goniôn* [in Odef. 7 and R def. 7, and O2], makes better sense, not only because the latter is already rendered by another [still problematic] expression *zāwiya kathīra* (more angles), but also in the light of the more frequent occurrence of the former construction in Greek.[133] The premise that something [line ΚΛ] seen "under more visual rays" (ὑπὸ πλειόνων ὄψεων/ *pleionôn opseôn*) than something else [line ΓΔ] is seen more clearly *is* part of the Greek concluding statements of the present proposition [in both O and R], even if it applied to specific cases, and in a place other than the causal premise.[134] It seems that the Arabic sets out to provide a more general statement of this as *its* causal premise, rather than a mere reference to the last Euclidean definition [as in O].[135] An almost verbatum formulation becomes an added definition in T.[136]

---

132 The adopted reading of the ending as an adverb (*akthara*), rather than the more familiar form *al-shuʿāʿ al-akthar,* is consistent with the way it appears in all manuscript copies [i.e., without the *al-* in front of *akthar*]. The same form is also used in T.

133 Ver Eecke, *Euclide, l'Optique et la Catoptrique* p. 57, n. 2, states that πλειόνων ὄψεων is a common expression in the version attributed to Theon, as opposed to πλειόνων γωνιῶν used exceptionally in its last Euclidean definition.

134 A more complete demonstration would involve the following stated assumptions: (1) the closer of the two magnitudes makes a larger angle at the eye; (2) a larger angle contains more rays; (3) for a closer object subtended by a greater angle, a greater number of rays fall on it; (4) objects seen under greater/more angles are seen more clearly [def. 4]; (5) objects seen by more rays are seen more clearly .

135 The statement of the Greek definition that "objects seen under greater/more angles (πλειόνων γωνιῶν) are seen more clearly" is, by itself, devoid of the reasoning that

Historical Context: The manner in which the Arabic version of the second proposition has come down to us reflects complex problems in transmission, many occasioned by the presence of linguistic and orthographic ambiguities in both Greek and Arabic. The cases of the Greek πλειόνων/*pleionôn* (greater/more) [employed here and in the last Euclidean definition], of the many forms of its even more problematic Arabic counterpart *akthar* (more in number/size, and clarity), and of the term *kathīr* [from the same root] and its orthographic counterpart *kabīr* (great), have all been noted here and elsewhere [see Summary and Commenray: Definitions 1 and 4, p. 8, p. 14, pp. 19–20]. In this proposition, in addition to the complications of the first part, *mā waqa'a 'alayhi al-shu'ā' akthara* (than upon which more of the ray falls), there is also the case of the part *ru'yatuhu aṣdaq* (its vision is more accurate).

The transition in the terminology of visual clarity from *ẓuhūr kathīr* (clear appearance) [corresponding to the ἀκριβέστερον φαίνεσθαι/*akribesteron phainesthai* of the last Euclidean definition],[137] to the alternative language of *ṣidq al-ru'ya* (accuracy of vision) [now used for the ἀκριβέστερον φαίνεσθαι/*akribesteron phainesthai* of Proposition 2], can hardly be explained in isolation from the many problems of transmission involved, including those of the terminological apparatus [for the relevant portion of the Greek and Arabic definition and proposition, see Variation Sample 2, pp. xl–xli]. The use and indeed prevalence of the adopted form [*aṣdaqu ru'yatan* (is seen more accurately)] over the abondened form [*ẓahara kathīran* (appears clear)] can be explained partly in terms of linguistic factors [the flexible nature of the root *K^{a}TH^{a}R^a*] and partly in terms of orthographic elements [the interchangeable transcription of *kathīr* (more)/*kabīr* (great) and *ru'ya* (vision)/*zāwiya* (angle) in

---

relates an angle's size with the number of rays in it, such a reasoning being present only in explicit statements such as the one in Ptolemy's *Optics* that "the greater of the angles of a visual pyramid contains a greater number of rays;" see bk. II, sec. 48, ed. Lejeune, *L'Optique de Claude Ptolémée*, p. 35.

136 "That upon which more of the ray falls (*al-shu'ā' akthara*) is seen more accurately (*aṣdaqu ru'yatan*) than that with less (*aqall*) [rays];" see Supplementary Passages, v. 1, p. 246. T's causal premise of the proposition has a slightly different formulation: "Since one angle is larger (*a'ẓam*) than the other, more (*akthar*) of the ray falls upon the line subtended by it than the other, so its vision is more accurate" (*ru'yatuhu aṣdaq*). The term "parallel," lacking in the enunciation of both M and T, is added in t2 and t7.

137 The Arabic definition reads: "What is seen (*ubṣira*) under greater/more angles (*zawāyā kathīra*) appears (*ẓahara*) more clearly (*kathīran*);" see Mdef. 4-3, Edition and Translation, v. 1, p. 3, and Summary and Commentary: Definition 4, pp 19–20.

Arabic manuscripts without proper or consistent dotting].[138] But the matter must be futher viewed in the context of the close association between the second and the fifth Euclidean proposition].[139] The concern of the two propositions with the apparent clarity and size of equal and unequally distant magnitudes had led to their close association in the Greek tradition. The association was to become further strengthened by their particular formulation in the Arabic tradition.[140]

The acquired centrality of the subject of visual clarity in the Arabic tradition may itself not be unrelated to the close affinity between the respective propositions on clarity of appearance [Proposition 2], discussed in a single Euclidean proposition, and size of appearance [Proposition 5], the subject of a large number of Euclidean propositions. A good measure of the centrality of the subject of visual clarity in the Arabic tradition is the abundance of not only the treatments of the subject in almost every text composed on matters related to vision from early on but also of the language of "*ṣidq al-ru'ya*" (accuracy/correctness of vision) together with a quantified treatment of it in terms of *number* of rays involved. For al-Kindī, whose related arguments appear in several places of his *De aspectibus*,[141] what determines the effectiveness of the central visual region in terms of clarity is that it is illuminated in *many* parts (*plures partes*) and from all sides.[142] Where the original Arabic is available for direct examination, the overlap of discourses about the optimum clarity of the central core of the visual region with the wording of the second Arabic

---

138 These terms are transcribed in Arabic as كبير/كثير and زاويه/رؤية, respectively. The latter is the case in at least one of the variants (m1), in which the expression *ru'yatuhu aṣdaq* appears as *zāwiyatuhu aṣdaq*; see M2(9) and the critical apparatus, v. 1, p. 7.

139 In proving that the closer of two equal and parallel objects is seen more accurately, the second Arabic proposition shows that clearer appearance is a logical consequence of the fact that the larger angle of the one closer to the eye contains more rays than the other. In the fifth proposition, however, it is the greater [and not clearer] appearance of a closer object which is to be proved by a similar consideration of the visual angle.

140 Ver Eecke has this footnote to the fifth proposition: "Une note de scoliaste dans un des manuscrits fait remarquer la correspondance de cette proposition avec la proposition II;" see *Euclide, l'Optique et la Catoptrique*, p. 4, n. 2. See also the long passage in Ptolemy's *Optica*, where he complains about those who relate the appearance of an object's magnitude to the number of rays striking it; bk. II, sec. 48, ed. Lejeune, pp. 35–36. In the case of the Arabic tradition, the marginal note on Proposition 2 in one variant (m1), fol. 59b, sums up the problem: next to "that upon which more (*akthar*) of the ray falls, its vision is clearer (*aṣdaq*), so the vision of GD [the closer magnitude] is clearer than AB [the further one]," is written: "The magnitude (*miqdār*) of GD is seen as greater (*akthar*) [i.e., in size, rather than in clarity] than AB, by [amount] GZ and HD."

141 Sec. 22, ed. Björnbo pp. 37–39; sec. 12, ed. Björnbo, pp. 17–19. It is maintained that the clarity of a close object is due to the strength of its lumination, which is optimized with proximity to the visual axis.

142 Sec. 14, ed. Björnbo, p. 24.

proposition is more immediately detectable. Ibn ʿĪsā demonstrates geometrically that the priviledged position of the visual axis in terms of "more accurate vision" (*aṣdaq ruʾyatan*) is due to the involvement of "more of the ray" (*mā waqaʿa min al-shuʿāʿ akthara*).[143] And Ṭūsī continues to speaks in a comparable tone when he adds yet another definition that "whatever has the visual cone's axis (*sahm al-makhrūṭ al-shuʿāʿī*) on it is seen more accurately (*aṣdaq ruʾyatan*) than its vicinity (*ḥawl*), as the ray falling on it is more (*al-shuʿāʿ al-wāqiʿ ʿalayhā akthara*)."[144]

Proposition 3: Limits of Visibility – Same Object, Different Distances

Synopsis: The enunciation is that every visible object has a certain limit of distance [from the observation point] beyond which it is no longer seen. The illustration is that of a straight line (AB) at a certain distance from the eye (G). The demonstration (*burhān*) begins with the geometrical construction of ray[s] (GA, GB) extending to the visible object (AB), itself displaced to ZH, where DE is drawn upon it [equal to AB]. Then follows, AB is seen because the ray[s] [GA, GB in m1; GAZ, GBH in m2–m5] fall on it, but DE is not, because no ray falls on it [def. 3]. The implicit assumption is that DE and AB are the same. The conclusion is that [because DE and AB are the same, it is as if] once AB reaches a certain limit (DE), it is no longer seen.

Textual Correspondence:[145] This is another proposition with a notable distinction from the Greek [close in O and R]. Besides minor differences – two letters, D, E marking the displaced object [the Greek has one (K)] and ZH being its [imaginary] limit of extension [the Greek has none] – there is a major difference in the demonstration. As in the first proposition, the Arabic bypasses the notion of "intervals between visual rays" (ἐν τῷ μεταξὺ διαστήματι τῶν ὄψεων/ . . . *diastêmati ton opseôn*),[146] while appearance [and disappearance] of objects is still explained in terms of the falling of the rays on objects [def. 3].[147] The original Euclidean proposition sets out to prove that an object placed within

143 The relevant passage in Ibn ʿĪsā's Ḥ [see List II] appears in fols. 8b–9a of ḥ1. See Sabra's related discussions in *The Optics of Ibn al-Haytham*, v. 2, pp. 25–26. Another passage in Ḥ contains the interesting combination that "whatever seen by a large angle is seen as larger (*aʿẓam*) and its vision is more accurate (*aṣdaq ruʾyatan*);" see ḥ1, fol. 79b.

144 " . . . and in a more condensed form (*ashaddu tarākuman*), so what is closer is more accurate than what is farther." In *Shuʿāʿ* [see SH in List II, v. 1, p. xlvii], there is an "overlap (*tadākhul*)" of cones of ray for a more "intense (*shadīd*) appearance (*ẓuhūr*)."

145 M3 compared to O3, R3, and T3: M3, v. 1, pp. 8–9; O3, ed. Heiberg, pp. 4–6, tr. Ver Eecke, p. 3; R3, ed. Heiberg, pp. 156–158, tr. Ver Eecke, p. 58; T3, edn. Hyderabad, p. 3.

146 See ed. Heiberg, p. 6. Burton, "The Optics of Euclid," p. 357, translates this as "divergence of visual rays;" see also Summary and Commentary: Proposition 1, p. 25.

147 A link between this proposition and Proposition 1 and Definition 3 [see Brownson, "Euclid's Optics and Its Compatibility with Linear Perspective"] is not made in Arabic.

the "interval" between visual rays ceases to be visible at some distance because as it is moved further away, its enclosing visual rays lose contact with its extremities. The Arabic proof does away with the presupposition of the discontinuous distribution of visual rays within the visual field, explaining visibility and invisibility simply in terms of what of the ray "falls" (*yaqa'u*) and "does not fall" (*lā yaqa'u*) on the object [def. 3] at certain distances.[148] And in contrast to the first definition and proposition where there is similarly no correspondence to the *diastêma* of the Greek [see Summary and Commentary: Definition 1 and Proposition 1, pp. 12–13 and p. 25, respectively], and indeed to all other parts of the text, for the third Euclidean proposition there is yet another formulation in Arabic with a striking and revealing set of expressions. Preserved as an alternative proof in the margins of a manuscript of the early version (m1) and integrated into the proof of the proposition's later recension (T), the second Arabic proof is further removed from that of the Greek and Arabic propositions in having been formulated in terms of the effect of the increasing distance on the decreasing size of the visual angle, successively expressed as becoming small (*taṣghuru*), narrow (*taḍīqu*), and vanished (*ghāba*).[149]

Historical Context: At least from the mid-3rd/9th century onwards, discussions devoted to the subject are placed in the framework of the diminishing angle. In the optical section of Qusṭā ibn Lūqā's treatise, the questions of the appearance and disappearance of objects and of one and the same object having different sizes as it approaches or moves away from the eye are given in terms of the changing size of the object's visual angle: "What is the reasoning" (*'illa*), it is asked, that "when [something] is continually distanced (*bu'd dā'im*) from the eye it becomes concealed (*khafiya*) and so not seen;" the reasoning is offered in the familiar language of the narrowing (*ḍīq*) of the visual angle until it reaches a point where it becomes a line with no width, while the object's smallness (*ṣighar*) reaches a magnitude too small to be perceived by visual sensation.[150]

---

148 Since the proposition involves the same object in two different locations, a discussion in terms of the closer object blocking the farther one would not be relevant, the idea being that the object at this limit of distance falls within the so-called "interval" between visual rays and becomes no longer reachable by them.

149 The reasoning offered in m1 [the variant with the most problematic figure] is as follows: "Whatever becomes distant from the eye gets smaller, because the angle by which it is seen becomes small (*taṣghuru*) [becomes narrow (*ḍīq*) in the other two marginal notes], until it moves away to a limit where it [the object] is not seen from there, and as long as the angle is not present (*ghā'ib*), it [the object] is not seen." The reasoning, together with the language of angular narrowing (*ḍīq*), reappears as the alternative reasoning (*'illa*) of Ṭūsī's *Taḥrīr* of this proposition.

150 "Whatever becomes distant (*ba'uda*) from the eye, the angle by which it is seen is continually narrowing, and when the narrowness (*ḍīq*) reaches the form of a line with no width to it, the object at the [angle's] base is not seen due to its narrowness," *Fī 'Ilal*, L [see List II, v. 1, p. xlvi], p. 6. The question appears as the fifth chapter heading on p. 4.

The expressions in Ṭūsī's *Taḥrīr* are comparable: "The reasoning" (*'illa*) is not what he [Euclid] mentioned, it is rather the case that [visual] angle AGB narrows (*taḍīqu*) until its two sides are as one (*ḍil'āhu ka-l-muttaḥidayn*), so the object becomes extremely small (*ghāyat al-ṣighar*), nonexistent (*mun'adim*) to the eye.

The notions of the smallness of an object's visual angle at far distances, its imperceptible size as the two rays join, and its invisibility when there is no length to subtend any angle also stand out in the second book of Ptolemy's *Optica*.[151] Such expressions as the "narrowing" (*ḍīq*) of the visual cone's angle and the "smallness" (*ṣighar*) of the enclosed portion of the eye in Ibn al-Haytham's related treatments point to the Arabic text of the relevant Ptolemaic passages as a possible source for the proposition's alternative formulation.[152] Enough of this same terminology is used in the margins of the Euclidean text, and in the "reasoning" (*'illa*) of Ibn Lūqā and Ṭūsī to suggest such a common source, even in the absence of any evidence for the full availibility of Ptolemaic *Optics* to Arabic authors other than Ibn al-Haytham.

Ibn al-Haytham's own lasting contributions to the subject of limits of visibility included the distinction between distance [from the eye] (*bu'd/remotio*) and interval (*masāfa/spatium*) [as a segment of it] and the establishment of a moderate range (*'arḍ i'tidāl*) for correct (*ṣaḥīḥ*) vision.[153] According to him, "sight cannot truly (*ṣaḥīḥ*) conceive an extremely small (*fī ghāyat al-ṣighar*) object in which minute features and distinct parts exist;" there is only a "range," which he terms the "moderate range" (*'arḍ i'tidāl*), within which an object can be perceived as it is.[154] Later treatments of the subject of limits of visibility in terms of least sensible magnitudes or a quantified relation between an object's size and its distance from the eye seem similarly independent of the Arabic Euclidean tradition.[155]

---

151 See Ptolemy, *Optica*, bk. II, sec. 47, 50, ed. Lejeune, p. 35 and p. 37, respectively. Elsewhere, Ptolemy writes of a given size being imperceptible at a considerable distance with reference to the smallness of the visual angle; see sec. 26, ed. Lejeune, pp. 26–27.

152 For the overlaps with the margins of m1, see critical apparatus in v. 1, p. 9. The terminology in *Kitāb al-Manāẓir*, bk. II, sec. 142 is "as the object recedes from the eye, the cone surrounding it will become smaller (*ḍāq*), the angle of the cone narrower (*ḍāq*), and the portion of the eye containing the form, smaller (*ṣaghura*);" ed. Sabra, p. 276, tr. Sabra, v.1, p. 175; see also Summary and Commentary: Proposition 10, p. 48.

153 *Kitāb al-Manāẓir*, bk. II, 3, sec. 157, ed. Sabra, pp. 285–286; tr. Sabra, v. 1, p. 183. On *masāfa* and *bu'd*, see Sabra, *The Optics of Ibn al-Haytham*, v. 2, p. 96.

154 Ibid., bk. III, sec. 9 and 15, ed. Sabra, p. 376 and 379; tr. Sabra, v. 1, pp. 253–255.

155 Theisen, *The Mediaeval Tradition*, pp. 281–282, includes a discussion of the related treatments of Pecham and Witelo and of the minimum angle of resolution of modern optical instruments. The exact ratio [1 to 530] for an object's diameter to its distance is specified in the late Persian version of the Euclidean proposition [see D in List II, p. lii], still in terms of the narrowing (*taṣāghur*) and nonexistence (*in'idām*) of the visual angle .

Propositions 4–10: Appearances Involving Sizes and Distances

Propositions 4–8: Relative Appearance of Equal Magnitudes

Proposition 4: The Case of Equal Magnitudes Aligned Along the Same Line

Synopsis: The enunciation is that when equal magnitudes are aligned [successively] on the same line, the magnitude with the longest radial path toward it is seen as the smallest. The illustration is that of three equal lines (AB, BG, GD) on a horizontal extension (AD) and a perpendicular line (EA) from the eye (E) to that extension. The demonstration (*burhān*) begins with the geometrical construction of the [visual] ray drawn as rays EA, EB, EG, ED and a line (BW) extending from one of the points (B) parallel to line GE to reach EA. Then [through the similarity of the two resulting triangles (AGE, ABW), inequality of two angles (WEB, WBE) and equality of two others (WBE, BEG)] follows that the opening of the ray (*makhraj al-shu'ā'*) toward (*ilā*) AB [i.e., angle WEB] is greater than that of BG [angle BEG]. The conclusion is that [because the opening for AB is greater than for BG], it is seen as greater [def. 4], and similarly, BG is seen as greater than GD.

Textual Correspondence:[156] In this proposition, where O and R differ noticeably in the demonstration, figure, and lettering, M is closer to O: in the figure, where line BW is drawn [O has an additional line not in R, and R has letters within angles not in O]; in the lettering [point Z in O appears as W, which has a close trancription in Arabic]; and in the notably different demonstration [O has a longer one to which M is close except for the absence of the ending]. The Arabic differs from the Greek in containing, in all variants of M, the uncommon expression "the opening of the ray" (*makhraj al-shu'ā'*) [supposed to determine the size of an object's appearance]. This expression, which is used in a premise [M4(16)] with no exact correspondence in O and R, is not to be found elsewhere in M or in the corresponding demonstration in T, which otherwise follows M closely. T differs slightly from M in the figure, specifically in having the letter Z instead of W at the midpoint of EA [as in O].

Proposition 5: The Case of Equal Magnitudes at Unequal Distances

Synopsis: The enunciation is that when equal magnitudes are unequally distant from the eye, they are seen as unequal, and the one closer to the eye is seen as greater. The illustration is that of two equal lines (AB, GD), where AB is closer to the point representing the eye (E). The demonstration (*burhān*) begins with the geometrical construction of the [visual] ray drawn as rays EA, EB, EG, ED.

---

[156] M4 compared to O4, R4, and T4: M4: v. 1, pp.10–13; O4: ed. Heiberg, pp. 6–8, tr. Ver Eecke, pp. 3–4; R4: ed. Heiberg, p. 158, tr. Ver Eecke, p. 59; T4: edn. Hyderabad, p. 4.

Then follows the statement that angle AEB is greater than angle GED. The conclusion is that because whatever is seen (*ubṣira*) under a large angle is seen as larger (*yurā a'ẓam*) and whatever is seen (*ubṣira*) under a small angle is seen as smaller (*yurā aṣghar*) [def. 4 has "appears as great (*ẓahara 'aẓīm*) and small (*ẓahara ṣaghīr*)"], then AB is seen as greater than GD.

Textual Correspondence:[157] In this proposition, where O and R differ slightly in the illustration and the lettering of the figure, M is closer to O: in the illustration, where it is stated let there be two equal magnitudes [as in O; R has it as one magnitude being equal to another]; and in the lettering, where AB is closer to the eye than GD [as in O; in R, it is not specified which is the closer one]. T's abridged version follows M closely, except for a slightly different formulation of the enunciation as "the closer of unequally distant magnitudes is seen as the greatest of them" (*yurā a'ẓamuhā*) ["seen as the greatest" (*yurā a'ẓama*) in M].

Propositions 6–7: The Case of Equal Magnitudes Separating Parallel Lines

Proposition 6: When Viewed from a Distance on the Same Plane

Synopsis: The enunciation is that parallel lines are seen from a distance as having unequal [separating] widths. The illustration is that of two parallel lines (AB and GD), and a point (E) [representing the eye]. The demonstration (*burhān*) begins with the geometrical construction of the [visual] ray drawn as rays EB, EW, EZ, ED, EH, ETand the lines of width (*khuṭūṭ al-'arḍ*) (BD, WH, ZT) [i.e., the distance between the two parallel lines at different points]. Then follows that these separating lines are seen as unequal [greatest, next greatest, and least in magnitude, according to the size of the angle they subtend] [def. 4]. The conclusion is that because the separating lines are seen as unequal, the two parallel lines (AB, GD) are seen as unequal in width [i.e., not parallel] [def. 4] .

Textual Correspondence:[158] This proposition represents major structural differences with the Greek. The sixth Euclidean proposition is presented as two independent propositions, where M6 corresponds to the first half of its Greek counterpart and M7 to the second half. In M6, where O and R differ slightly in figure and lettering, M is closer to O: in the illustration, where the subject is two parallel magnitudes [as in O, R has one magnitude parallel to the other]; and in the lettering, where AB and GD represent the two magnitudes and E the eye [other letters differ]. T follows M closely. The point for the qualification "seen from a distance" in the enunciation of both M and T is not made clear.

---

157 M5 compared to O5, R5, and T5: M5: v. 1 pp. 14–15; O5: ed. Heiberg, p. 8, tr. Ver Eecke, p. 4; R5: ed. Heiberg, pp. 158–160, tr. Ver Eecke, p. 59; T5: edn. Hyderabad, p. 4.

158 M6 compared to O6-1, R6-1 and T6: M6: v. 1, pp. 16–17; O6: ed. Heiberg, pp. 8–10, tr. Ver Eecke, pp. 4–5; R6: ed. Heiberg, p. 160, tr. Ver Eecke, pp. 59–60; T6: edn. Hyderabad, p. 4–5.

Proposition 7: When Viewed from a Distance on a Different Plane

Synopsis: The enunciation is that parallel lines on a plane below the eye are seen from a vertical distance (*fī al-sumk*) as having unequal [separating] widths. The illustration is that of two parallel lines (BG, DE), their separating lines (WZ, HT, IK) [also parallel], and a point (A) representing the eye at a vertical distance. The demonstration (*burhān*) begins with the geometrical construction of the [visual] ray drawn as rays AW, AZ, AH, AT, AK, AI, a perpendicular line (AL) to plane BGDE, another line (LMNS) extending from its foot (L) to the separating lines (WZ, HT, IK), and from their meeting points (M, N, S) lines to point A (AM, AN, AS). Then [through a few geometrical identities] follows that the widths (WZ, HT, IK) are seen as unequal [greatest, next greatest, and least in magnitude, according to the size of the angle they subtend] [def. 4]. The [implicit] conclusion is that [because the widths are seen as unequal], the two parallel lines (AB, GD) are seen as unequal in width [def. 4] [i.e., not parallel].

Textual Correspondence:[159] This proposition, which presents the second case of the sixth Euclidean proposition separately, differs from the corresponding Greek part as follows: it contains a stated enunciation [reconstructable from other parts of O and R] in which it specifies the two lines as being "parallel" [this is implicit in O and R], "below the eye" (*munkhafiḍa*) [the eye is elevated in O and R], and "seen from a vertical distance" (*fī al-sumk min al-bu'd*) [T has "a far vertical distance" (*fī al-sumk min ba'īd*)]. The Arabic proposition lacks the concluding statement repeating the enunciation [close in O and R]. The wording and lettering of the Arabic proposition do not exactly correspond to either of the two Greek versions, which differ slightly in the figure, lettering, length, and phraseology. T follows M closely, including the lack of concluding statements, but its lettering is slightly different from it [still not the same as either O or R].

Proposition 8: The Case of Equal Magnitudes Aligned at Different Ends of the Same Line

Synopsis: The enunciation is that when equal magnitudes are positioned at different places of the same line, they are seen as unequal in size. The illustration is that of two straight lines (AB, GD) [aligned horizontally] on the same line (AD) at unequal distances from the eye (E) and ray[s] EA, ED [EA longer than ED]. The demonstration [without the corresponding term *burhān*] begins with the geometrical construction of a circle, AED, around a triangle with the same letters, and from the meeting points of rays EB, EG with AD (B, G), extension lines BZ and GH and perpendicular lines (BT, GI). Then [through few

---

159 M7 compared to O6-2, R6-2, and T7: M7, v. 1, pp. 18–21; O6-2, ed. Heiberg, p. 10, tr. Ver Eecke, p. 5; R6-2, ed. Heiberg, pp. 160–162, tr. Ver Eecke, p. 60; T7, edn. Hyderabad, p. 5.

identities and inequalities regarding arcs and angles] follows the statement that magnitudes AB and AG are seen from angles AEB and GED. The conclusion is that [because what is seen from a greater angle appears as greater (def. 4)], AB is seen as greater than GD [with no mention of BG].

Textual Correspondence:[160] In this proposition, where O and R differ considerably in enunciation, illustration, figure, lettering, and demonstration, M corresponds to O closely. The Arabic proposition marks the first appearance of the expressions *'alayhā* or *'alayhumā*, as in "we draw a circle marked *upon it*" (*'alayhā*), AED: the English translation given here is "we draw a circle, AED". The proposition also marks the first absence of expressions used throughout the preceding propositions: besides "*burhānuhu*" (its demonstration), the curious construction "*nukhriju al-shu'ā' shu'ā'āt* . . . " ("we draw the ray, as rays . . . ") at the beginning of all previous geometrical constructions. In the illustration of M, the ray[s] are represented as EA, ED [as in O], and later as EB, EG [in O the second set is not specified as rays]. In T, which follows M closely, the latter are presented with three letter (EBZ, EGH).

Propositions 9–10: Appearances vs. Distances

Proposition 9: Ratio of Sizes vs. Distances – Equal and Parallel Magnitudes

Synopsis: The enunciation is that when equal and parallel magnitudes are unequally distant from the eye, the difference of their aspects (*ikhtilāfu manāẓirihā*) is not seen according to the magnitudes of their distances [i.e., the ratio of their sizes is not seen to be the same as the ratio of their distances]. The illustration is that of two straight lines (AB, GD) [with unequal distances from the eye (E)]. The demonstration begins with the geometrical construction of ray[s] AE, EG [to the uppermost points of the two magnitudes] and the arc of a circle (HZT) with E as center and EZ as radius [Z is where line GD meets ray EA]. Then [through a set of geometrical identities and inequalities] follows the statement that [the ratio of] BE to ED [or rather EB to ED as the ratio of the distances from the eye] is greater than [the ratio of] angle HET to ZET and hence greater than [the ratio of] lines GD to AB [subtending those angles]. The implicit assumption is that objects viewed under greater angles are seen as greater [def. 4-1]. The conclusion is that for the two magnitudes, the difference of aspects [i.e., the ratio of the angles or magnitudes subtending them] is not seen according to [i.e., the same as] the ratio of their distances.

---

160 M8 compared to O7, R7, and T8: M8, v. 1, pp. 22–25; O7, ed. Heiberg, p. 12, tr. Ver Eecke, pp. 5–6; R7, ed. Heiberg, p. 162, tr. Ver Eecke, p. 61; T8, edn. Hyderabad, pp. 5–6.

Textual Correspondence:[161] In this proposition, slightly different in O and R, the Arabic, which differs from the Greek in the critical parts, represents elements from both O and R: in containing the term "parallel" [not in R] for the two magnitudes, it is closer to O [m4 has *mutawāziya* written over *mutasāwiya*, as if introduced later]; in lacking the part corresponding to the Greek τοῦ ὄμματος/*tou ommatos* [only in O], it is closer to R. And there are parts in the long demonstration where the Arabic includes features with no exact correspondence in the Greek. One is the single appearance of the expression *ikhtilāf al-manāẓir* [in the title of m1, m3, m5] within the Arabic text [M9(3, 29); T has *ikhtilāfuhā fī al-ru'ya*]; another is two rather unique constructions "*fī al-manẓar*" and "*fī al-zāwiya*" [M9(7) and M9(27); T has *fī al-ru'ya* for both], which judging from the context may not be entirely independent.[162] The case of the corresponding Arabic expression for the Greek ἀναλόγως/*analogôs* (in proportion) [in both O and R] is most problematic.[163] In both enunciation and conclusion, the ratio of the apparent size of two equal and parallel magnitudes (*ikhtilāf al-manāẓir*) is stated and proved to be seen *not* according to (*'alā*) [short for *'alā nisba* (in proportion), as in T] the magnitude/magnitudes (*qadr*/*aqdār*) of their distances, a sense in clear contrast to the Greek[164] [see Variation Sample 3, p. xlii].

---

161 M9 compared to O8, R8, and T9: M9, v. 1, pp, 26–29; O8, ed. Heiberg, pp.14–16, tr. Ver Eecke, pp. 6—7; R8, ed. Heiberg, pp. 164–166, tr. Ver Eecke, pp. 61–62; T9, edn. Hyderabad, p. 6.

162 These two constructions [only in one variant (m1) in contrast to the occurrence of *ikhtilāf al-manāẓir* in all copies] are used in the premises preceding the geometrical construction and final conclusion, where the ratio of distances are stated to be different from the ratio of the magnitudes "in aspect" (*fī al-manẓar*) and "by angle" (*fī al-zāwiya*), respectively. The inclusion of both in the later recension suggests that they were integrated as part of the text; the appearance in T of the same expression (*fī al-ru'ya*) points to a more standard form of application later.

163 On the term *analogôs*, see Heath, *The Thirteen Books of Euclid's Elements*, v. 2, bk. V, def. 3 and 6, p. 116 and p. 129. The uncommon expression "*fī al-manẓar*" following the first appearance of the concept of ratio [see M9(7) and crticial apparatus, v. 1, p. 27] seems to stress the *appearance* of the relationship in question: "The amount GD to AB [ratio of the size of the two magnitudes] in appearance (*fī al-manẓar*) [i.e., appear to be] not the same as the amount (*ka qadr*) of BE over DE [ratio of their distances]." So does its equivalent in Ṭūsī's *Taḥrīr*, where *fī al-ru'ya* is used in place of *fī al-zāwiya* (in angle) [M9(27)]; the use of *ikhtilāf fī al-ru'ya* (difference in vision) in T for *ikhtilāf al-manāẓir*, and *fī al-ru'ya* (in vision) for both *fī al-manẓar* and *fī al-zāwiya* in this same proposition could be, at least in the latter case, related to the close transcription of *ru'ya* and *zāwiya* in Arabic; see Summary and Commentary: Proposition 2, p. 32, n. 138.

164 For various accounts of the Greek proof using modern notation, see Ver Eecke, *Euclide, l'Optique et la Catoptrique*, pp. 6–7, n. 1; see also Brownson, "Euclid's *Optics* and Its Compatibility with Linear Perspective," p. 182, and Theisen, *The Mediaeval Tradition*, p. 286, where the proof of this proposition is tied to two others from Euclid's *Elements*, VI. 1 and 33; see Heath, *The Thirteenth Books of Euclid's Elements*, v. 2, p. 191 and p. 273, respectively].

Historical Context: The most critical aspect of the transmission of this proposition concerns the terminology of the concept of proportion, where an abbreviated, nontechnical Arabic term passing for Greek *analogôs* changes – in fact reverses – the original sense of the proposition altogether. The Greek proposition, itself much discussed with reference to its possible compatibility with linear perspective,[165] teaches that the appearance of one magnitude to another equal and parallel to it is not in the same proportion [i. e. its ratio is not the same] as their respective distances from the eye.[166] But in both the Arabic enunciation and conclusion, the term corresponding to the Greek *analogôs* is not the standard expression *'alā qadr nisba*, widely used in early mathematical texts,[167] but just *'alā qadr/aqdār* (*according to* the magnitude/magnitudes] in all copies. This shorter, nontechnical, form carries with it a different sense from the original which, once applied in context, represents one ratio [of appearances] as not having a dependent *relation* on another [of distances].[168]

---

165 For a thorough treatment of the proposition in that context, see Brownson, "Euclid's *Optics* and Its Compatibility with Linear Perspective;" for more recent treatements, see Tobin, "Ancient Perspective and Euclid's *Optics*," with a discussion of differences in the Greek versions, and Knorr, "On the Principle of Linear Perspective in Euclid's *Optics*."

166 The discussion of the intended "mathematical meaning" of proportion in this proposition, as opposed to the modern construction "one quantity having a dependent relation to another" in the well-known formula by Jean Perelin (Viator), is in Brownson, "Euclid's *Optics* and Its Compatibility with Linear Perspective," pp. 183–184. The author, who argues in favor of the compatibility of "natural perspective" [i.e., ancient and medieval theories on the perception of the visible world] and "artificial perspective" [the Renaissance representation of it] against Panofsky and his followers, relates the problems to a "misunderstanding of the mathematical meaning of proportion" in this proposition.

167 The longer and more standard form is part of the vocabulary of the Arabic tradition of the fifth book of Euclid's *Elements*, where the definitions of ratio and proportion are given; see Plooij, *Euclid's Concept of Ratio and His Definition of Proportional Magnitudes as Criticized by Arabian Commentators*, p. 19. The extant Arabic text of Ptolemy's *Planetary Hypotheses* includes the close variations, ratios (*'alā nisab*) and proportion (*'alā tanāsub*); see Goldstein "The Arabic Version of Ptolemy's *Planetary Hypotheses*," p. 34, and Sabra, "Ptolemy vs. Alhazen," appendix, p. 244 (transcription), and pp. 221–223 (translation and commentary). The disregard of the standard form in the the *Optics* suggests the inaccessibility of the these particular works, which may well be due to the very early date of its translation; see The Arabic Versions, v. 1, p. xix.

168 The longer expression *'alā nisba* (in proportion) appears in *Taḥrīr al-Manāẓir* and *Tajrīd al-Manāẓir* [which relates the proposition to Theodosius' *Sphaerica*, III, 11; see T and J, respectively, in List I, v. 1, p. xxxvi and p. xliv] with the sense of a fixed ratio.

The employment of the short form *'alā qadr* (according to), rather than *'alā qadr nisba* (in proportion), makes the statement of the Arabic enunciation and conclusion one according to which "the ratio of the appearance of the size of two magnitudes [to which the expression *ikhtilāf al-manāẓir* corresponds] is not seen *according to* the amount (*qadr*) of their distances. This is a conception that is, in effect, the reverse of the modern one, according to which one ratio is a dependent function of the other. The Arabic wording affects not just the meaning of *ikhtilāf al-manāẓir,* the expression for the optical principle in its single appearance in an Arabic text with that same title; it affects the message of the proposition, and the terms of the critical relationship between appearance and distance, which were to become so central to both optics and perspective.[169]

The absence of an equal numerical relation between the ratio of the appearance of two equal magnitudes and their respective distances from the same observation point in the original Greek proposition does not imply that distance *cannot be* a factor for determining appearances. The passage in Ptolemy's *Optica*, where one and the same object is considered at different locations [as the mathematical equivalent to the two equal but unequally distant objects of the present proposition], and where distance, as well as angle and position, is a determining factor for the appearance of an object's size,[170] is not in conflict with the formulation of the Euclidean proposition. The same may not be said of the Arabic Euclidean proposition, where a literal reading of it accords better with a nondependent – not merely nonequal – relation between the ratio of appearances and distances. A notable feature of the Arabic tradition is the absence of any developments in the line of Ptolemaic treatments[171] as extended by Ibn al-Haytham[172] or the medieval and Renaissance perspectivists. The

---

169 This topic is the subject of Kim Veltman's unpublished dissertation, *Optics and Perspective: A Study in the Problems of Size and Distance*.

170 See Ptolemy, *Optica*, bk. II, sec. 53–54, ed. Lejeune, pp. 38–39; see also Sabra, *The Optics of Ibn al-Haytham*, v. 2, pp. 93–94. The astronomical passage in the *Planetary Hypotheses*, where Ptolemy applies the optical principle under discussion to the case of celestial bodies at great distances, is a more direct indication of his position insofar as he explicitly states that the estimated magnitude of these bodies varies according to the distance, though at a smaller ratio; see Goldstein, "The Arabic Version of Ptolemy's *Planetary Hypotheses*," p. 7.

171 That a direct relationship between appearance and distance in the Ptolemaic treatment did not reach or inspire Arabic authors predating Ibn al-Haytham is suggested by the absence of any evidence to that effect, either in the textual tradition or in the marginal commentaries of the manuscript corpus.

172 Ibn al-Haytham discusses the problem in at least two works. A passage in his *Solution of Difficulties in the Almagest* refers explicitly to the related passage of Ptolemy's *Optica* in this connection: "Ptolemy has shown in the *Optics*, in the course of his discussion of size, that size is not perceived according to the angle alone, but according to the magnitude of the angle and of the [object's] distance and according to whether the object is inclined or frontally situated," *Ḥall shukūk fī al-Majisṭī yashakkiku*

peripheral nature of any developments in the line of linear perspective in particular must be further viewed in the context of other related problems in transmission, specifically that two Euclidean definitions most relevant to it (fifth and sixth in Heiberg's edition) were not directly transmitted as part of the Arabic definitions [see Structural Correspondence, pp. xxix–xxx, and n. 36].

Proposition 10: Right-angled Figures Viewed from a Far Distance

Synopsis: The enunciation is that right-angled figures when seen (*ubṣirat*) from a distance are seen as (*turā*) circular. The illustration is that of a right-angled figure (ABGD) [with no point representing the eye]. The demonstration begins [with no geometrical constructions] with the assertion that the eye, when at a distance, does not pause at a single point but shifts, so that what is seen is figure EZHTKLMN but not what is in between [i.e., corners]. The assumptions are that any object has a certain limit of distance beyond which it is no longer seen [prop. 3], and that the smallest parts of an object are the first to disappear from a distance [external assumption: margin of m1]. The concluding statement is that because what is hidden from sight [i.e., angles] is the part of the figure with least magnitude and, any figure with no angles is circular, then all figures are seen from a far distance as circular and so is the right-angled figure, ABCD.

Textual Correspondence:[173] This is another case where the Arabic proposition is considerably different from the Greek. And where O and R differ [in the illustration and the figure], M has elements from both. None of the variants of M have the "right-angled figure" (*al-shakl al-qā'im al-zawāyā*) of the illustration [corresponding more closely to the ὀρθογώνια μεγέθη/*orthogônia megethê* (orthogonal magnitudes) of the enunciation][174] on an elevation [in O and at least

---

fihā ba'ḍu ahl al-'ilm; see Sabra and Shehaby (eds.), *Alhazen's al-Shukūk 'alā Baṭlamyūs;* for the translation and discussion of this passage, see Sabra, "*Psychology vs. Mathematics*," p. 232. See also the second book of Ibn al-Haytham's own *Optics,* bk. II, sec. 135–143, ed. Sabra, p. 273–278, tr. v. 1, pp. 173–177, esp. sec. 140–143, pp. 174–176: "If . . . sight perceived the sizes of visible objects only through their estimation by means of the angles produced by the objects at the center of the eye it would not perceive . . . the same object from unequal distances to be of the same magnitude (provided those unequal distances are moderate), nor ever perceive equal objects from unequal but moderate distances to be equal . . . . The faculty of judgment, when discerning the object's size, will not take into account the angle alone but will consider both angle and the distance, because it has been established in it that the angle varies with the distance."

173 M10 compared to O9, R9, and T10: M10, v. 1, pp. 30–33; O9, ed. Heiberg, p. 16, tr. Ver Eecke, p. 8; R9, ed. Heiberg, p. 166, tr. Ver Eecke, p. 62; T10; edn. Hyderabad, pp. 6–7.

174 The modern corresponding terms are *quadrilateral* or *rectangular* [see Burton, "The Optics of Euclid," p. 359, and Ver Eecke, *Euclide, l'Optique et la Catoptrique*, p. 8 and p. 62, respectively] which are distinct characetrizations of the figure's sides and angles.

one MS of R].[175] As for the figure, copies slightly differ: one variant (m1) has a circle contained in a six-sided figure, itself bound by a rectangular figure [R has a circle bound by a rectangular figure]; the rest (m2–m5) [and following them T] are closer to O in having a six-sided figure within a rectangular figure. The accompanying Arabic demonstration is different from both Greek versions: it is not strictly based on the third Euclidean proposition [as in O and R], where disappearance at a limit of distance is treated in terms of the visual rays losing contanct with the object [see Summary and Commentary: Proposition 3, p. 34], but on the first proposition, insofar as the phenomenon is explained not just in terms of the movement of sight [see Summary and Commentary: Proposition 1, p. 24] skipping the corners of an object from a far distance, but also in terms of the nonsimultaneous nature of disappearance. This is in fact the reverse of the first proposition: There the central core of the visual field holds a prominent place for being seen first; here it is the corners which are the first to be "concealed (*yakhfā*) from sight (*baṣar*)." Such reasoning is presented in only one of the variants (m1); the others (m2–m5) go directly from the claim that the points [representing corners] are not seen to the concluding statements on the circular appearance of the right-angled figure. The unique treatment in m1 contains a causal premise within the text [beginning with "*li-anna*" (because)], as well as a marginal commentary [that the angles are the first to disappear (*yaghīb*) for being the smallest (*aṣghara*) part of an object; see critical apparatus, v. 1, p. 31]. The margin of Proposition 3 in m1, is once again close to that in Ṭūsī's alternative proof (*'illa*) of Proposition 10,[176] this time without the exact same terminology.[177]

---

[175] This part is bracketed in Heiberg's edition of R; see *Opticorum recensio Theonis*, p. 166. Ver Eecke, *Euclide, l'Optique et la Catoptrique*, p. 8, and p. 62, n. 2, also provides reference to the manuscript containing this part.

[176] "I say the reasoning (*'illa*) is not this [i.e., the eye moves in such a way that points such as angles are not seen] but rather that the chords (*awtār*) like ZH, subtending angles are smaller than the diameters of the figure, and that which is smaller dies out (*yafūtu*) in a shorter distance (*bu'd aqall*) than what is larger, so when the distance becomes such that [the magnitude of] angles are concealed from the eye but the figure's diameters are not, then it [the figure] is seen as one without angles;" see T, edn. Hyderabad, p. 7. On another case of overlap between m1 and T, see Summary and Commentary: Proposition 3, p. 36.

[177] The apparant shape is *mustadīr* (circular) in m1, but *ghayru dhī zawāyā* (without angles) in T; the verb for that which disappears is *yaghību* in m1, *yafūtu* in T. Relative disappearance of chords (*awtār*) and diameters (*aqṭār*) and the relationship between size and distance in disappearance in T is comparable to Ibn al-Haytham's, specifically, in stating that the disappearance of smaller objects occurs in less distance (*bu'd aqall*).

Historical Context: The case of the present proposition is distinct from the previous ones in some ways. In contrast to the third Euclidean proposition [to which it has close ties], the alternative Arabic proof is contained *within* the text as well as in the margins of one variant (m1); in comparison to this and the first Euclidean proposition [to which the Arabic is particularly linked], the shift in the proof seems to be independent of linguistic matters [see Summary and Commentary: Proposition 1 and 3, p. 25 and p. 34, respectively]. The occurrence of the alternative proof is hard to explain, insofar as the assumptions of the Greek [Euclidean def. 3], apparently abandoned, seem less problematic than the assumptions of the Arabic. The context of the inspirations or reactions associated with the problems involved only provides some direction.

The subject of the circular appearance of rectangular shapes takes on extensive and diverse forms beginning with Greek sources. The author of the passage in the pseudo-Aristotelian work *Problemata* treats the case of close and far objects at once, when he states that "a square (τετράγωνον/*tetragônon*) [i.e., a four-sided figure] appears to have sundry angles, but if we stand farther off it looks like a circle."[178] In contrast to the assumption of the uneven distribution of visual rays within the cone of vision, the notion of "cut-off" angles at great distances is one that runs through many early discussions of the visual effect in question. The assumption is at the base of the Euclidean proof that the visual phenomenon is due to the angles no longer being seen beyond a certain distance (in its Ptolemaic formulation, the visual angle is diminished to a point of imperceptibility at exceedingly great distances).[179] The invisibility of the object's corners in producing the same effect also forms the core of various Arabic treatments, including the alternative reasoning of the Euclidean proposition itself, where sight (*baṣar*), in moving (*yantaqil*) from one point on the object's outline to another, misses some points in between (*mā bayn*).[180] The clarifying statements in the marginal passage in m1 and in Ṭūsī's recension of that proposition act as an alternative causal premise that this is "because (*li-anna*) the smallest parts of an object [namely angles] are the first to disappear (*yaghīb*) from sight (*'an al-baṣar*) at a far distance."

---

178 The full passage is as follows: "Similar to this is the phenomenon that a square appears to have sundry angles, but if we stand farther off it looks like a circle. For as the fall of the rays is in the form of a cone, when the figure is removed to a distance, those rays that are at the angles are cut off and do not see anything because they are weak and few, when the distance grows greater, but those that fall on the centre persist because they are collected together and strong. When the figure is near they can see also the parts at the angles, but when the distance becomes greater they cannot," *Problems*, bk. XV, 911b19–21, Loeb edn., p. 335.

179 See *Optica*, bk. II, sec. 26, ed. Lejeune, p. 25, for a statement to the effect that for objects at exceedingly great distances, their angles are so diminished that they become imperceptible.

180 Only one variant (m1) is explicit about the critical premise that "a figure for which there are no angles is a circle;" see M10(14), v. 1, p. 33.

Aḥmad ibn ʻĪṣā's treatment of this proposition is of particular interest on many levels. Citing Euclid by name (*Uqlīdis*) and his book by its old Arabic title (*Ikhtilāf al-manāẓir*), he voices next to an almost verbatim reproduction of the Arabic proof of the Euclidean proposition, responses to the problems and solutions involved. This he presents not only through a different demonstration based on different assumptions but in a different framework and language:[181] On the one hand, he presents a critique of what "Euclid said" (*qāla Uqlīdis*) in that proposition on both physical and epistemological grounds, calling the assumption of the leap (*ṭafra*) of sight [the Arabic proposition does not contain this term], together with the inability to perceive the figure's real shape (*al-shakl al-ḥaqīqī*), among "the most amazing of amazements" (*min aʻjab al-ʻajab*);[182] on the other hand, he offers geometrical demonstrations for a variety of angled figures, which now [to the amazement of *his* readers], treat the "circle" no longer as a figure having no angles, but as a figure being equidistant from its own center.[183]

---

181 He speaks of the round (*mudawwar*) appearance of "angled shapes (*al-ashkāl dhawāt al-zawāyā*)" such as a quadrilateral (*murabbaʻ*) from a far distance (*masāfatin baʻīdatin*) next to an almost verbatim phraseology of the Arabic Euclidean proof, ḥ1 [see Ḥ, List II, v. 1, p. xlvi], fols. 89a–90a. Note that the term *murabbaʻ*, which has come to mean a square, refers to any quadrilateral figure here.

182 "Euclid said in his book *Ikhtilāf al-manāẓir* that figures having angles, like a quadrilateral (*murabbaʻ*), are seen from a certain distance as circular (*mustadīr*), so if they become distant from the eye they are seen as round (*mudawwar*)." Repeating the lines of the alternative Arabic proof closely, the argument follows that this is "the most amazing of amazements" (*aʻjab al-ʻajab*) because if it is in the nature (*ṭabʻ*) of sight (*baṣar*) to make a leap (*ṭafra*) and see a quadrilateral object at a distance as round, then as a result of that leap the object's real (*ḥaqīqī*) shape is not seen, and this would be the case for round (*mudawwar*) and triangular (*muthallath*) objects alike, these being seen other than (*ghayr*) their real shapes, whether it is roundness (*tadwīr*) or triangularity (*tathlīh*). But sensation (*al-ḥiss*) shows us the opposite of this, except in the case of largeness and smallness [of magnitudes]. If, on the other hand, this [property] is particular to a quadrilateral figure (*murabbaʻ*), then the eye would make leaps even before looking at it from a distant . . . . The reasoning offered follows in a much clearer text: "We, however, say that the cause (*ʻilla*) in this [case] is that when a quadrilateral object is moved to a [far] distance (*masāfatin*), the difference (*ikhtilāf*) between the distance of its center to its angles and to the middle of its sides is not of a perceptible magnitude [and so] the rectangular shape (*al-shakl al-murabbaʻ al-zawāyā*) is seen as circular," ḥ1 [see Ḥ in List II, v. 1, p. xlvi], fols. 89a–90a.

183 In the geometrical demonstration, it is shown that if from the center of the quadrilateral figure a line is drawn perpendicularly such that there is a point from which the excess (*faḍl*) of lines connecting the figure's center to its far corner and middle of its sides is not a sensible magnitude (*qadr maḥsūs*), then from that point the figure is considered to be seen as circular. Therefore, it is clear that while all other explanations of this visual effect depend on the characterization of a circle as a figure with no angles, this

The transmission of this proposition is of much historical interest, not just in terms of what was or was not available to such early authors as Aḥmad ibn 'Īsā, but especially, of the identity and milieu of this particular author. In a book entitled *Kitāb al-Tarbī' wa al-tadwīr* (*Rectangularity and Circularity*), composed by al-Jāḥiẓ, the famous ninth-century literary figure, an objection addressed to a figure comparable to Aḥmad ibn 'Īsā, seems to be made with reference to this same proposition in the ironic statement that: "if we say a quadrilateral structure is seen from a distance only as circular, then perhaps the sun is polygonal (*muḍalla'a*) and the stars are quadrilateral (*murabba'a*)."[184]

Of the comparable language of Ibn 'Īsā's text of that same proposition, little seems to have been passed on directly. The same cannot be said of the Arabic Euclidean proposition with which it otherwise shares many elements, or of the more frequently encoutered vocabulary of its commentary on the margins of m1, which may itself be an insertion *from* as well as an integration *into* later Arabic texts. In Ibn al-Haytham's third book of the *Optics,* where visual illusions are treated, the explanation of the problem of "why a polygonal (*muḍalla'*) figure is perceived to be circular (*mustadīr*)" is treated with the terminology of the concealment (*khafy*) of the angles due to their relative smallness (*ṣighar*) at a distance (*bu'd*).[185] As in the case of the third Euclidean proposition, this may well reflect the vocabulary of the corresponding passage in the Arabic text of Ptolemy's second book, whether or not it has acted as a common source for the inserted commentary on the Arabic proposition, and the integrated proof of Ṭūsī's recension [see also Summary and Commentary: Proposition 3, p. 36 and n. 152].

---

proof is based on a more technical description of a circle as a figure having all its points equidistant from a center. In Euclid's *Elements* a circle is defined as "a plane figure contained by one line such that all the straight lines falling upon it from one point among those lying within the figure are equal to one another;" see Heath, *The Thirteen Books of Euclid's Elements*, bk. I, v. 1, p. 153.

184 The term *muḍalla'a* (مضلعة) makes better sense than *muṣallaba* (مصلبة = cross-shaped) in Pellat's edition [see *Kitāb al-Tarbī' wa al-tadwīr*, ed. Pellat, p. 91], without changing the argument itself]. The text opens with a satirical note describing a certain Aḥmad ibn 'Abd al-Wahhāb as being quadrilateral (*murabba'*) in form, but estimated as round (*mudawwar*) due to his size! [for the editor's English translation of the opening passage, see *The Life and Works of Jāḥiẓ*, ed. Pellat, p. 126, where the title of the book is given as *The Circle and the Square*]. Note that the two terms *murabba'* and *mudawwar* used by al-Jāḥiẓ for the geometrical figures are exactly the same as those in Ibn 'Īsā's text [ḥ1, fol. 89a], who himself uses these as alternative terms to *al-ashkāl qa'imat al-zawāyā* (right-angled figures) and *mustadīr* (circular) of the Arabic Euclidean proposition [M10(2), v. 1, p. 30], and whose name, dates [before 250/864, according to Krause, "Stambuler Handschriften," p. 513], and possibly intellectual orientation [see entry under a Aḥmad ibn 'Īsā (d. 247/861) in *EI2, Supplement,* pp. 48–49] are comparable to the target of al-Jāḥiẓ in this piece.

185 See *Kitāb al-Manāẓir*, bk. III, sec. 9, ed. Sabra, p. 416; tr. Sabra, v. 1, p. 281.

Propositions 11–18: Appearances Involving Orientations

Proposition 11–15: Appearances Involving Elevations and Directions

Proposition 11: Relative Appearance of Depths of Surfaces Below the Eye

Synopsis: The enunciation is that the most remote of surfaces below the eye is seen as the highest. The illustration is that of a point (A) [representing the eye] higher than a line BEDG [the surface] and four rays (AE, AB, AD, AG) falling [on the surface]. The demonstration begins with the geometrical construction of a line (ZH) from the midpoint of BE (Z) at right angles [upwards]. Then follows, since the visual rays fall on ZH before ZG, H is higher than G, T higher than D, and I higher than E; the rays by which GD, DE and EB are seen [AH, AT; AT, AI and AI, AB respectively] [def. 3] are each higher than the next. The conclusion is that [because whatever is seen by higher rays is seen as higher, def. 5, not in Arabic], GD is seen as higher than DE and DE higher than EB.

Textual Correspondence:[186] This proposition follows the Greek closely, with a lettering closer to O [otherwise close in O and R]. A few minor differences between the Greek and Arabic are the following: In M, the plane is marked with four letters [in O, with three letters, in R, two letters]; the ray AB is not specified as being perpendicular to that plane below [the corresponding line is in both O and R]; and the perpendicular line ZH is erected from the midpoint (Z) of the magnitude closest to the eye (BE) [for both O and R, this is an arbitrary point on the corresponding line]. No references are made to the assumption regarding appearance of higher rays [Euclidean def. 5] underlying the demonstrations, either within or in the margins of the Arabic text [that does not contain this definition] or in the short demonstration of T [where an additional definition is included regarding higher or lower rays, or those to the right and to the left].

Proposition 12: Relative Appearance of Heights of Surfaces Above the Eye

Synopsis: The enunciation is that the most remote of surfaces above the eye is seen as the lowest. The illustration is that of a point (A) [representing the eye] lower than a line BG [representing the surface] and four rays (AB, AD, AE, AG) falling [on that surface]. The demonstration begins [without a geometrical construction] with the statement that ray AG is lower than AE, AE lower than AD, and AD lower than AB [referring to the previous proposition]. The conclusion is that [because whatever is seen by lower rays is seen as lower, def. 5, not in Arabic], GE [seen by EA, AG] is seen as lower than DE [seen by DE, DA], and DE is seen as lower than DB [seen by DA, AB].

---

[186] M11 compared to O10, R10, and T11: M11, v. 1, pp. 34–37; O10, ed. Heiberg, pp. 16–18, tr. Ver Eecke, pp. 8–9; R10, ed. Heiberg, pp. 166–168, tr. Ver Eecke, pp. 62–63; T11, edn. Hyderabad, p. 7.

Textual Correspondence:[187] In this proposition, where the figures of the two Greek versions differ considerably, the Arabic has a figure and lettering closer to O. The demonstration steps of the Greek [close in O and R] are matched closely in M and again condensed in T, though neither specifies the ray AB as being perpendicular to the plane above [as in the corresponding line in both O and R] or contains any comments about the assumption regarding appearance of lower rays [Euclidean def. 5, not in Arabic] underlying the demonstration.

Proposition 13: Relative Appearance of Directions of Magnitudes at a Distance Facing the Eye

Synopsis: The enunciation is that for magnitudes at some distance from the eye and facing it, those to the right are seen to the left, and vice versa. The illustration is that of two lines (BD, AG), and a point (E) [representing the eye]. The demonstration begins with the geometrical construction of six visual rays (ET, EI, ED, EZ, EH, EG). Then follows, EZ is more to the right than EH, and so is EH compared to EG [def. 6, not in Arabic]. The conclusion is that EG is thought to be (*yuẓunnu*) more to the left of EH and EH more to the left of EZ, and so ED, EI, and ET are thought to be moved to the right.

Textual Correspondence:[188] In this proposition, where O and R differ in the demonstration and lettering, M is closer to O in both cases. Its demonstration is closer to O in discussing the relative appearance of rays (EZ, EH, EG), rather than points [in R]. The Arabic differs from both Greek versions in describing the two magnitudes facing the eye as being "far from the eye" (*mutabā'id 'an al-baṣar*).

Proposition 14: Relative Appearance of Heights of Magnitudes on the Same Side Below the Eye

Synopsis: The enunciation is that the more remote of equal magnitudes below the eye on the same side is seen as higher than the one below it. The illustration is that of three lines (AB, GD, EZ) [representing the magnitudes] and a point (H) [the eye] above them. The demonstration begins with the geometrical construction of three rays (HA, HG, HE). Then follows, AH is higher than HG, and HG higher than HE, their end points (A, G, E) terminating magnitudes AB, GD, EZ. The implicit assumption is that if point A on object AB is seen as higher than point G on object GD, then object AB is seen as higher than GD.

[187] M12 compared to O11, R11, and T12: M12, v. 1, pp. 38–39; O11, ed. Heiberg, pp. 18–20, tr. Ver Eecke, p. 9; R11, ed. Heiberg, pp. 168–170, tr. Ver Eecke, pp. 63–64; T12, edn. Hyderabad, p. 7.

[188] M13 compared to O12, R112, and T13: M13, v. 1, pp. 40–41; O12, ed. Heiberg, p. 20, tr. Ver Eecke, pp. 9–10; R12, ed. Heiberg, p. 170, tr. Ver Eecke, p. 64; T13, edn. Hyderabad, p. 7.

The conclusion is that [because for the highest ray, its end point is highest, and because the endpoint is on the furthest object, and the furthest objects appear as highest] AB is seen as higher than GD and GD higher than EZ.

Textual Correspondence:[189] In this proposition, where O and R differ in both demonstration and lettering, M is closer to O in both cases. Its demonstration is closer to O in discussing the relative appearance of rays (AB, DG, EZ) through all three points (A, G, E) that are simultaneously on these and the three visual rays (EA, EG, EA) [as in O], as opposed to one point [the furthest in R], extended to the other two cases.

Proposition 15: Relative Appearance of Heights of Magnitudes on the Same Side Above the Eye

Synopsis: The enunciation is that the most remote of equal magnitudes above the eye is seen as lowest. The illustration is that of three lines (AB, GD, EZ) above a point (H) [representing the eye]. The demonstration begins with the geometrical construction of three visual rays (HB, HD, HZ). Then follows, HB is lower than HD, and HD, lower than HZ, their end points (B, D, Z) terminating magnitudes AB, GD, EZ. The implicit assumption is that if point B on object AB is seen as lower than point D on object GD, then object AB is seen as lower than GD. The conclusion is that [because for the lowest ray, its end point is lowest, and because this endpoint is on the farthest object], AB is seen as lower than GD and GD lower than EZ.

Textual Correspondence:[190] In this proposition, where O and R differ in the length of demonstration and lettering of the figure, M is closer to O. In T the proposition is limited to the enunciation and illustration, followed by a statement that "the demonstration is just as the case of the previous one."

---

[189] M14 compared to O13, R13, T14: M14, v. 1, pp. 42–43; O13, ed. Heiberg, pp. 20–22, tr. Ver Eecke, p. 10; R13, ed. Heiberg, p. 170, tr. Ver Eecke, p. 64; T14, edn. Hyderabad, p. 8.

[190] M15 compared to O14, R14, and T15: M15, v. 1, pp. 44–45; O14, ed. Heiberg, p. 22, tr. Ver Eecke, pp. 10–11; R14, ed. Heiberg, pp. 170–172, tr. Ver Eecke, pp. 64–65; T15, edn. Hyderabad, p. 8.

Propositions 16–18: Relative Appearance of Unequal Magnitudes Viewed Simultaneously

Proposition 16: The Case of Unequal Magnitudes Below the Eye

Synopsis: The enunciation is that when two magnitudes, one greater than the other, are below the eye, the part seen of the greater one while seeing the smaller one, is less [in magnitude] than when the eye is lowered [i.e., when the eye is looking from above the smaller one (E), less of the greater one is seen (BZ) than when the eye is lowered (H), where more of the greater one is seen (TB)]. The illustration is that of two lines (AB, GD), where AB is greater than GD, and two points (E, H) [representing the high and low positions of the eye]. The demonstration involves the geometrical construction of a ray (EZ) from the eye (E), and another ray (HT) from the lower position of the eye (H), both through the tip of the smaller magnitude (G). Then follows, the amount seen by the visual ray (EZ) of the greater magnitude (AB) while seeing the smaller one (GD) is BZ [def. 3], so AB is seen as greater than GD by the amount AZ; that seen by the lower visual ray (HT) in the same situation is TB [equal to GD itself] [def. 3], so AB is seen as greater than GD by the amount AT. The conclusion is that [because AT is less than AZ], the part seen of the greater one while seeing the smaller is less than when the eye is lowered.

Textual Correspondence:[191] In this proposition, where O and R differ in enunciation, figure, lettering, and wording, M is closer to O in all but the figure, where the two positions of the eye are not placed on the same horizontal extension [as in R], yet are not connected by a [superfluous] line [in both O and R]. The Arabic differs from the Greek in both enunciation, and [similarly incomprehensible] demonstration. There are no references to the magnitudes BZ, GD as being *below* the visual ray EZ or as seen *by the same eye* [as in the corresponding lines in O and R respectively]. T lacks any reference to the visual ray EZ. Its enunciation specifies the greater magnitude to be the further one, and its terminology is markedly different [*mar'ī* for *mubṣar* (visual object), the verbal root *nazila* for *khafiḍa* (lowering) and the term *ḥayna'idh* for the simultaneous appearance of the two magnitudes].

Proposition 17: The Case of Unequal Magnitudes Above the Eye

Synopsis: The enunciation is that when two magnitudes, one greater than the other, are above the eye, the part seen of the greater one while seeing the smaller one is more [in magnitude] than is the case when the eye is raised [i.e., when the eye is looking from below (E) at the smaller one, more of the greater one is seen (BZ) than when the eye is raised (H)]. The illustration is that of two lines (AB,

---

[191] M16 compared to O15, R15, and T16: M16, v. 1, pp. 46–49; O15, ed. Heiberg, pp. 22–24, tr. Ver Eecke, p. 11; R15, ed. Heiberg, p. 172, tr. Ver Eecke, p. 65; T16: edn. Hyderabad, p. 8.

GD), where AB is greater than GD, and two points (E, H) [representing the low and high positions of the eye]. The demonstration involves the geometrical construction of a ray (EZ) from the eye (E) and another ray (HT) from the higher position of the eye (H), both through the tip of the smaller object (G). Then follows, the amount seen by the visual ray (EZ) of the greater magnitude (AB) while seeing the smaller one (GD) is ZB [def. 3], so AB is seen as greater than GD by the amount AZ; that seen by the higher visual ray (HT) in the same situation is TB [def. 3], so AB is seen as greater than GD by the amount AT. The conclusion is that [because ZB is greater than TB], the part seen of the greater one while seeing the smaller is more than the case when the eye is raised.

Textual Correspondence:[192] In this proposition, where O and R differ in enunciation, demonstration wording , and lettering [though not in the figure], M is again closer to O, still differing from it and from T [which refers to the previous demonstration] in the same way as in the preceding proposition.

Proposition 18: The Case of Unequal Magnitudes Leveled with the Eye

Synopsis: The enunciation is that when two unequal bodies [with unequal upright alignment] are on the same line, and the smaller one is closer to the eye, they are seen by the same difference [in magnitude] from far and near. The illustration is that of two lines (AB, GD), where AB is greater than GD [the one closer to the eye], and two points (Z, H) [representing the far and near positions of the eye]. The demonstration begins with the geometrical construction of a visual ray ZGE, and from a further point (H), the visual ray HZZGGE. Then follows, AB is seen as greater than GD by the amount AEin both cases. The conclusion is that [because the apparent difference in size between the two objects is the same amount (AE) in both cases] the two objects are seen by the same difference from far and near.

Textual Correspondence:[193] In this proposition, where O and R differ in the length of the demonstration and the lettering of the figure, M is closer to O in all cases, still differing from it in phraseology, most noticeably in the enunciation. The Arabic differs from both Greek versions in its designation of the two visual rays drawn from the two different positions of the eye in the [uncommon] form of three letters [ZGE] and six letters [HZZGGE]. T's unusually short version does not contain any references to the visual rays.

---

[192] M17 compared to O16, R16, and T17: M17, v. 1, pp. 50–53; O16: ed. Heiberg, pp. 24–26, tr. Ver Eecke, pp. 11–12; R16, ed. Heiberg, pp. 172–174, tr. Ver Eecke, pp. 65–66; T17, edn. Hyderabad, p. 8.

[193] M18 compared to O17, R17, and T18: M18, v. 1, pp. 54–55; O17, ed. Heiberg, p. 26, tr. Ver Eecke, pp. 12–13; R17, ed. Heiberg, p. 174, tr. Ver Eecke, p. 66; T18, edn. Hyderabad, pp. 8–9.

Propositions 19–22: Determination of Heights, Depths, and Lengths by Means of Rays

This set of four propositions [O18–21/R18–21] is commonly known as the "surveying propositions." The first two concern the determination of a given height [one in the presence of sunshine, the other by means of a plane mirror], the following two, that of a certain depth and length, respectively. The wording "to know how great a given height, depth, or length is," together with the method of determining an unknown value from three known ones through proportional relations between sides of two similar triangles, can be contrasted to the rest of the Euclidean propositions insofar as they involve the problem of the *determination* rather than the *appearance* of magnitudes. But even in their application of the geometrical laws of vision, the four cases maintain the purely demonstrative character of a geometrical proposition: they all begin with the statement of their respective problem as enunciation, proceed with a geometrical demonstration [to demonstrate that there are means to determine magnitudes such as heights, depths or lengths], and end with the conclusion "That is what we wished to demonstrate;" none contains instructions as to how to measure the three known values or calculate the unknown value from them.

Proposition 19–20: Determination of Heights by Means of Rays

Proposition 19: Determination of Heights by Means of Sun Rays

Synopsis: The problem is to determine the height of an object by means of the sun (*bi al-shams*). The illustration is that of a vertical line (AB) [representing an upright object], a slanted line (GD) [representing the solar ray] falling on the object's upper point (A) and the ground level (D), and a horizontal line (BD) [the object's shadow on the ground]. The demonstration begins by instructions on placing an object that is known [in height], EZ, parallel to AB such that GD falls on its upper point at Z, and it proceeds with the statement of the similarity of the two resulting triangles (ABD, ZED). The conclusion [through some geometrical identities] is that because the ratio of DB to BA is known [from the measurable values of EZ and DE], then the magnitude of AB can be determined.

Textual Correspondence:[194] In this first of the four so-called "surveying" propositions, where the enunciation, figure, and lettering differ in O and R, M is closer to O, though in the enunciation the term corresponding to sunshine [in O but not R] appears only in m1 and m5 [T has "sun ray" (*shu'ā' al-shams*]. The Arabic differs from the Greek as none of the copies of M [or T] has a statement corresponding to the one [again in O but not R] that line EZ "is not illuminated

[194] M19 compared to O18, R18, and T19: M19, v. 1, pp. 56–57; O18, ed. Heiberg, pp. 26–28, tr. Ver Eecke, p. 13; R18, ed.. Heiberg, pp. 174–176, tr. Ver Eecke, pp. 66–67; T19, edn. Hyderabad, p. 9.

by the [sun] ray GD" when it meets it. The Arabic [in both M and T] is also explicit about the similarity of the two triangles.

Proposition 20: Determination of Heights by Means of Reflecting Visual Rays

Synopsis: The problem is to determine the height of an object by means other than the sun. The illustration is that of a line [representing an upright object (AB)], and a point [representing the eye (G)]. The demonstration begins by instructions on placing a plane mirror (ED) [horizontally on the ground]. Then follows the geometrical construction of line EB [from the extension of ED to point B], of a visual ray GH, such that it is deflected (*in'aṭafa*) [i.e., reflected] at H to fall on the farthest point of AB (A), and of a line [extended in the opposite direction] from H to T [where a perpendicular from the eye (G) meets this line]. Then the deflection (*in'iṭāf*) of ray GH at H to reach A is stated to be at equal angles, so the two resulting triangles [AHB and GHT] are similar. The conclusion [through some geometrical identities] is that because the ratio of HB to AB is known [from the measurable values of GT and TH], the magnitude of AB can be determined.

Textual Correspondence:[195] In this proposition, the Arabic is distinctly different from the Greek [close in O and R] in more than one place. It lacks reference to a *Catoptrics* [in both O and R] immediately following the statement of the principle of the equality of angles for reflection at a plane mirror. The principle itself is expressed as *al-in'iṭāf* [commonly meaning refraction, here meaning reflection], the verb *in'aṭafa* (deflects) corresponding to ἀντανακεκλάσθω/*antanakeklasthô* in O and ἀνακεκλάσθω/*anakeklasthô* in R. In the rest of the proposition, where O and R differ slightly in enunciation, figure, lettering, and length of the demonstration, M is closer to O in the last two cases and closer to R in the first two. The enunciation of M has the expression "without the sun" (*min ghayr al-shams*) [as in R], rather than "when the sun is not present" [in O, m3 lacks *ghayr*, T has "by a mirror" (*bi al-mir'āh*)]. The shape of the mirror, which is represented in the edited text of O as a rectangle, and in the edited text of R as a rectangle with a semicircle as its upper side, appears differently in the Arabic variants: m5 has it as a rectangle [as in O], m1 and m3 have just a semicircle, and m2 and m4 only a straight line.

Historical Context: In this second of the set of two propositions on height determination, which treats the problem by means of reflecting visual rays, the transmission of the principle of reflection at equal angles takes on a curious and misdirected course. The terminology rendering the principle of reflection is not the standard form, *in'ikās*, but rather *in'iṭāf*, which became a common

---

195 M20 compared to O19, R19, and T20: M20, v. 1, pp. 58–61; O19, ed. Heiberg, pp. 28–30, tr. Ver Eecke, pp. 13–14; R19, ed. Heiberg, p. 176, tr. Ver Eecke, p. 67; T20, Hyderabad, p. 9.

expression for refraction.[196] The "slip", whether or not conceptually driven (resulting from the confusion of one optical principle with another), is likely to have involved a terminological bend, either as a misapplication of the principles' respective expressions or a mistranscription of their relatively close orthography.[197] Not only is the intended optical principle linguistically twisted, the misnomer finds its way into many later sources, including Ṭūsī's recension, where the single occurrence of the principle in a marginal proposition is raised to the status of an independent definition: "When a ray encounters a polished surface such as a mirror, it is "refracted" (*in'aṭāfa*) [meant to be reflected] at two equal angles, one called angle of incidence (*zāwiyat al-shu'ā'*) and the other "angle of refraction" (*zāwiyat al-in'iṭāf*) [meant to be angle of reflection].[198]

As marginal as the slight variation may seem, having entered into all copies of the Arabic proposition, and further imported into Ṭūsī's much more widely circulated recension, the effect on the subsequent development of the problem of indirect vision seems to have been far from marginal.[199] The problematic transmission of the principle of reflection, coupled with the practical "nontransmission" of the treatments of refraction, came to be added up not to one but a series of "puzzling" treatments including Ṭūsī's own, treatments in which the interchangeable and inconsistent use of the terms *in'ikās* and *in'iṭāf* for the long acknowledged optical phenomena are particularly striking.[200]

---

196 On the corresponding Greek and terms for the bending of the rays, see Variation Sample 4, p. xliv. References to sources containing the more common Arabic form appear on pp. 57–58, n. 205–206.

197 In a hand transcription, the orthography of the pairs: انعطاف and انعكاس or انعطف and انعكس may be close.

198 See *Taḥrīr al-Manāẓir*, *Majmū' al-rasā'il*, p. 2; *Revue de l'Institut des Manuscrits Arabes* (*RIMA*), p. 251, and Supplementary Passages, v. 1, p. 246; see also Variation Sample 4, p. xlv, which includes the Greek and Arabic corresponding texts. On Ṭūsī's use of *al-in'iṭāf* as refraction in I [see List II, v. 1, p. xlviii], see p. 59, n. 208.

199 Far from being a single slip of the pen, the statement appears in many manuscript copies, from all 5 copies of the early Arabic version to more than 50 manuscript copies of its 13th-century recension, covering a wide range of dates and places, including those linked to the copy of Quṭb al-Dīn al-Shīrāzī (d. 711/1311) [t36, t41, t51, t56, in List I, v. 1, pp. xlii–xliv, and p. liii].

200 For example, the treatments of Shīrāzī in three of his astronomical works, chronologically in this order: *Nihāyat al-idrāk fī dirāyat al-aflāk*, *a-Tuḥfa al-shāhiyya*, in Arabic and *Ikhtiyārāt Muẓaffarī* in Persian. For the relevant passages in the first two, see Sabra, *The Optics of Ibn al-Haytham*, v. 2, p. lxx, n. 110; the third work, *Ikhtiyārāt*, contains, as the Persian summary of *Nihāya*, the interchangeable terminology, not for the astronomical problem of the appearance of stars at the horizon, but for the optical problem of the appearance of a piece of grape in a bowl of water, Tehran University MS Mishkat 469, fol. 8a, see Dānishpazhūh, *Mishkāt*, v. 3, part 2, p. 822.

With various formulations of at least the principle of reflection in ancient sources, from optical and catoptrical works[201] to Platonic and Aristotelian passages,[202] the most immediate task is the identification of the sources of confusion. While early Arabic translations of relevant Greek works[203] present us with a real challenge in terms of records of accessibility,[204] extant Arabic texts well demonstrate the terminological inconsistency. The more standard term *in'ikās* is used as early as in Ḥunayn ibn Isḥāq's *Ten Treatises on the Eye*, in the Arabic versions of the Pseudo-Euclidean *De speculis*, and Aetius' *Placita philosophorum*,[205] and as late as some works of Ṭūsī and his commentators.[206]

---

201 On the standard Greek literature, see Lejeune, *Euclide et Ptolémée,* 1956, pp. 98–99; and Lejeune, *Recherches sur la catoptrique grecque*, Brussels, 1957, pp. 34–35; For more recent literature, see Smith, "Ptolemy's Search for a Law of Refraction: *Saving Catoptrics*;" Knorr, "Archimedes and Pseudo-Euclidean Catoptrics: Early Stages in the Ancient Geometric Theory of Mirrors;" Jones, "On Some Borrowed and Misunderstood Problems in Greek Catoptrics." On Arabic optics (*manāẓir*) and catoptrics (*marāyā*) in the context of the Greek tradition, see Sabra, *The Optics of Ibn al-Haytham*, v. 2, p. lvi.

202 The relevant passages from Plato's *Timaeus*, Aristotle's *De sensu, Meteorologica*, and *Posterior Analytics*, as well as the Peripatetic *Problemata*, are included in Boyer "Aristotelian References to the Law of Reflection," where a case is made for the familiarity of the Aristotelian school with equality of angles in reflection at plane mirrors.

203 For the transmission of the Aristotelian corpus, see Peters, *Aristoteles Arabus, The Oriental Translations and Commentaries on the Aristotelian Corpus*, p. 17, 39, 45, and 66. See also Petraitis, *The Arabic Version of Aristotle's Meteorology*, and the report of the Arabic version of the pseudo-Euclidean *De speculis*, in Sabra, "A Note on Codex Biblioteca Medicea-Laurenziana Or. 152."

204 For many of the relevant Greek treatments, such as the proposition in the [pseudo-] Euclidean *Catoptrics*, the works Hero of Alexandria and Archimedes under the same title, and the related parts of Ptolemy's *Optica*, extant Arabic texts have not been documented. On the first of these works, see Ver Eecke, *Euclide, l'Optique et la Catoptrique*, p. 100, and Takahashi, *The Medieval Latin Traditions of Euclid's Catoptrica*, with a historical treatment of the law of reflection. On the treatments of Hero and Archimedes, see Cohen and Drabkin, *A Source Book in Greek Science*, pp. 261–268, especially p. 265, n. 1, and Knorr, "Archimedes and the Pseudo-Euclidean *Catoptrics*." On Ptolemy's treatments, see Lejeune's *Euclide et Ptolémée*, pp. 98–99, and *Recherches sur la catoptrique grecque*, pp. 34–35. An Arabic manuscript of an Aristotelian work on mirrors (*al-Mir'āt)* is reported in Ahlwardt, *Königlichen Bibliothek*, p. 355, number 6022/1 [no longer in Berlin]. For ancient reports on an optical work by Aristotle, see Boyer, "Aristotelian References to the Law of Reflection," p. 93, n. 14.

205 On the first of these, see Max Meyerhof's edition, p. 109, lines 7–8 [reference from Sabra, *The Optics of Ibn al-Haytham*, v. 2, p. lviii, n. 80], where *in'ikās* is used together with *inkisār*, in terms of the "turning back" (*rujū'*) of the visual rays (*manāẓir*); in the Arabic manuscript of *De speculis*, the verb *yan'akis* is used [fol. 104b, the Latin has *convertitur*; see *Tractatus* [pseudo-] *Euclidis De speculis*, Björnbo and Vogl, *Alkindi, Tideus und Pseudo- Euklid*, p. 100 [reference from Theisen, *The Mediaeval Tradition, p.* 294 and n. 54]. For the occurrence of *in'ikās* in Qusṭā ibn Lūqā's Arabic translation of

As for the nonstandard terminology of *in'iṭāf* and its occasional confusion with the former, in Fārābī's long passage on the science of optics deflected (*mun'aṭifa*) rays appear alongside reversed (*mun'akisa*) or bent (*munkasira*).[207] Ṭūsī's application is both confusing *and* inconsistent: On the one hand, he employs *ini'kās* in his repeated references to and even proof of the principle of reflection in the two short treatises, while *ini'iṭāf* is the term in his *Taḥrīr al-Manāzir*, both in the Euclidean proposition involving a mirror and in his added definition restating the ancient phenomenon [see Variation Sample 4, p. xlv]; on the other hand, he uses the term *ini'iṭāf* in *In'ikās al-shu'ā'āt wa in'iṭāfuhā* for a form of refraction a cone of rays undergoes at the surface of such transparent

---

Aetius' *Placita philosophorum,* see Daiber, *Aetius Arabus, Die Vorsokratiker in arabischer Überlieferung,* p. 204.

206 In Ṭūsī's *Shu'ā'* and *In'ikās al-shu'ā'āt wa in'iṭāfuhā* [see SH and I in List II , v. 1, pp. xlvii–xlix], the expression is used for the deflection of the cones of ray at equal angles when encountered by polished surfaces. In the case of Ṭūsī's commentators, there is Badr al-Dīn al-Ṭabarī's short Persian treatise, *Irtifā'* [see E in List II, v. 1, p. l], with two chapters devoted to the problem of height determination by means of a plane mirror. The author, "recalling" another method by which the height of tall objects can be made known (*ma'lūm*) by means of a mirror placed at different locations on the ground, states that what the people of the art (*ahl-i ṣinā'at*) may have offered on this subject "has not reached him, or has been seen by him anywhere." The commentator of *Sī faṣl* (*Thirty Chapters*) attributed to Ṭūsī and of Euclid's *Elements* expresses the equality of visual angle (*zāwiyat al-shu'ā'ī*) and the angle of reflection (*zāwiyat al-in'ikāsī*) in the more common formulation and explicitly associates the principle with the science of optics (*'ilm al-manāẓir*).

207 "Rays that pass through transparent bodies to a visible object are either straight (*mustaqīma*) or deflected (*mun'aṭifa*) or reversed (*mun'akisa*) or bent (*munkasira*) . . . Straight rays are those that, having issued from the eye, extend rectilinearly on the line of sight until they weaken and come to an end . . . Deflected rays are those that, having passed out of the eye, meet on their way, and before they weaken, a mirror that precludes them from passing through in a straight line, thereby causing them to be deflected (*tan'aṭif*) and turned to one side of the mirror. They then extend in the direction into which they have turned towards the beholder. Reversed rays are those that return from the mirror on the path they traversed at first, until they fall on the body of the beholder from whose eyes they have issued, and it is by means of this [kind of] ray that the beholder sees himself. Bent rays are those that return from the mirror towards the beholder from whose eyes they have issued, but extend obliquely beside him until they fall on something else behind the beholder or on his right or his left or above him, and it is thus that we see what lies behind or beside us;" see ed. Amīn, pp. 79–83, and English translation, Sabra, *The Optics of Ibn al-Haytham*, v. 2, pp. lvi–lvii, where the passage is viewed in the context of sources from Greek antiquity.

bodies as still waters.[208] There, "*zāwiyat al-in'iṭāf*" is no longer Ṭūsī's "angle of deflection" at the surface of polished bodies such as mirrors, as in the last definition of his *Taḥrīr*, but the angle that the refracted end of the visual cone makes with the refracting surface. What is more, the so-called "angle of refraction" (*zāwiyat al-in'iṭāf*) is still held to be equal with the angle of incidence, called in both cases "*zāwiyat al-shu'ā'*" (angle of the visual ray).[209]

The treatment of Ṭūsī may have indeed been among the primary sources for the "perplexity" (*taḥayyur*) that according to the introduction in Kamāl al-Dīn al-Fārisī's *Tanqīḥ al-Manāẓir*, acted as the starting point of his own optical researches. As he recalls in a revealing passage in that introduction, he had read in "the statements of some leading philosophers, and in more than one of them, that light shines from a luminous object in straight lines, and when it encounters a surface such as the surface of water, four equal angles are produced, named angles of direct radiation (*istiqāma*), reflection (*in'ikās*), penetration (*nufūdh*), and refraction (*in'iṭāf*)." The troubling concept of the equal size of angles of incidence and refraction in particular, whether or not directly rooted in the "misnomer" of the Arabic Euclidean texts [of the "shortcomings" of which Fārisī explicitly speaks of] had, according to Fārisī's testimony, the fateful consequence of his lifetime devotion to the science of optics.[210]

---

208 See English translation, Winter and 'Arafat, "A Statement on Optical Reflection and 'Refraction' Attributed to Naṣīr ud-dīn aṭ-Ṭūsī," p. 141; partial German translation, Wiedemann, "Über die Reflection und Umbiegung des Lichtes von Naṣīr al Dīn al Ṭūsī."

209 In Shu'ā' [SH in List II, v. 1, p. xlvii], the angle that falls between these two angles is called *zāwīyih-i avval* (the first angle). In having the refracted ray lie along the extension of the reflected ray, the model follows the pre-Ptolemaic ones offered by Ibn 'Īsā and Ibn al-Haytham's *Commentary on the Almagest* as described in Sabra, *The Optics of Ibn al-Haytham*, v. 2, p. lix, and n. 85.

210 The expression for "puzzle" (*taḥayyur*) together with Fārisī's reasons to take up the initial course of his optical researches appear in the introduction of his important commentary on Ibn al-Haytham's *Kitāb al-Manāẓir*, entitled *Tanqīḥ al-Manāẓir*, ed. Hyderabad, p. 6: "There was a period of time during my research in matters related to optics when I was extremely interested in the explanation of the manner of the perception of sensible forms, especially by refraction (*in'iṭāf*), because I had seen that the appearance of visible objects in the water or from behind glass is in an unfamiliar way and other than [their appearance] in a direct manner in the air, and [because of] the shortcomings of Euclid's book [i.e., *Optics*] on the object of my search, and [also] because I saw in the statements of some leading philosophers, and in more than one of them, that light shines from a luminous object in straight lines, and when it encounters a surface such as the surface of water, it is reflected from it at angles equal to their opposite and penetrates (*yanfūdh*) into it on the extension of illumination, and is refracted in the extension of reflection, and from this four equal angles are produced, named angles of direct radiation, reflection, penetration, and refraction (*zāwāyā al-istiqāmah, al-in'ikās, al-nufūdh, wa al-in'iṭāf*). So I became puzzled (*taḥayyartu*) by these principles (*aḥkām*)." Ṭūsī's treatment in "*Risāla fī in'ikās al-shu'ā'āt wa in'iṭafihā*" [Winter and 'Arafat, "A Statement on Optical Reflection and 'Refraction' Attributed to Naṣīr ud-dīn aṭ-Ṭūsī"] has

Proposition 21: Determination of Depths by Means of Visual Rays

Synopsis: The problem is to determine the depth of an object. The illustration is that of a vertical line (AD) [representing the depth of the object] and a point (E) [representing the eye]. The demonstration begins with the geometrical construction of a few lines [ED, BZ, EZ, but not AB]. Then follows the statement of the similarity of the two resulting triangles (ABD, EZD). The conclusion [through geometrical identities] is that because the ratio of DA to AB is known [from the measurable values of EZ and ZB], the magnitude of AD can be determined.

Textual Correspondence:[211] In this proposition, where the difference between O and R is most immediately clear from their figures, M is closer to R in the figure [where the desired magnitude is the side of a square] but not in the text [it lacks the corresponding part of the text in R]. It is closer to O, where line EZ is drawn perpendicular to BZ rather than parallel to AD [in R]. But the term used for the Greek ἀκτίς/*aktis* [commonly translated as "solar rays"], represented by line ED passing through the ground level [m1 has "the sphere's suface" (*basīṭ al-kura*)], is *shu'ā' al-baṣar* (visual ray[s]). The construction is closer to R, though the ray is represented with two letters rather than three [as in R]. T has "radial line" (*al-khaṭṭ al-shu'ā'ī*) EBD [with three letters].

Proposition 22: Determination of Lengths by Means of Visual Rays

Synopsis: The problem is to determine the length of an object on the ground surface. The illustration is that of a horizontal line (AB) [representing a "long" (*tawīl*) object] and a point above it (G) [representing the eye]. The demonstration begins with the geometrical construction of two rays (GA, GB) and a point close to the eye (D), extended as a line parallel to AB (DE). Then [through similarity of triangles ABG and GDE, and some geometrical identities] follows the statement that because the ratio of GA to AB is known [from the measurable magnitudes GD and DE], the magnitude of AB can be determined.

---

been cited by Sabra in this connection; see *The Optics of Ibn al-Haytham*, v. 2, pp. lxix–lxxi, and n. 112.

[211] M21 compared to O20, R20, and T21: M21, v. 1, pp. 62–65; O20, ed. Heiberg, p. 30–32, tr. Ver Eecke, pp. 14–15; R20, ed. Heiberg, p. 178, tr. Ver Eecke, pp. 67–68; T21, edn. Hyderabad, p. 9.

Textual Correspondence:[212] In this proposition, where O and R differ in some places [besides the figure and lettering, in the position of the parallel line within the triangle formed by the two visual rays and the desired length, drawn from a point *close to the eye* in O, but arbitrarily in R], M is closer to O in most cases. In M the object is described as "long" (*ṭawīl*), a description not in the Greek. The description *qarīban min mawḍiʿ al-baṣar* (close to the eye's position) for the point from which the line parallel to the object is drawn is in the Greek [in O but not in R]. Both descriptions are absent in T, where the proposition describes AB as an upright height (*irtifāʿ*) [rather than a length (*ṭūl*) on a plane] with a largely different figure and procedure. The departure from the original proposition may be related to the absence of the term *ṭūl* in the enunciation of some manuscript copies of M, as in variant m5, where it appears as an addition above the line.

Propositions 23–64: Back to Appearances

In the rest of the propositions it is once again appearance itself, not the application of its laws as in the four preceding cases, that is the main subject of treatement. The next set of propositions treats appearances in two and three dimensions (M23–42) and, with the exception of one (M63), goes back to appearances involving one-dimensional magnitudes (M43–64). Those propositions treating the appearance of flat figures include the appearance of curvatures (M23) and circles (M36–42); those treating the appearance of solid figures (M24–35) involve the appearance of spheres (M24–28), cylinders (M29–31), cones (M32–35), and other three-dimentional shapes (M63).

It is with these higher-numbered propositions that a direct correspondence between the Arabic and Greek further breaks down [see Correspondence Table 2, p. xxxiv]. There are cases where an Arabic proposition corresponds only to parts of the Greek [M23 to O22-1/R22]; there are also cases where more than one proposition in Arabic correspond to parts of the same proposition in the Greek [M29, M30 to O28-1, O28-2/R 28; M45, M46, M47 to O40-1/R 40-1, R 40-2, R40-3]. And there are also cases where two successive propositions in Arabic are in the reverse order of the Greek [M53–54 to O46, O45/R46], or else two propositions far apart in Arabic correspond to two successive ones in the Greek [M43, M48 to O37, O38/R41–R42]. This can be explained either by the transmission of more than one Greek version [on M36–42, see Textual Transmission, v. 1, p. lxi, and p. lxiii, n. 58], or by an attempt for a better match of subjects in the Arabic [e.g., M54 matching M55 better than in the Greek order]. With few exceptions [e.g., M48 matching M43 better], the arrangement of the propositions seems far from arbitrary, and at least in the case of the very next proposition, with a consequence that seems far from negligible.

---

212 M22 compared to O21, R21, and T22: M22, v. 1, pp, 66–69; O21, ed. Heiberg, p. 32, tr. Ver Eecke, p. 15; R21, ed. Heiberg, p. 178, tr. Ver Eecke, p. 68; T22, edn. Hyderabad, pp. 9–10.

Proposition 23: Appearance of Curvatures – Viewed from the Same Plane

Synopsis: The enunciation is that when a segment of a circle is placed below the eye and in the same plane with it, it is seen as a straight line. The illustration is that of a circle's segment (GB), with a center (K) placed in the same plane below the eye (A). The demonstration begins with the geometrical construction of ray[s] (AB, AD, AE, AZ, AH, AT, AG) [but not line BG, drawn only in the figure], and lines (KB, KD, KE, KZ, KH, KT, KG). Then follows the statement that such lines are seen [from point A] as having different magnitudes according to the size of visual angles [def. 4], the furthest [relative to the perpendicular line (KZ)] being seen as the greatest, such that segment (BG) is seen as line BG.

Textual Correspondence:[213] This proposition corresponds to the first of three treatments in the Greek [O22-1, with no correspondence in R, itself containing the same enunciation, but only the demonstration corresponding to O22-3].[214] In the enunciation [same in O and R], M differs from the Greek in having the subject, namely "a segment of the circle" (*qiṭ'atan min al-dā'ira*) [the equivalent term for an arc would have been *qaws*], placed "under the eye" (*waqa'a taḥt al-baṣar*) [equivalent to a convex shape facing the eye]. The demonstration, figure and much of the lettering follows O22-1 primarily,[215] with a reasoning based on appearance through angles [def. 4], rather than nonsimultaneous nature of appearance [prop. 1], as in O22-3 and R22. In the Arabic demonstration, the number of the visual rays and the segment's radii are seven [as in O], rather than three [in R], though the figure in M slightly differs from O insofar as line BG, representing the appearance of segment AB, is drawn in the figure of all but one variant (m2) [in contrast to O22-1]. Neither of the two additions [or interpolations][216] in the form of alternative treatments [O22-2, O22-3 (R22)] are reflected directly in M or in T.

Historical Context: The transmitted Arabic proposition, which contains the main [i.e., the first] Euclidean demonstration to the exclusion of the two alternative ones [i.e., the next two], has apparently not raised problems of consistency,

---

[213] M23 compared to O22, R22, and T23: M23, v. 1, pp. 70–71; O22, ed. Heiberg, pp. 32–36, tr. Ver Eecke, pp. 15–17; R22, ed. Heiberg, p. 180, tr. Ver Eecke, pp. 68–69; T23, edn. Hyderabad, p. 10.

[214] Of the three demonstrations appearing in O, the first two are not in R, the last two are not in any variants of M.

[215] It should be noted that the perpendicularity of the line mentioned in the second Euclidean trearment is reflected in the Arabic. See the diagram for this proposition, which relates only to the equivalent of a convex arc, as in O22-1.

[216] Theisen, *The Mediaeval Tradition*, p. 297, briefly discusses these in the Latin tradition, with reference to Heiberg's considerations of possible interpolations.

complexity, or veracity to its Arabic interpreters, as it has to its modern commentators.[217] The Arabic proposition follows the first Euclidean proof closely in showing that the straight-line effect of circular segments placed edgewise before the eye is produced by a certain order of appearance for points on their circumference. This order, following from the unequal appearance of the radial lines of the curvature, is such that their end points on the segment give the *impression* of a straight line. No comments are made about the impression amounting to the appearance of a curve *or* a shape other than a straight line.[218] There are no forms of reasoning in terms of nonsimultaneous appearance of different parts of the same object [prop. 1], as in R, and in the third part of O; Neither are there any signs of the impact of the varying lengths of the rays or of far distances on the stated impression of a straight line. This is the case with the second part of O [with no correspondence in R or M], in which the first treatment is illustrated by the appearance of ropes viewed from below and the second, by shadows of rings lying in the same plane with the source of light.

In this proposition, as in those discussed already in historical context [see Summary and Commentary: Propositions 1, 2, 3, 9, 10, 20, pp. 7–60], the nature of transmission directly affects the course of developments. The absence of the two additional treatments in the Arabic which, at least in the case of the first, teaches that distance from the eye *is* a factor in determining appearances, may be viewed as another instance [in addition to the third and eighth Euclidean propositions; see Summary and Commentary: Propositions 3 and 9] of a missed opportunity for the critical development of a direct relationship between distance and appearance.[219] What does get transmitted, on the other hand, provides an occasion for such notable developments as al-Kindī's challenge to the intromission hypotheses in all their manifestations.[220] The argument, resting on the refutation of the conception of an object's form as a total unit, is meaningful only in the light of the geometrical model of the first treatment, which is the only part of the proposition transmitted into Arabic. That treatment proceeds to demonstrate that an arc looks like a line to an eye in the same plane with it,

---

217 It is considered "erroneous" by Heiberg [*Litterargeschichtliche Studien über Euklid*, p. 136], "confusing" by Theisen [quoting Heiberg in *The Mediaeval Tradition*, pp. 296–297], and "inconsistent" in terms of the relevance of the proof to the enunciation by Brownson ["Euclid's Optics and Its Compatibility with Linear Perspective," p. 177].

218 See the full argument in Brownson, "Euclid's Optics and Its Compatibility with Linear Perspective," p. 177.

219 This relationship has been fully treated by Veltman, *Optics and Perspective: A Study in the Problems of Size and Distance.*

220 In Lindberg, *Theories of Vision*, reference is made to al-Kindī's use of this proposition for his refutation of the intromission theory: "If sight occurred through intromission of the forms of visible thing he [i.e., al-Kindī] argues, a circle situated edgewise before the eye would impress its form in the eye and consequently would be perceived in its full circularity;" see p. 23; for the context of the discussion and many relevant references, see also pp. 22–24, and pp. 223–224, n. 23–27.

because points farther away from its belly, being end points of those radial lines that "appear longer" due to the size of the visual angles subtending them, look closer [and farther away in the case of a concave arc], such that the total form gives the impression of a straight line. The inability of the intromission model to break out of the totality of a form [here, circular] to be subjected to geometrical analysis comes to form the core of al-Kindī's argument against intromission hypotheses.[221] Aḥmad Ibn 'Īsā's extensive treatment of the proposition, whose extant Arabic text reveals its close overlap with the Arabic proposition, occupies a different place in his work:[222] The treatment of the straight-line appearance of an arc [the term he uses is *qaws*], surpassing the Arabic proposition in having both the arc and its chord (*watar*) bisected by a perpendicular (*'amūd*) at the outset, has no better place in his version of the text than where it is situated for its specific purpose. Immediately following the case of the appearance of surfaces [from quadrilateral figures (*al-ashkāl al-murabba'a*) to higher-numbered sides and angles (*wa mā fawqahā min dhwāt al-aḍlā' wa al-zawāyā*); see Summary and Commentary: Proposition 10, p. 47], and just preceding the appearance of solid figures [from the spheres to the cylinders of the originally higher numbered propositions], his treatment of the whole set seems part of an attempt to provide geometrical reasoning for appearance of shapes *other than what they really are.*

The approach of Ibn al-Haytham is to treat the problem neither as an epistemological tool for discussing veracity of shape perception [as in Ibn 'Īsā] nor as a methodological tool for refuting another theory [as in al-Kindī].[223] His is a psychological approach in which a distinction is made not only between a moderate distance (*bu'd mu'tadil*) and an excessively great one (*bu'd mutafāwit*) but also between familar objects (*al-mubṣarāt al-ma'lūfa*) and unfamiliar ones (*al-mubṣarāt al-gharība*). Ibn al-Haytham's treatment of the problem in the language of "convexity" (*taḥdīb*) or "concavity" (*taq'īr*) and under perception rather than visual illusions seems to follow Ptolemy's extension of the problem to the case of far distances.[224]

---

221 Al-Kindī, *De aspectibus*, sec. 7, ed. Björnbo, pp. 9–10.

222 The treatment appears on fols. 93b–96a of ḥ1 [see Ḥ in List II, v. 1, p. xlvi]. That Ibn 'Īsā, in his extensive treatment of this proposition and inclusion of the appearance of the arc's inner part [*bāṭin al-qaws* as in M], has been working from the transmitted portion of the Euclidean proof is suggested by the rather superfluous use of the perpendicular line to the arc's middle point [in both cases].

223 For the details of Ibn al-Haytham's treatment, see Sabra, "Form in Ibn al-Haytham's Theory of Vision," pp. 121–123.

224 "When the eye looks at a very remote body in which there is a certain convexity (*taḥdīb*) or concavity (*taq'īr*) . . . sight will not perceive its convexity or concavity, but rather perceive it as plane and even; see *Kitāb al-Manāẓir*, bk. II, sec. 119, ed. Sabra, pp. 265–266, tr. Sabra, v. 1, p. 167. There is a corresponding passage in Ptolemy's *Optics*, where it is stated that when the proportion of size to distance is imperceptible at a very far distance, then the convex and concave shapes of both lines and surfaces will appear as

The language of convexity and concavity having been applied in the Arabic translation of a later Euclidean proposition [O57/M63] with which the present one [O22-1/M23] is often compared, the question of the absence of the later technical vocabulary in the earlier one is rather curious.[225] The higher numbered proposition, where such expressions are applied to figures with aligned middle and outer parts as appearing "sometimes outward and sometimes inward,"[226] has been in fact expressed in contradictory terms with the one here in which the arc,[227] being in the same plane with the eye, is supposed to be always seen as a straight line.

As for the Euclidean proposition on outward and inward appearances itself [M63/O57/R56], while it did deal with "relative appearances" as the main theme of the Euclidean text, it did not receive much attention in the Arabic tradition. Neither did the many other propositions on the appearance of geometric entities other than those already treated, namely the opening propositions on relative time, clarity, and limits of appearance [M1–3], propositions on the circular appearance of quadrilateral shapes from far away [M10] and linear appearance of curvatures viewed from the same plane [M23]. This is true not only for the solid figures of the next twelve propositions [M24–35], including the subset dealing with the popular subject of appearances involving illusions [M25, M31, M33],[228] but also in the case of the appearance of flat figures immediately following [M36–64].

---

straight and level; *Optica*, bk. II, sec. 69, ed. Lejeune, p. 48. There is also a corresponding passage in the alternative proof in the margins of some copies of Ṭūsī's proposition 23, which shares some of these features: "The reasoning (*'illa*) is not what is considered, rather it is that sight, at the same time as perceiving convexity (*taḥaddub*) and concavity (*taqa''ur*), perceives the difference between the longest and shortest paths (*sumūt*) and settles with one of them, and this [situation] does not occur unless the eye is in a moderate range (*al-ab'ād al-mu'tadila*), that is neither extremely close nor at a distance which would be considered exceptionally far (*mutafāwit*), and if it is in this far [distance], then it would not perceive the difference between its paths, and so the section (*qiṭ'a*) would be seen as a straight line."

225 An exception is the marginal proofs on Ṭūsī's *Taḥrīr*, where the terms *taḥaddub* (convexity) and *taqa''ur* (concavity) appear with reference to Proposition 23 [the corresponding term for the latter in M and T is "inside of the arc" (*bāṭin al-qaws*)].

226 These are sometimes alternatively termed as *muḥaddab* and *qā'ir* .

227 On the inconsistency between the conclusion of the earlier proposition [convexity and concavity appearing as linear] and the later one [perceptibility of these properties], see Ver Eecke, *Euclide, l'Optique et La Catoptrique*, p. 50, n. 1. Theisen, *The Mediaeval Tradition*, p. 298, n. 69–70, includes references to Witelo and Pecham who further treat the case of spheres appearing as planes at a distance.

228 Appearances involving illusions are here expressed by the verbal form *yuẓannu* for Greek δοκεῖ/*dokei* (thinking).

Propositions 24–35: Appearance of Solid Figures

Propositions 24–28: Appearance of Spheres

Proposition 24: The General Case of a Sphere Viewed from Any Angle

Synopsis: The enunciation is that when a sphere is viewed from any point whatsoever, what is seen of it is less than a hemisphere and enclosed by a circle. The illustration is that of a sphere centered at A viewed from a certain point (B) [representing the eye]. The demonstration begins with the geometrical construction of a line (BA) [from the position of the eye to the sphere's center], a plane (AB) through it, the great circle (BGD) [without specifying how], four lines (BG, BD, AD, AG) [the first two visual rays, without such specification], and a line (HT) parallel to line GD [drawn only in figure]. Then [through a few other constructions] follows, GB and BD are tangents to the sphere [at points G and D where the angles are right] and the only part of the sphere seen is a [circular] line (*khaṭṭ!*) [i.e., segment (*qiṭʻa*) or arc (*qaws*)]. The conclusion is that because the part of the sphere seen by the rays BG, BD [i.e., arc GD] [def. 3] is less than a semicircle [i.e., arc HGDT], what is seen of the sphere is less than a hemisphere.

Textual Correspondence:[229] In this proposition, where O and R differ in the enunciation, construction of the demonstration, and lettering of the figures, M has elements of both: In the lettering as well as in the geometrical construction [four lines as in O; R has five including the equivalent of GD], it is closer to O, though in the enunciation, the part seen of the sphere in M is stated as both "less than a hemisphere" [only in O] and "enclosed by a circle" [only in R]. The Arabic differs from the Greek: in the enunciation – it lacks any explicit reference to the presence of just one eye [in both O and R] – and in the geometrical construction – it lacks [in all copies] the case of line GD [drawn in both O nad R], as well as any reference to the common section (GDTH) [Ver Eecke, *Euclide: l'Optique et la Catoptrique*, p. 17, n. 2 and p. 69, n. 2, refers to relevant parts in Euclid's *Elements* and Theodosius's *Spherics*]. T follows M in the case of enunciation, but not geometric construction [it joins GD and refers to GHTD].

Proposition 25: The Case of a Sphere Viewed by an Approaching Eye

Synopsis: The enunciation is that when the eye approaches a sphere, what is seen of it is less [than before] but thought (*yuẓannu*) through vision (*fī al-ruʼya*) to be more. The illustration is that of a sphere with center A viewed from a certain point (B) [the eye]. The demonstration begins with a set of geometrical

229 M24 compared to O23, R23, and T24: M24, v. 1, pp. 72–75; O23, ed. Heiberg, pp. 36–38, tr. Ver Eecke, pp. 17–18; R23, ed. Heiberg, pp. 180–182, tr. Ver Eecke, pp. 69–70; T24, edn. Hyderabad, p. 10.

constructions [as in the previous propositions] and proceeds to the case where the eye is closer to the sphere [at point T]. The conclusion is that in the latter case what is seen of the sphere in reality (*bi al-ḥaqīqa*) [i.e., KL] is less than GD [the case where the eye was at B], but it is thought through vision to be more, because KL is viewed through angle KTL, which is larger than GBD [def. 4-1].

Textual Correspondence:[230] In this proposition, where O and R differ in the demonstration steps as well as in the figure and its lettering, M is closer to O in lettering, in the geometrical construction [five lines including GD as in O; R has four, followed by the equivalent of GD] and in the demonstration, through the presence of a causal statement that the visual angle at the closer distance (KTL) is greater than that at the farther distance (GBD) [as in O]. The Arabic differs from the Greek primarily in its linguistic emphasis on the sense of illusion [besides *yuẓannu* for the Greek δοκεῖ/*dokei*, the contrast between "seen in reality" (*yurā bi al-ḥaqīqa*) [not in the Greek] and "thought through vision" (*yuẓannu fī al-ru'ya*)]. T does not contain the latter expression here, and its causal premise is not the last statement in all of its copies.

Proposition 26–28: Three Cases of Spheres Viewed by Both Eyes

Proposition 26: When the Distance Between the Eyes Is Equal to the Circle's Diameter

Synopsis: The enunciation is that when a sphere is seen by both eyes such that the distance between the two eyes is equal to the sphere's diameter, then what is seen of the sphere is its entire half. The illustration is that of a sphere with center A, a circle on it (GB) [points G and B not specified], and a diameter (GB). The demonstration begins with the geometrical construction of two lines, one from B and the other from G to two points (E, D) [representing the two eyes], another line (DE) [representing the distance between the eyes], and a line (AZ) from the center (A) parallel to lines EG and DB. Then [through further constructions: parallelograms BZ and GZ revolving about AZ so that B and G interchange their position] follows the statement that a full semicircle [BG] is seen. The conclusion is that because the part seen of the sphere by the two eyes [through visual rays DG, EB] [def. 3] is a full semicircle, then what is seen of the sphere [BG] is a full hemisphere.

---

230 M25 compared to O24, R24 and T25: M25, v. 1 pp. 76–79; O24, ed. Heiberg, pp. 38–40, tr. Ver Eecke, p. 18; R24, ed. Heiberg, pp. 182–184, tr. Ver Eecke, p. 70; T25, edn. Hyderabad, p. 11.

Textual Correspondence:[231] In this proposition, where O and R differ in the length of the demonstration and the lettering of the figure, M is closer to O in lettering as well as in the illustration, where reference is made to the sphere, its center, the circle in it, and its diameter [all in O; only the sphere and its diameter in R]. T's shorter demonstration deviates from M: it contains an alternative terminology: *'ayn* (eye: organ) for *baṣar* (eye: sight) and *al-mar'ī* (visible) for the part seen of the sphere. This is followed by an alternative treatment, with an additional figure [starting with "I say, that is not correct and right (*ṣaḥīḥ wa ṣawāb*)"] that demonstrates the visible surface to be smaller.

Proposition 27: When the Distance Between the Eyes Is Greater than the Circle's Diameter

Synopsis: The enunciation is that when the distance between the two eyes is greater than the sphere's diameter, then what is seen of the sphere is more than a hemisphere. The illustration is that of a sphere with center A, a circle in it (EDH), and two points (B, G) representing the two eyes. The demonstration begins with the geometrical construction of the distance between the two eyes (BG), and two visual rays (BE, GD) extended to meet at a point (Z). Then follows, because the two lines ZE, ZD fall on the circle [i.e., are tangent to it] from a point outside it, segment ED is less than a semicircle, while segment EHD is more than a semicircle. The conclusion is that because the part seen of the sphere by the two eyes B and G [through visual rays BE, GD] [def. 3] is more than a semicircle, while from point Z segment ED [less than a semicircle] would have been seen, then what is seen of it [EDH] is more than a hemisphere.

Textual Correspondence:[232] In this proposition, where O and R differ in phraseology, length of the demonstration, the figure and its lettering, M has elements from both: in the lettering and in the demonstration, where there is no reference to tangents [in R], it is closer to O. But in the illustration, where the circle is marked with three points rather than four [in O], it is closer to R. The Arabic differs from both Greek versions in the figure and in parts of the text [the statement in M27(18) that "from the intersection point, Z, the smaller segment ED would have been seen"]. T does not include this statement, but it similarly lacks the term *mumās* (tangent) in this proposition, even though it uses it in the previous alternative demonstration.

---

[231] M26 compared to O25, R25 and T26: M26, v. 1 pp. 80–83; O25, ed. Heiberg, pp. 40–42, tr. Ver Eecke, p. 19; R25, ed. Heiberg, pp. 184–186; tr. Ver Eecke, p. 71; T26, edn. Hyderabad, p. 11.

[232] M27 compared to O26, R26 and T27: M27, v. 1, pp. 84–87; O26, ed. Heiberg, pp. 42–44, tr. Ver Eecke, pp. 19–20; R26, ed. Heiberg, p. 186, tr. Ver Eecke, p. 71; T27, edn. Hyderabad, pp. 11–12.

Proposition 28: When the Distance Between the Eyes Is Smaller than the Circle's Diameter

Synopsis: The enunciation is that when the distance between the two eyes is less than the sphere's diameter, then what is seen of the sphere is less than a hemisphere. The illustration is that of a sphere with center A, a circle in it (BG), and the distance (DE) between the two eyes (D, E). The demonstration begins with the geometrical construction of two lines (DB, EG), specified both as "tangents" (*mumās*) to the circle and as visual rays (*shu'ā'*) extended on a straight line to meet at a point (Z). Then follows, because the two lines ZG, ZB fall on the circle [i.e., tangent to it] from a single point [i.e., "from a point outside"], segment BG is less than a semicircle. The conclusion is that because the part seen of the sphere as contained by the two visual rays DB, EG [def. 3] is less than a semicircle, then what is seen of it [BG] is less than a hemisphere.

Textual Correspondence:[233] In this proposition, where O and R differ in the steps of the demonstration, the figure, and its lettering, M is closer to O in the figure and lettering and in the geometrical construction, where the circle is specified with two letters [as in O, in contrast to three in R] and the two tangents [in both O and R] are specified as visual rays [as in O, and in contrast to R]. The Arabic differs from the Greek where the line joining the tangent points (BG) is not drawn in either text or figure [in contrast to both O and R], and the visible surface of the sphere (GB) is marked with two letters [in contrast to three letters in both O and R]. T's abridged demonstration ends with the statement that there is a defect (*khilal*) in the two previous propositions resulting from the reasoning (*qīyās*) of the proposition preceding both.

Propositions 29–31: Appearance of Cylinders and Pillars

Proposition 29: The General Case of a Cylinder Viewed from Any Angle

Synopsis: The enunciation is that when a cylinder is viewed from any point whatsoever, what is seen of it is less than its half. The illustration is that of a cylinder, its base centered at point A. The demonstration begins with the geometrical construction of a circle (GB) around point A, a point (D) in the same plane as the cylinder's base, a line (DA) from the eye's position to the circle's center, two visual rays (DB, DG) tangent to the circle, and from their contact points (B, G) two lines on the surface of the cylinder (BE, GZ) at right angles to the base, and (through those lines and visual rays) two planes [DBBE, GZDG], likewise tangent to the cylinder. The conclusion is that the part seen of the cylinder contained by the two visual rays DB, DG [def. 3] is less than a semicircle, and the same amount is seen of the cylinder [i.e., less than its half].

233 M28 compared to O27, R27 and T28: M28, v. 1, pp. 88–91; O27, ed. Heiberg, pp. 44–46, tr. Ver Eecke, p. 20; R27, ed. Heiberg, pp. 186–188, tr. Ver Eecke, p. 72; T28, edn. Hyderabad, p. 12.

Textual Correspondence:[234] In this proposition, where O and R differ in figure, lettering, and demonstration steps, M [which corresponds to the first demonstration in O and only to the enunciation in R], is closer to O: in lettering, figure [R has a figure closer to O28-2], and demonstration. The Arabic differs from the Greek in the enunciation, where it does not specify that the cylinder be viewed by one eye [as in both O and R] and in the conclusion, where it does not not add the case of both eyes or refer to the case of the sphere [only at the end of O28-1]. T is a much shorter version of M.

Proposition 30: The Case of Right Pillars and Cylinders

Synopsis: The proposition begins [without an enunciation] with the geometrical construction of a circle centered at A and from this point a line (AZ) to a fixed position for the eye (Z), a line perpendicular to this as the circle's diameter (GD) and around it (AZ) a circle (ABZE), four lines (AB, BZ, ZE, EA), two of which (BZ, ZE) are tangents to the circle [because of the right angles at B, E]. Then [in the manner of the previous proposition] follows the statement that segment BE contained by the two visual rays BZ, ZE is less than a semicircle. Next comes the geometrical construction of the sides of right pillars or cylinders [for which the proposition is stated to have been set up] and the assertion that the region (*nāḥiya*) BGDE is not seen [def. 3]. The conclusion is that because the part seen of right pillars and cylinders [segment BE] is less than a semicircle [GBED], then what is seen [BE] is less than their half.

Textual Correspondence:[235] This proposition represents structural differences with respect to both O and R. The two different treatments in O28 are presented as two independent propositions, where M29 corresponds to the first treatment and only R's enunciation, and M30 to the second and only R's demonstration. The Arabic follows the Greek closely, but where it states that "we have set up this proposition for right pillars *(al-'umud)* and cylinders," the Greek has "cones (*konôs*) and cylinders." T has "cones and cylinders" (*al-makhrūṭāt wa al-asāṭīn*).

Proposition 31: The Case of a Cylinder Viewed by an Approaching Eye

Synopsis: The enunciation is that when the eye approaches a cylinder, what is seen of it is less [than when the eye was far] in reality (*bi al-ḥaqīqa*) but thought (*yuẓannu*) [through vision] to be greater. The illustration is that of a cylinder, its

---

234 M29 compared to O28-1, R28 and T29: M29, v. 1, pp. 92–95; O28-1, ed. Heiberg, pp. 46–48, tr. Ver Eecke, pp. 20–21; R28, ed. Heiberg, pp. 188–190, tr. Ver Eecke, p. 72; T29, edn. Hyderabad, p. 12.

235 M30 compared to O28-2 and T30: M30, v. 1 pp. 96–99; O28-2, ed. Heiberg, pp. 48–50, tr. Ver Eecke, pp. 21–22; R28, Heiberg, pp. 188–190, tr. Ver Eecke, p. 72; T30, edn. Hyderabad, p. 12.

base circle BG with center A, and a point (E) [the position of the eye]. The demonstration begins with the geometrical construction of a line connecting the two points (EA), two visual rays (EB, EG), and two lines (BZ, GH) perpendicular to the cylinder's [plane] surface. Then follows the statement that according to the previous demonstration, [surface] BZGH [viewed from point E] is less than half of the cylinder. This is all repeated for the case where the eye is moved closer to the cylinder at point T [geometrical construction of visual rays TL and TK and perpendicular lines KM, LN, and the assertion that segment MKLN is contained by the two rays TL, TK]. The conclusion is that what is seen of the cylinder when the eye gets closer (KLMN) is [really] less than before (BZGH), but by eyesight (*fī mir'āt al-'ayn*) it is thought [wrongly] to be greater, because the angle through which KLMN is viewed (T) is larger than that through which BZGH is viewed (E) [def. 4-1].

Textual Correspondence:[236] In this proposition, where O and R differ in phraseology and length of the demonstration, in the figure and in the lettering [R has additional circles and lines but a similar demonstration], M is closer to O, in all cases including in the problematic expression "at right angles with the cylinder's surface" (*ma'a basīṭ al-usṭuwāna*) in all copies [see also, Ver Eecke, *Euclide: l'Optique et la Catoptrique*, p. 22, n. 2]. T's much shorter version has "perpendiculars to the cylinder's plane" (*'amūd fī saṭḥ al-usṭuwāna*).

Propositios 32–35: Appearance of Cones

Proposition 32: The General Case of a Circular Cone Viewed from Any Angle

Synopsis: The enunciation is that when a circular-based cone is viewed [from any point], then what is seen of it is less than its half. The illustration is that of a cone with a circular base (BG), apex A, a point (D) [representing the position of the eye], and two visual rays DB, DG as tangents to the cone. The demonstration begins with the statement that segment BG is less than a semicircle according to the preceding demonstration. Then follows the geometrical construction of the sides of the cone (AB, AG) and the claim that the part of the cone seen as contained by these and BG [def. 3] is less than half the cone. The conclusion is that because what is seen of the circular cone [as contained by DB, DG and segment BG] is less than a semicircle, what is seen of it is less than half.

---

236 M31 compared to O29, R29, and T3: M31, v. 1, pp. 100–103; O29, ed. Heiberg, pp. 50–52, tr. Ver Eecke, p. 22; R29, ed. Heiberg, p. 190, tr. Ver Eecke, p. 73; T31, edn. Hyderabad, pp. 12–13.

Textual Correspondence:[237] In this proposition, where O and R differ in enunciation, phraseology, and length of the demonstration, in the figure, and in the lettering [again R has additional circles and lines but a similar demonstration], M is closer to O, though in its enunciation it lacks reference to the right angle at the cone's apex [R also lacks this]. The Arabic differs from the Greek in not specifying the cone to be viewed by one eye [as in both O and R]. T's abridged demonstration has the standard expression *makhrūṭ al-mustadīr* for the circular cone, instead of "a cone with a circular base" in its enunciation.

Proposition 33: The Case of a Circular Cone Viewed by an Approaching Eye

Synopsis: The enunciation is that when the eye approaches a circular cone whose base is in the same plane with it, then that which is seen of the cone is smaller [when closer] but it is thought *(yuẓannu)* to be larger. The illustration is that of a circular cone, whose base is circle AB with center K and its apex is point G. The initial position of the eye is point D. The demonstration begins with the geometrical construction of a line connecting the center of the cone's base to the eye (DK), ray[s] DA, DB, and the two sides of the cone (AG, BG) and the statement that from point D, what is seen of the cone by visual rays DA, DB is part ABG, which is less than half [with no reference to previous demonstrations]. This is repeated for the eye at the closer point E [with geometrical construction of visual rays EZ, EH, and the sides of the cone ZG, HZ], and the statement that [this time] what is seen of the cone is part ZGH, which is less than ABG. The conclusion is that what is seen of the cone when the eye gets closer (ZGH) is less than before (ABG) [i.e., in reality], but it is thought through vision *(yuẓannu fī ru'ya)* to be greater, because the angle through which ZGH is viewed (ZEH) is larger than the angle through which ABG is viewed (ADB) [def. 4-1].

Textual Correspondence:[238] In this proposition, where O and R differ in enunciation, demonstration, figure, and lettering, M is closer to O in most cases. The enunciation states the eye to be in the same plane as the cone's base in all copies [as in O, and in contrast to R which lacks reference to "the cone's base"]; the illustration of M includes the reference to the cone's apex [as in O, and in contrast to R]. The rest of the demonstration, the figure, and the lettering also follow O more closely [R has additional circles and lines], except at the very end where M lacks the reference to the previous cases of the sphere and cycinder [only in O; R similarly lacks this part]. The Arabic differs from the Greek in the figure, where the center of the cone's base (K) and its apex (G) do not coincide

---

237 M32 compared to O30, R30, T32: M32, v. 1 pp. 104–105; O30, ed. Heiberg, p. 52, tr. Ver Eecke, p. 23; R30, ed. Heiberg, p. 192, Ver Eecke, pp. 73–74; T32, edn. Hyderabad, p. 13.

238 M33 compared to O31, R31, and T3: M33, v. 1, pp. 106–107; O31, ed. Heiberg, p. 54, tr. Ver Eecke, pp. 23–24; R31, ed. Heiberg, pp. 192–194, tr. Ver Eecke, p. 74; T33, edn. Hyderabad, p. 13.

and R], and in the conclusion, where reference to the visual angle appears in the form of three letters [in O it is by the two visual rays, and in R by one letter]. T is a condensed version of M.

Proposition 34: The Case of a Circular Cone Viewed by an Eye Positioned on a Specific Line

Synopsis: The enunciation is that in the case of a circular cone, when the eye is anywhere on the intersection line between two specific planes [those passing through two visual rays tangent to its base and two generators reaching the tangent points], then what is seen of the cone is always the same. The illustration is that of a circular cone, with base circle BG and apex point A. The eye's initial position is point D. The demonstration begins with the geometrical construction of two rays (DZ, DG), and from the points of contact with the cone's base (Z, G) two lines to its apex (ZA, GA), and from the intersection of two planes passing through these (GDA, AZD) the common line (AED), along which the eye is to be moved. Then [after the construction of ET, EH] follows, because the eye's movement along this particular line always keeps its plane parallel to the one in the original position, the visual angle remains the same in each case. The conclusion is that because what is seen through equal angles is seen as equal [def. 4-4], then the appearance of the cone [through these two equal angles] remains the same.

Textual Correspondence:[239] In this proposition, where O and R differ in enunciation, demonstration, figure, and lettering, M is closer to O in all cases, except for the opening line of the enunciation. The figure is similar in structure and lettering [with minor differences, such as BG drawn as the diameter in all copies of M]. The demonstration is also rather close, including the crucial parts [in O, but not R] about the absurdity of "bending" (*yan'aṭifān*) visual rays and the final statement of the conclusion. T is an abridged version of M without any of these features but with the added comment that the conclusion can be extended to other cases [i.e., other positions of the eye along the same line].

Proposition 35: The Case of a Circular Cone Viewed by an Eye Positioned on a Line Parallel to a Generator of the Cone

Synopsis: The enunciation is that when the eye is [on a line] equidistant from a cone (*mutabā'id 'an al-makhrūṭ bi ustuwā'*) [see other variants], more of the cone is seen when the eye is at a higher position [on that line] than when it is lower. The illustration is that of a circular cone, with base circle BG, apex point A, and sides AB, AG. The demonstration begins with the geometrical construction of a line connecting the end points (B, G), and extending line BG

[239] M34 compared to O32, R32, T34: M34, v. 1 pp. 108–113; O32, ed. Heiberg, pp. 56–58, tr. Ver Eecke, pp. 24–25; R32, ed. Heiberg, pp. 194–196, tr. Ver Eecke, p. 75; T34, edn. Hyderabad, pp. 13–14.

[not specified as the circle's diameter] to point H, from point T a line (TK) parallel to AB, and extending lines AT and AK to points H and L, respectively. Then follows the statement that what is seen of the cone is different from points L and H but the same from points T and H or K and L according to the previous demonstration [prop. 34]. The conclusion is that because what is seen of the cone when viewed from point H [same as from point T] is greater than that which is viewed from point L [same as point K], then what is seen of the cone from the higher point (T) is greater than what is seen of it from the lower point (K).

Textual Correspondence:[240] In this proposition, where O and R differ considerably in enunciation, demonstration, figure, and lettering, M is closer to O in most cases. In the case of the enunciation, O states that from a higher point more of a cone is seen [though in appearance this is less] than when viewed from a lower point, while R's much shorter statement specifies the eye to be equidistant from the cone. In the enunciation of M, the eye is stated in m1, m3, and m5 as being *mutabāʿid ʿan al-makhrūṭ bi ustuwāʾ* [on a line] equidistant from the cone [as in R] but in m2 and m4, the corresponding expression is *mutabāʿid ʿan al-makhrūṭ al-mustawan* (far from the right cone) [a sense not in either O or R]. This part of the enunciation, where the even-numbered manuscripts contain elements absent from both Greek versions and the odd-numbered manuscripts present a close correspondence to the Greek version R, is of some significance for providing another striking case against any assumption of a consistent correlation between the two distinct variants of Greek and Arabic [see also Summary and Commentary: Definition 1, p. 7, Definition 2, p. 18]. The rest of the enunciation [the same in all variants of M] is closer to O, except that it lacks the part "though in appearance this is less." The demonstration also matches O more closely, including the reference to the previous demonstration [not in R, though added in Ver Eecke, *Euclide: l'Optique et la Catoptrique*, p. 76, n. 2]. In all but one of the variants of M (m5), the figure appears with BG as the diameter [in contrast to both O and R]. T's figure also has this line through the center, though its text has features not found in any of the variants of M: its enunciation has a slightly different construction for the eye's equidistant condition (*ʿalā al-buʿd mutasāwī*); its demonstration lacks reference to the previous demonstration, and its conclusion includes the distinction between the actual (*bi al-ḥaqīqa*) and apparent (*bi al-naẓar*) size [that part of the enunciation in O lacking in all variants of M].

---

240 M35 compared to O33, R33, and T35: M35, v. 1, pp. 114–117; O33, ed. Heiberg, pp. 58–60, tr. Ver Eecke, p. 25; R33, ed. Heiberg, pp. 196–198, tr. Ver Eecke, p. 76; T35, edn. Hyderabad, p. 14.

Propositions 36–42: Appearance of Circles

The following set of propositions on the appearance of circles can be divided into three groups: the appearance of circles as circles (M36–37), circles not as circles (M38–41), and circles sometimes as circles and sometimes not (M42) [these correspond to O34–36, R34–37, and Pappus's *Collection*, VI (C50, C42, C44, C51), ed. Hultsch, pp. 568–586; tr. Ver Eecke, v. 2, pp. 436–448, see Knorr, "Pseudo-Euclidean Reflections in Ancient Optics," pp. 35–39, and "When Circles Don't Look Like Circles," pp. 292–294]. The demonstrations are based on the equal and nonequal appearance of two of the circle's diameters.

Propositions 36–37: Cases for the Appearance of Circles as Circles

Proposition 36: The Case Where the Eye Is Positioned on a Line Passing through the Circle's Center at Right Angles to the Circle's Plane

Synopsis: The enunciation is that from any point on a line perpendicular to a circle's plane, drawn from its center upwards, the diameters of that circle are seen as equal [so the figure is seen as a circle]. The illustration is that of a circle with center A from which line AB is drawn perpendicuarly upward. The position of the eye is point B. The demonstration begins with the construction of the circle's diameters (GD, EZ) meeting at point A [the center] and the joining lines [i.e., rays] BG, BE, BD, BZ. Then [through equality of triangles] follows the statement that the angles (DBG, EBZ) subtending the two diameters are equal. The conclusion is that because objects seen under equal angles are seen as equal [def. 4-4], then GD and EZ [the diameters] are equal [i.e., seen as equal].

Textual Correspondence:[241] In this proposition, where O and R differ slightly in enunciation, figure, lettering, and demonstration, M is closer to O in the enunciation, where the line is stated to be "drawn upward" [as in O], in the illustration, where the diameters are "drawn in the circle's plane" [as in O; in R, C and T, they are just the circle's diameters], and in the figure's lettering, where M follows O [R and C have different lettering; in R angles are further marked with letters inside]. The figure itself needs some explanation: the "upright" line is drawn sideways in m1, m3, and m5 [as in O] but straight up in m2 [as in R and C] [m4 has an incomplete figure]. The Arabic demonstration differs slightly from the Greek: toward the end, where the angles are represented by three letters and not with reference to the sides enclosing them [in O, R, and C] and in the concluding statement, where the two diameters are proved to be equal and not *seen as* equal [in O, R, and C]. T's abridged version has it right. The case introduced at the very end of O, where the line is not perpendicular to the

[241] M36 compared to O34-1, R34, C50-1, and T36: M36, v. 1, pp. 118–121; O34-1, ed. Heiberg, pp. 60–62, Ver Eecke, p. 26; R34, ed. Heiberg, pp. 198–200, tr. Ver Eecke, p. 76; C50-1, ed. Hultsch, pp. 580–581, tr. Ver Eecke, pp. 445–446; T36, edn. Hyderabad, pp. 14–15.

circle's plane but equal in magnitude to half of its diameter, does not appear as part of M36 [or T36], as it does in O and C, but as an independent proposition (M37, T37) [as in R].

Proposition 37: The Case Where the Eye Is Positioned on a Line Passing through the Circle's Center Not at Right Angles to the Circle's Plane but Equal to the Circle's Radius

Synopsis: The enunciation is that when the line drawn [upward] from the circle's center is not at right angles to the circle's plane but equal to half its diameter [i.e., its radius], its diameters are [still] seen as equal [and so the figure is seen as a circle]. The illustration is that of a circle (ABGD) and a line (ZE) equal to its radius. The demonstration begins with the construction of two diameters at right angles to each other (AB, GD) and the ray[s] ZA, ZB, ZC, ZD [from point Z, not specified as the eye]. Then follows, because BE, EA, and EZ are equal, a semicircle drawn on line AB [diameter of the circle] will pass through point Z, making angles AZB and GZD right. The conclusion is that because objects seen under equal angles are seen as equal [def. 4-4], then AB and AD [seen under angles AZB and GZD, both being right angles and so equal] are seen as equal. T's abridged demonstration contains the expression "nothing but equal" (*lā ghayr mutasāwiyatan*) and a statement that this condition holds for all other points on line AB.

Textual Correspondence:[242] This proposition represents structural differences from the Greek, insofar as it corresponds to the second part of the previous proposition in O [and C] and only to the first part of what is introduced as a separate proposition in R [R35-1]. Where O and R differ in their overlapping parts [phraseology, steps of demonstartion, and lettering of figure], M is closer to O in the lettering and in the demonstration, where M holds radius BE equal to EA and to line EZ before establishing the equality of the three lines [in O, BE and EA each equal EZ; in R, all three are equal], and where it contains the part of the geometrical construction of a semicircle on the diameter AB [only in O, and lacking in R]. But M is closer to R in the illustration, where Z is not specified as the eye's position [as in R, and in contrast to O], though the geometrical construction of rays ZA, ZB, ZA, and ZD presupposes this. The Arabic differs from the Greek in the enunciation of M, where the expression for the size of the erected line as "equal to half the diameter" [i.e., radius] is different from the Greek [same in O and R, C has "circle's radius"], and in the geometrical construction, where the diameters are specified as being "at right angles" [in contrast to O, R, and C].

---

[242] M37 compared to O34-2, R35, C50-2, and T37: M37, v. 1, pp. 122–125; O34-2, ed. Heiberg, pp. 62–64, tr. Ver Eecke, pp. 26–27; R35-1, ed. Heiberg, pp. 200–202, tr. Ver Eecke, p. 77; C50-2, ed. Hultsch, pp. 580–581, tr. Ver Eecke, pp. 445–446; T37, edn. Hyderabad, p. 15.

Note: A part missing in the Arabic versions is the case where the eye is positioned on a line passing through the circle's center, neither at right angles to the circle's plane nor equal to its radius, but inclined to its diameter, where the raised angles are equal [O 34-3, R35-2, not in C or V] and where the diameters are still seen as equal [so the figure is seen as a circle]. There is no correspondence to such a case in any of the variants of M or T [in only one variant of M (m4), the previous Arabic proposition is repeated verbatim in its place still as M37, but so is the fourth proposition in the same variant (m4)]. According to the expected order of propositions, the next Arabic proposition (M38) should have been the one to correspond to what is yet another case for the equal appearance of diameters [i.e., appearance of circles as circles], as in the third part of Proposition 34 in O and the second part of Proposition 35 in R. But such a part, or proposition, is absent from the Arabic text altogether [as it is in C], and the next Arabic proposition (M38) is the first of the few propositions on the unequal appearance of diameters [i.e., appearance of circles not as circles].

Propositions 38–41: Cases for the Appearance of Circles Not as Circles

Proposition 38: The Case Where the Eye Is Positioned on a Line Passing through the Circle's Center Neither at Right Angles to the Circle's Plane Nor Equal to its Radius Nor Inclined to Its Diameter Where the Raised Angles Are Equal

Synopsis: The enunciation is that when the line drawn from the eye to the circle's center is not at right angles to its plane, and is neither equal to half its diameter [i.e., its radius], nor inclined to one of them where the raised angles would be equal, its diameters are [in contrast to the previous cases] not seen as equal [so the figure is no longer seen as a circle]. The illustration is that of a circle (ABGD). The demonstration begins with the geometrical construction of two diameters not at right angles to each other (AB, GD) [though still equal, as the circle's diameters] and from point E [not specified as the eye] a line (EZ) not perpendicular to the circle's plane or equal to the circle's radius, such that neither angles GEZ, ZEB nor AEZ, DEZ are equal, and then joining lines [i.e., visual rays] ZG, ZD, ZA, ZB. Then follows [after a set of premises on unequal angles], angle AZB is either greater than GZD or smaller [i.e., unequal to it]. The conclusion is that [because magnitudes seen under unequal angles are seen as unequal] [reverse of def. 4-4], the [equal] diameters [AB, GD] seen under [unequal] angles AZB, GZD are seen as unequal.

Textual Correspondence:[243] In this proposition, which corresponds to the first part of a long proposition [O35, R36; in the case of the much shorter C42, only

---

[243] M38 compared to O35-1, R36-1, C42, and T38: M38, v. 1, pp. 126–129; O35-1, ed. Heiberg, pp. 64–66, tr. Ver Eecke, pp. 27-28; R36-1, ed. Heiberg, pp. 202–204, tr. Ver Eecke, pp. 78–79; C42, ed. Hultsch, pp. 568–569, tr. Ver Eecke, pp. 436–437; T38, edn. Hyderabad, p. 15.

in enunciation] where O and R differ [enunciation, phraseology and lettering], M is closer to O in the illustration, where the circle is marked by four letters, not just by its center [in R], and in the geometrical construction, where there are two diameters intersecting, and not one diameter intersecting with another [in R]. But M clearly reflects elements from R in the enunciation, where there is a statement after the raised equal angles "the greater to the greater and the smaller to the smaller" ["greater or smaller than the circle's radius" in R, itself also apparently misplaced in the enunciation; C42 has this part but lacks the part preceding it]. The Arabic, which devotes the whole proposition to this case, also differs from the Greek in the illustration, where the two diameters are specified as "*not* intersecting at right angles" [M38(10)]. In the text of both O and R, and in the figure of O, the two diameters *are* at right angles, though in R's figure they are not drawn as such. The Arabic also lacks any reference to the propositions following this one [in both O and R, though not in the same place]. T's abridged demonstration, with notably different variants, does not always specify the figure as a circle [ed. *RIMA*, p. 274] or that the diameters are at right angles. Yet it states that they do not have equal inclination (*maylan mutasāwīyan*), using standard terms for acute and obtuse angles (*ḥādda, munfarija*). T also does contain a statement at the end that the manner of this (*al-ḥāl fī dhālika*) [i.e., the diameters not appearing as equal] will be demonstrated in the next proposition and the ones following it.

Proposition 39: The Case Where the Eye Is Positioned on a Line at Right Angles to the Circle's Plane but Not at Its Center Rather Outside It

Synopsis: This proposition [not marked as a lemma] begins, without an enunciation or illustration, with the construction of a circle, centered at A, a point (B) as the position of the eye, a perpendicular (BG) [to the circle] not falling on center A but rather outside it, two lines AG, and also (*ayḍan*) AB joining A to G and B, respectively, and three other lines – a diameter (DE) [without such specification], a perpendicular to it from G (GZ), and a line (BZ) joining B to Z. There follow a set of inequalities between lines and angles and the assertion that angle GAB is smaller than angle ZAB [as *remaining* angles in their respective triangles]. The conclusion is that because BAG is smaller than BAZ [following a set of premises involving the pythagorean relation of right triangles], it is smaller than any angle formed by the lines passing through the center A and at right angles with line AB.

Textual Correspondence:[244] This proposition represents yet more structural differences from the Greek. It appears as an independent proposition, in contrast

[244] M39 compared to O35-2, R36-2, C44, T39: M39, v. 1, pp. 130–133; O35-2, ed. Heiberg, pp. 66–70, tr. Ver Eecke, pp. 28–30; R36-2, ed. Heiberg, pp. 204–206, tr. Ver Eecke, pp. 79–80; C44: ed. Hultsch, pp. 570–572, tr. Ver Eecke, pp. 438–439; T39, edn. Hyderabad, p. 15.

to O and R, where this is a lemma attached to the first case of the unequal appearance of the circle's diameter [O35-2, R36-2, also corresponding to C44]. And where O and R differ [phraseology, length of demonstration, figure, and lettering], M is now closer to R in the figure and its lettering [O involves many more lines, C has no circle], in the geometrical construction, where GA is drawn and then "also" (*ayḍan*) AB [only in R] and in the demonstration, where side (*ḍil'*) AG is greater than AZ; angle ZAB is further specified as the remaining (*bāqiya*) angle. But M clearly reflects elements from O in the enunciation, where the subject is the *smallest angle* [and not *a line making the smallest angle* as in R] and in the demonstration, where it contains the qualifying statement that the greater side subtends the greater angle [only in O]. The Arabic differs from O and R in concluding with the Pythagorean premises [as in C42, lacking in O, R, and T] before the conclusion that angle BAG is smaller than BAZ.

Proposition 40: The Case Where the Eye Is Positioned on a Line at Right Angles to the Circle's Plane, Longer Than Its Radius, Such That It Makes a Right Angle with One of the Two Diameters and an Arbitrary Angle with the Other Diameter, and at Right Angles with the First One

Synopsis: This proposition begins [without an enunciation or illustration] with the geometrical construction of a circle (ABGD), two diameters at right angles (AB, GD), a point (E) for the eye's position, from it two lines – a line (EZ) longer than the circle's radius, such that it makes a right angle with one of the diameters (GD) and an arbitrary angle with the other (AB) [the claim that "diameters AB and GD appear unequal" is lacking here] and a perpendicular line (EK) to line AB in the circle's plane – and the circle's diameter (HT). Then follow a few other constructions [Fig. 40.2]: a line (LM) equal to the circle's radius, and from its midpoint (N), a perpendicular line (NS) equal to EZ, two other lines (NO, NF) equal to EZ such that angles LNO and LNF are equal to angles HZE and BZE, and three circle segments (LSM, LOM, LFM) on diameter LM, passing through the end points S, O, and F. There follow a set of equalities and inequalities and the assertion that angle S [equal to GED] is greater than angle O [equal to HET]. The conclusion is that because diameter GD [under angle GED] is seen as greater than HT [under angle HET] [def. 4-1], HT is seen as greater than AB [and so they are seen as unequal], and GD is seen as the greatest and AB the smallest of all the lines passing through point Z and line EZ.

Textual Correspondence:[245] This proposition also represents structural differences from the Greek. As a special case, where the line is greater than the

[245] M40 compared to O35-3, R36-4, C51-1, and T40: M40, v. 1, pp. 134–139; O35-3, ed. Heiberg, pp. 70–76, tr. Ver Eecke, pp. 30–33; R36-4, ed. Heiberg, pp. 208–212, tr. Ver Eecke, pp. 81–82; C51-1, ed. Hultsch, pp. 582–584, tr. Ver Eecke, pp. 446–448; T40, Hyderabad, p. 16.

radius, it appears as an independent proposition, in contrast to O and R also C, where this is a part of another proposition [O35-2, R36-2, C51]. Where O and R differ [enunciation, illustration, phraseology, figure, lettering], M is closer to O in both figure and lettering [especially variants m1, m5, where EZ is drawn smaller than the radius, as in O], in the illustration, where M similarly lacks the part in R about exterior angles [an interpolation according to Ver Eecke, *Euclide: l'Optique et la Catoptrique*, p. 82, n. 1], and in the conclusion, where GD and AB are specified as the greatest and smallest of all the lines passing through point Z and line EZ. But M shares with R the length and phraseology of the demonstration. It also lacks parts in O, which are similarly lacking in R: in the illustration, the part "I say that the two diameters . . . are seen as unequal" [present in O, not present in R] is lacking. This part is not lacking in T.

Proposition 41: The Case Where the Eye Is Positioned on a Line through the Circle's Center, Shorter than the Circle's Radius

Synopsis: The enunciation is that when the line drawn from the eye through the circle's center is not greater than the radius but smaller, the opposite of what preceded occurs; namely the [previously] greater diameter would be seen as the smaller, and vice versa. The illustration is that of a circle (ABGD), on which are constructed two diameters at right angles (AB, GD), another, EZ, at an arbtrary angle, a point (E) for the eye's position, and from it a line (TH) shorter than the circle's radius. Then follow other constructions (Fig. 41. 2): a line (KL) equal to the circle's radius from its midpoint (M) a perpendicular line (MN) equal to TH, two lines (MO, MS) equal to HT, such that angles KMS, KMO are equal to angles ZHT, AHT, and then three circle segments (KNL, KOL, KSL) on diameter KL, passing through the end points S, N, and O. The demonstration ends [through a set of equalities and inequalities] with the conclusion that diameter AB is seen as greater than EZ [def. 4-1], and that AB is the greatest and EZ the smallest of all the lines passing through point Z and line EZ.

Textual Correspondence:[246] In this proposition, an independent treatment for the case where the line is less than the radius, where O and R differ, M is closer to O in the figures and their lettering [the figures in both R and C are considerably different]. And M again shares a feature with R besides the length and phraseology of the demonstration: In the opening line of the enunciation, the line is described as being not larger but rather smaller [in O, it is just smaller]. T's demonstration is a condensed version of M.

[246] M41 compared to O35-4, R36-5, C51-2, T41: M41, v. 1 pp. 140–145; O35-4, ed. Heiberg, pp. 76–80, tr. Ver Eecke, pp. 33–34; R36-5, Heiberg, pp. 212–214, tr. Ver Eecke, pp. 82–84; C51-2, ed. Hultsch, p. 584, tr. Ver Eecke, p. 448; T41, edn. Hyderabad, pp. 16–17.

Proposition 42: The Case of the Varying Appearance of Circles

Synopsis: The enunciation is that wheels of chariots are seen sometimes as circular and sometimes as distorted. The illustration is that of a circle (ABGD) whose plane is parallel to that of the eye. The demonstration begins with the construction of two diameters (AB, GD) at right angles at a point (E) [E should be the position of the eye on a parallel plane]. Then follows, when the line from the eye through the circle's center is either perpendicular to its plane or equal to the radius, the diameters are seen as equal [Propositions 36–37, no specific reference] and it is demonstrated (*tabayyana*) that the wheel is seen as circular, but if that line is neither perpendicular to the plane of the circle nor equal to the radius, then the diameters are seen as unequal [Propositions 38–40, no specific reference], so it is demonstrated that [in these cases] the wheel is seen as distorted.

Textual Correspondence:[247] In this proposition, where O and R differ slightly, M is closer to O: in the illustration, where the circle is specified with four letters [as in O; R does not refer to the circle], and where the eye is placed "on a plane parallel" to that of the circle [in O it is just "not in the same plane," R lacks any such reference], and in the figure, where the center is marked [m2 is closer to R in lacking this].

Propositions 43–56: Appearances Involving Loci

Propositions 43–47: Cases Where the Eye's Locus Is Fixed While the Viewed Object Is Being Displaced

This set of five propositions addresses three cases. The first proposition (M43) deals with the case where the object is located and is displaced in the same plane as the eye; the other four (M44–47) deal with the case where the object is located and is displaced in a plane different from that of the eye. M44 deals with the case where the object is located and displaced in a plane at right angles to that of the eye; M45–47 deal with cases where the object is located and displaced in a plane not at right angles to that of the eye. M45 deals with the case where the object's magnitude is greater than the circle's radius, M46 with the case where the object's magnitude is equal to the circle's radius, and M47 with the case where the object's magnitude is smaller than the circle's radius.

---

[247] M42 compared to O36, R37, and T42: M42, v. 1, pp. 146–147; O36, ed. Heiberg, p. 80, tr. Ver Eecke, p. 34; R37, ed. Heiberg, p. 214, tr. Ver Eecke, p. 84; T42, Hyderabad, p. 17.

Proposition 43: The Case Where the Eye Is Positioned on the Circumference of a Circle, Along Which the Viewed Object Is Displaced [Followed by the Opposite Case, Where the Viewed Object Is Fixed as a Chord of the Circle Along the Circumference of Which the Eye Is Displaced]

Synopsis: The enunciation is that there is a fixed place (*mawḍiʿ*) [i.e., locus] for the eye from which a displaced object is always seen with the same magnitude [i.e., to demonstrate that there is such a locus for the eye]. The illustration is that of a point (A), representing the [initial] position of the eye, and a magnitude (BG), the viewed object. The demonstration begins with the geometrical construction of ray[s] AB and AG, a circle (ABG) around points A, B, G , and after the displacement of the magnitude BG to GD [within the same circle], also the ray AD. Then follow a series of equalities of arcs and angles and the assertion that because the viewed object in the two different positions (GB, GD) is seen under two equal angles (BAG, DAG), it is seen with the same magnitude] [def. 4-4]. The demonstration now switches to the case of a fixed object (diameter AG) and different positions of the eye (B, D). Then because angles ABG and ADG are equal [as right angles, when AG is a diameter, and as angles within two equal triangles, when AG is not the diameter], the [fixed] object is seen with the same magnitude [from the two different positions on the circle].

Textual Correspondence:[248] This proposition is considerably different from both O and R. M43 does not contain the last parts in O [on the case of the eye having a fixed position at the circle's center, while the object is being displaced on the circle's circumference]; it does not even contain a specific reference to the circle's circumference [also in the text of R]; neither is it followed immediately by a separate proposition [as in both O and R] regarding the case of the fixed position of the object and displaced position of the eye [such a case becomes M48, apparently in a conscious effort to group it with a series of later propositions (M49–56) on fixed objects and displaced eye positions]. Still, the case of a fixed object [i.e., diameter] and a displaced eye position [on the circle's arc] is treated as part of this same proposition and figure (M43), again in contrast to both O and R. As for the first part of the proposition [the only part corresponding to O37, R41], where O and R differ, M has elements from both: in the demonstration steps, it is closer to R [the part on the falling of the rays from the eye (A); see M43(7); on a possible copying error in O and the absence of the last part from it, see Ver Eecke, *Euclide: l'Optique et la Catoptrique*, p. 35, n. 1]; in the lettering, M is closer to O in containing three-letter presentation of angles [in O] as opposed to letters within angles [in R]. The two-part demonstration in T follows M closely, but one of the published versions has a figure that differs considerably, even from m2 and m4. The treatment of the

---

[248] M43 compared to O37, R41, and T43: M43, v. 1, pp. 148–151; O37, ed. Heiberg, pp. 80–82, tr. Ver Eecke, pp. 34–35; R41, ed. Heiberg, p. 226, tr. Ver Eecke, pp. 89–90; T43, edn. Hyderabad, p. 17.

opposite case is presented in T's enunciation by the expression *wa bi al-'aks* (vice versa) and with a specific reference to Proposition 48 at the end.

Propositions 44–47: Cases Where the Eye's Fixed Position Is the Center of a Circle Around Which an Upright Magnitude Is Being Displaced

Proposition 44: The Case Where the Magnitude Is Erected at Right Angles to the Circle's Plane

Synopsis: The enunciation is that if a magnitude is raised perpendicularly above a plane and displaced around a circle with the eye at its center, then it will be seen always as equal [i.e., with the same magnitude]. The illustration is that of a line (AB) [representing the magnitude's height] and a point (G) [the eye]. The demonstration begins with the geometrical construction of a line (GB) and a circle (BD) centered at G. Then follows, for all [visual] rays drawn from the center G to the magnitude (GB), their [visual] angles are equal [without any mention of equality of the upright triangles and the fixed amount of the visual angle at each position]. The conclusion is that [because the visual angle is the same for any position of AB around the circle] the magnitude is seen always as equal [i.e., with the same height] [def. 4-4]. The proposition ends with a statement extending the validity of the conclusion to the case where the eye is placed on any point on a perpendicular line erected from the center of the circle parallel to the object.

Textual Correspondence:[249] In this proposition, where O and R differ slightly in enunciation and figure [lettering is the same], M is closer to R in all cases. M's enunciation does not contain the parts on the position of the eye at the circle's center or a position for the object parallel to its original position [as in R, and in contrast to O, which has both]. Also in M's figure only the radius GB is drawn [as in R; O has the diameter BD; variants m2, m5 lack point D on the circle]. The demonstration in M lacks [as in both O and R] the explanation why the visual angle is the same at each position of the magnitude around the circle. The figure in one of the published copies of T (t3) [see List I, v. 1, p. xxxvii] represents the final case of the eye's locus being on a line parallel to the object.

Propositions 45–47: Cases Where the Object Is Not Erected at Right Angles to the Circle's Plane

These cases are treated in the form of three separate propositions: The first is the case when the magnitude of the object is greater than the circle's radius (M45), the second, when it is equal to the circle's radius (M46), and the third, when it is smaller than the circle's radius (M47). The three Arabic propositions represent

[249] M44 compared to O39, R38, T44: M44, v. 1 pp. 152–155; O39, ed. Heiberg, p. 84, tr. Ver Eecke, p. 36; R38, ed. Heiberg, pp. 216–218, tr. Ver Eecke, pp. 84–85; T44, edn. Hyderabad, pp. 17–18.

major structural differeces from the Greek. In O, only one case is treated, that of the magnitude of the nonperpendicular object being equal to the circle's radius (O40). In R, where all three cases are included, they are treated as one proposition (R40). T follows the structure of M, though with a reversed order, starting from the case where the object's magnitude is less than the circle's radius.

Proposition 45: The Case Where the Magnitude of the Nonperpendicular Object Is Greater than the Circle's Radius

Synopsis: The proposition begins [without an enunciation] with the geometrical construction of a circle (AD) from a point on which (D) a line (DZ) is erected such that it is not at right angles to the circle's plane, and from whose center (point E) the object is to be viewed. The illustration, from which the enunciation can be reconstructed, is that if line DZ is displaced around the circumference of the circle, it is seen sometimes as equal [to itself], sometimes as greater, and sometimes as smaller [it should be "it is seen always as unequal," as in the enunciation of R, since this is what is demonstrated in each case]. Then follows the layout of the three cases [i.e., that line DZ is either equal to or else greater or smaller than the circle's radius] and the introduction of the case to be treated in the proposition. The demonstration for the first case (M45) begins [again without an enunciation] with the geometrical construction of some lines, including EG and AB equal and parallel to DZ, and the assertion that line AB is seen smaller than all the lines moved along the circle's circumference. The long demonstration includes many steps: the construction of two parallelograms (ABGE, DZGE) and, in a second figure (45.2), two other parallelograms (FMNS, QONS) with two angles (SNF, SNQ) in one set equal to two others (AEB, DEZ) in the other set; a set of premises including an explicit reference to Propostion 36 [in *abjad* form] and the specification of the smallest angle formed. The conclusion is that because angle SNF is smaller than angle SNQ [and so AEB is smaller than DEZ as their equivalent pair], AB appears smaller than DZ [def. 4-2] [so in this case the same magnitude is seen as unequal in different positions].

Textual Correspondence:[250] In this proposition [i.e., in the first of three cases], where O and R differ considerably in enunciation, figures, and demonstration, M is closer to R in the illustration, where DZ is not specified as being equal to the circle's radius [as in R, and in contrast to O], and in the demonstration and figures, where the three cases are laid out through two figures for each case [O has one figure for one case]. But M is closer to O in the part of the illustration where reference is made to the "greater, smaller and equal appearance of DZ" [as in O, R has only greater and smaller (its enunciation has it as always

---

[250] M45 compared to O40-1, R40-1, T45: M45, v. 1, pp. 156–163; O40-1, ed. Heiberg, pp. 86–88, tr. Ver Eecke, pp. 36–37; R40-1, ed. Heiberg, pp. 218–222, tr. Ver Eecke, pp. 85–86; T45, edn. Hyderabad, pp. 18–19.

unequal)]. The Arabic differs from the Greek in not having a separate enunciation [in contrast to both O and R], in referring to a specific proposition (M36) [no number is given in O, R, or even T], and in lacking the part after the conclusion [only in R] that this can be shown for the cases where DZ is equal to or smaller than the circle's radius. T follows M closely but treats the first and third cases in the reversed order.

Proposition 46: The Case Where the Magnitude of the Nonperpendicular Object Is Equal to the Circle's Radius, While Everything Else Remains the Same

Synopsis: The proposition begins [without an enunciation] with the illustration of a line (DZ) equal to the circle's radius and a semicircle equal to another (TKLM). The enunciation [if stated] would be that in this case AB appears smaller than DZ. The demonstration begins with the geometrical construction of two parallelograms (ABGE, DZGE) and, in a second figure (46. 2), two other parallelograms (KOTN, LSTN), where two angles (TNS, TNO) in one set are equal to two others (AEB, DEZ) in the other set. The conclusion is that because angle TNS is smaller than angle TNO [so AEB is smaller than DEZ], AB is seen as smaller than DZ [def. 4-2]; so the magnitude appears different once again.

Textual Correspondence:[251] In this proposition, which corresponds to the second part of R40 [O40 does not contain a second case; the case where DZ is equal to the circle's radius is its first and only case], M is close to R in containing no enunciation, two similar figures, and a shorter demonstration which [in contrast to R] is presented as a separate proposition .

Proposition 47: The Case Where the Magnitude of the Nonperpendicular Object Is Smaller than the Circle's Radius, While Everything Else Remains the Same

Synopsis: The proposition begins [without an enunciation] with the illustration of a line (DZ) smaller than the circle's radius and a semicircle equal to another (TKLM). The enunciation [if stated] would be that AB [still] appears smaller than DZ. The demonstration begins with the geometrical construction of two parallelograms (ABGE, DZGE) and, in a second figure (47.2), two other parallelograms (KFTN, LOSN), where two angles (SNO, SNF) in one set are equal to two others (AEB, DEZ) in the other set. The conclusion is that because angle SNO is smaller than angle SNF [so AEB is smaller than DEZ], AB appears smaller than DZ [def. 4-2] [so the magnitude appears different once more].

---

[251] M46 compared to R40-2, T46: M46, v. 1, pp. 164–167; R40-2, ed. Heiberg, pp. 222–224, tr. Ver Eecke, pp. 87–88; T46, Hyderabad, p. 19.

Textual Correspondence:[252] In this proposition, which corresponds to the third part of R [O40 does not contain a third case or include a treatment of the case of the object being smaller than the circle's radius in its only case], M is close to R, again containing no enunciation, two figures, and a shorter demonstration presented as a separate proposition.

Propositions 48–56: Cases Where the Viewed Object Is Fixed, While the Eye Is Displaced

Propositions 48–49: Cases Where the Eye's Locus Is a Circle

Proposition 48: The Case Where the Fixed Object Is in the Same Plane with the Eye

Synopsis: The enunciation is that there is a place (*mawḍiʿ*) [T has "stretch" (*masāfa*)] from which a fixed object appears always with the same magnitude [i.e., to demonstrate that there is such a locus for the eye]. The illustration is that of a magnitude (AB) [representing the object] and a point (G) [the eye's position]. The demonstration begins with the geometrical construction of the ray[s] GA, GB, a circle (GAB) around triangle AGB [representing the eye's plane, which is the same as that of the object], and from another position for the eye (D) [on the same circle, although this is not specified], the ray[s] DA, DB. Then follows the statement that angles E and Z, being in the same segment of the circle, are equal. The conclusion is that because angles E and Z are equal, the object is seen, when viewed from the respective points E and G [T has "in the two states" (*ʿalā ḥālatayn*)] and so from any point on the same segment [O and R have "on arc BΔΓ"], as being the same [i.e., in magnitude: def. 4-4].

Textual Correspondence:[253] In a strict structural correspondence with the Greek, this proposition would have followed a much earlier proposition (M43), which treats a fixed eye and a displaced object [as in O37–38 and R41–42]. But here it is grouped with the propositions on fixed objects and displaced eye positions (M49–56). In this, as in Proposition 47, where O and R differ [lettering of the angles, phraseology], M is still closer to R [one letter inside the angles in all copies but m5]. The Arabic differs from both Greek versions in figure and lettering: There is a full circle around the two triangles, not a segment [as in O and R]. T has a statement after the conclusion that refers to Proposition 43.

---

[252] M47 compared to R40-3, T47: M47, v. 1, pp. 168–171; R40-3, ed. Heiberg, pp. 224–226, tr. Ver Eecke, pp. 88–89; T47, edn. Hyderabad, p. 19.

[253] M48 as compared to O38, R42, and T48: M48, v. 1, pp. 172–173; O38, ed. Heiberg, p. 82, tr. Ver Eecke, p. 35; R42, ed. Heiberg, p. 228, tr. Ver Eecke, p. 90; T48, edn. Hyderabad, p. 19.

Proposition 49: The Case Where the Fixed Object Is Perpendicular to the Eye's Plane

Synopsis: The enunciation is that if an object is raised at right angles to the assumed plane (*al-saṭḥ al-mawḍū'*) [i.e., the eye's plane], then it is seen from any point on the enclosing circle always with the same magnitude. The illustration is that of a magnitude (AB) [representing the object], a point (G) [the eye], and a circle (GD) centered at B with radius GB [representing the eye's plane]. Then follows [without any geometrical constructions], because B is a right angle, all the rays from G to [the top of] A produce equal angles [no justification is given]. The conclusion is that because these angles [through which the object is seen from any point on the circle's circumference] are equal, the object is seen as being the same [i.e., in magnitude: def. 4-4].

Textual Correspondence:[254] In this proposition, where O and R differ slightly in enunciation, illustration and figure, M is closer to R in the enunciation, where it lacks the specification that the circle's center be where the object meets the plane [only in O], in the illustration, where it specifies the elevation of the object [only in R], and in the figure, where magnitude AB is drawn close to the circle's circumference [in variant m5 it is shorter than the radius as in O; it is longer than the radius in R; the figures in variants m2, m4, and m5 lack line BG]. T, with a figure closer to O in all published versions, refers to the similarity of triangles in establishing the equality of the raised angles.

Propositions 50–53: Cases Where the Eye's Locus Is a Line

Propositions 50–51: Cases Where a Fixed Object Is Seen as Having Different Magnitudes

Proposition 50: The Case Where the Eye's Locus Is a Line to the Side of the Object

Synopsis: The enunciation is that there is a place (*mawḍi'*) [i.e., locus] from which a fixed object is seen as having different magnitudes [i.e., to demonstrate that there is such a locus]. The illustration is that of a line (AB) [drawn horizontally to represent the object], a side line (GD) [representing the eye's locus], and two points (E, D) on it [respective positions of the eye]. The demonstration begins with the geometrical construction of an extension for line AB up to point G, and GE, the mean proportional of lines BG, GA, three lines (AE, EB, BD) [should be rays EA, EB, DB], a circle's segment (AEB) about triangle AEB [not specified as tangent to GD at E], and two other lines (AD, BZ) [should be ray DA and line BZ]. Then follows the statement that angle AEB

---

[254] M49 compared to O41, R39, T49: M49, v. 1, pp. 174–175; O41, ed. Heiberg, pp. 88–90, tr. Ver Eecke, p. 38; R39, ed. Heiberg, pp. 216–218, tr. Ver Eecke, p. 85; T49, ed. Hyderabad, p. 19.

is greater than angle ADB [because AZB, which is equal to it, is greater than ADB as an outside angle]. The conclusion is that because AB is seen from points E and D under unequal angles, it is seen [from those two points (or any other point on line GD)] as being different [i.e., in magnitude: def. 4-4].

Textual Correspondence:[255] In this propositon, where O and R differ considerably in structure, enunciation, demonstration, figures, and lettering, M is closer to R in structure, in having one case and one figure [O has an alternative treatment and figure], and in the enunciation, where the line in question is not specified as oblique [it is in O] and the object is said to be seen as unequal [O has sometimes equal, sometimes not for the two cases]. M's lettering, different from both O and R [the lettering in R is identical to that of the alternative treatment in O], is, in this case, closer to O [no single letters within angles as in R]. T's abridgement follows M, except for the figure, where line GD is not drawn as tangent to the circle's segment [as in M, O, and R].

Proposition 51: The Case Where the Eye's Locus Is a Line Parallel to the Object

Synopsis: This proposition begins [without an enunciation] with the illustration of a line (GD) [representing the eye's locus] parallel to the object (AB). The demonstration begins with the geometrical construction of a line bisecting the object at E and at right angles, a certain position of the eye (Z) on line GD, the ray[s] ZA, ZB, a circle's segment AZB [in m2 a full circle] around triangle AZB, a line (BH) [should be line AD to get point H on the circle, and then line BH, as in the corresponding case in R]. Then follows the statement that angle AZB is greater than angle ADB [because AHB (equal to AZB) is greater than ADB as an exterior angle]. The conclusion is that because AB is seen from points Z and D under unequal angles, it is seen [from those two points] as being different [i.e., in magnitude], though it is seen as equal from another point on that line such as G [def. 4-4].

Textual Correspondence:[256] In this proposition, where O and R differ considerably in demonstration, figure, and lettering, M is closer to O in the figure and lettering, including the absence of single letters within angles [as in R], in the demonstration, where the circle's segment is not specified as tangent to GD [it is in R], and in the conclusion, where the case of another point (G) from which the object is seen as equal to D is added [as in O]. However, M is closer to R in the illustration, where GD is specified as being parallel to AB [in

[255] M50 compared to O42-1, R43, T50: M50, v. 1, pp. 176–179; O42-1, ed. Heiberg, pp. 90–92, tr. Ver Eecke, p. 38; R43, ed. Heiberg, pp. 228–230, tr. Ver Eecke, pp. 90–91; T50, edn. Hyderabad, pp. 19–20.

[256] M51 compared to O43, R44, T51: M51, v. 1 pp. 180–183; O43, ed. Heiberg, pp. 94–96, tr. Ver Eecke, p. 40; R44, ed. Heiberg, p. 230, tr. Ver Eecke, pp. 91–92; T51, edn. Hyderabad, p. 20.

the enunciation of both O and R, and in the illustration of R; M has no enunciation]; and in the figure, where the fewer lines drawn match those in R more closely. T's abridged demonstration follows M closely.

Propositions 52–53: Cases Where Two Fixed Equal Objects Are Seen as Having Different Magnitudes

Proposition 52: The Case Where the Eye's Locus Is a Line Not at Right Angles to the Extension of the Two Objects

Synopsis: The enunciation is that there is a common place (*mawḍi' mushtarak*) for the eye from which equal magnitudes would be seen as unequal [i. e. to demonstrate that there is such a locus for the eye]. The illustration is that of two lines (AB, BZ) [drawn on the same vertical extension (AG) to represent the two equal magnitudes]. The demonstration begins with the geometrical construction of a line (BD) perpendicular to the extension of the lines (AG) and two points (D, E) [representing the eye's positions in the case of equal and unequal appearance of the two magnitudes respecively], three [visual] rays (AE, EB, EG) [the first should be EA], a circle (EADG) around triangle AGE, and a line (BZ) [an extension of line DB to point Z on the circle]. Then follows, because line GE is greater than line AE, magnitude AB is seen as greater than magnitude AG. The conclusion is that because magnitudes AB and BG are seen as equal from point D but unequal from point E [def. 4-4], there are places [any point on line DZ] from which AB and BG are seen as equal, and places [any point on line EH or even outside the circle] where they are seen as unequal.

Textual Correspondence:[257] In this proposition, which corresponds to the first part of O44, with no corresponding proposition in R, M does not always follow O closely. The enunciation in M does not specify the orientation of the magnitudes [the exact sense of the term *mushtarak* (common) is unclear; on the Greek case, see Ver Eecke, *Euclide: l'Optique et la Catoptrique*, p. 41, n. 2]; it does not include the case of equal appearance of the two equal magnitudes [as in O]. The illustration section and the lettering of the figure do not follow O closely, either [M's figure varies slightly among the variants: m2, m4 lack joining lines; the figure in m2 has been misplaced and should be switched with the one in M51]. The demonstration, which lacks some premises from O [equalities and inequalities of arcs] is not followed in M by an alternative treatment [as in O]. T's abridged demonstration follows M closely [the figure of one of the published versions (t3) lacks joining lines].

---

[257] M52 compared to O44-1, T52: M52, v. 1 pp. 184–187; O44-1, ed. Heiberg, pp. 96–98, tr. Ver Eecke, p. 41; T52, edn. Hyderabad, p. 20.

Proposition 53: The Case Where the Eye's Locus Is a Line Not Erected at Right Angles from the Midpoint of the Distance between Two Fixed Objects

Synopsis: This proposition begins [without an enunciation] by illustrating that for two equal magnitudes (AB, GD) perpendicular to a certain plane, there is a place [i.e., locus] for the eye from which the two magnitudes are seen sometimes as equal and sometimes as unequal. The demonstration begins with the geometrical construction of a line (BD) [the distance between the two fixed objects] and from its midpoint another line (EZ) perpendicular to it, and from Z [the position of the eye on this second line] the ray[s] ZA, ZB, ZG, ZD. Then follows a set of premises on equality of triangles BZA and DZG and the assertion that because angles BZA and DZG are equal, AB and GD are seen [from Z, or any point on EZ] as equal [def. 4-4]. The demonstration is then repeated for the second position of the eye (H) with the geometrical construction of the ray[s] HB, HA, HG, HD and a line (BT) on BH equal to HD to establish equality of triangles and angles BTA and GHD. The conclusion is that because angle BHA is smaller than GHD [equal to BTA], the two magnitudes are seen [from H or any point on HE] as being unequal [def. 4-4].

Textual Correspondence:[258] This proposition, which has no corresponding one in R, differs from O in at least one obvious sense: in M's demonstration of the second case, line HE [from any point of which AB and GD would be seen as unequal] is not drawn [either as part of the geometrical construction or in the figure], whether from the eye (H) to the midpoint E or vice versa [in one variant (m3) HA passes through E]. T's abridged demonstration follows M closely but lacks the statement [found in all variants of M though lacking in O] that "angles at bases with equal sides subtending them are equal."

Propositions 54–55: Cases Where the Eye's Locus Is a Single Point

Proposition 54: The Case Where Two Fixed Unequal Objects Are Seen as Having the Same Magnitude

Synopsis: The enunciation is that there is a place (*mawḍi'*) for the eye from which two unequal magnitudes would be seen as equal [i.e., to demonstrate that there is such a place]. The illustration is that of two unequal magnitudes [AB greater than BG, laid on the same line]. The demonstration begins with the construction of two similar segments of a circle (BDG, BDA) greater than a semicircle around lines BG and AB [only in m4 and m5 are they drawn as greater] and two lines (DA, DG) [no mention of D as the eye's position]. Then follows, because the segments ABD and BDG are similar, the angles ADB and GDB are equal. The conclusion is that because from point D magnitudes AB and BG are seen under equal angles, they are seen as equal [def. 4-4].

---

[258] M53 compared to O46 and T53: M53, v. 1, pp. 188–191; O46, ed. Heiberg, pp. 100–102, tr. Ver Eecke, pp. 43-44; T53, ed. Hyderabad, pp. 20–21.

Textual Correspondence:[259] In this proposition, where O and R are close in the enunciation, demonstration, figure, and lettering, but differ slightly in length [R lacks the part on the intersection of the two segments (at Z)], M is closer to R. But the Arabic differs from both Greek versions in lacking the statement toward the end that places the eye at the intersection of the two segments (Z) and in the lettering of the figure. T's abridged demonstration follows M closely but adds "so we have found that place" before concluding "that is what we wished . . . ."

Proposition 55: The Case Where a Combination of Two Unequal Objects Is Seen as Equal to Either One Separately

Synopsis: The enunciation is that there is a place for the eye from which the combination of two unequal magnitudes is seen as equal to either of them seen separately [i. e. to demonstrate that there is such a place]. The illustration is that of two unequal magnitudes (AB, BG) [placed adjacent on the same straight line, AB greater than BG]. The demonstration begins with the geometrical construction of three semicircles (ADB, BEG, AZG) around lines AB, BG, and AG, and four lines, two from the midpoint of arc AG at Z (ZA, ZG), and two joining lines (DB, BE). Then follows, angles ADB, BEG, being right angles, are equal. The conclusion is that because from points D and E magnitudes AB and BG are seen under equal angles, they are seen as equal [def. 4-4], and AG, seen under right angle AZG, is also seen as equal to each of them [def. 4-4].

Textual Correspondence:[260] In this proposition, where O and R differ slightly, M is closer to O: in the enuciation, it contains reference to the eye [as in O and in contrast to R]. The only other part where the two Greek versions differ [reference to the position of the eye upon the corresponding semicircles BAΔ, TZΔ in O, BAΔ, BKT, TZΔ in R], is absent from M altogether. In one variant (m1) the two magnitudes [and their semicircles] are drawn as equal; this is also the only variant that includes reference to BG in the corresponding text. T's abridged demonstration follows M but with a slightly expanded enunciation and a different phraseology for the demonstration.

Proposition 56: Cases Where the Eye's Locus Is a Circle's Circumference – the Case Where an Object Is Seen as Half, Quarter ,or Other Fraction of Its Actual Size

---

[259] M54 compared to O45, R46, T54: M54, v. 1, pp. 192–195; O45, ed. Heiberg, p. 100, tr. Ver Eecke, p. 42; R46, ed. Heiberg, pp. 232–234, tr. Ver Eecke, pp. 92–93; T54, edn. Hyderabad, p. 21.

[260] M55 compared to O47, R47, T55: M55, v. 1 pp. 196–197; O47, ed. Heiberg, p. 104, tr. Ver Eecke, p. 44; R47, ed. Heiberg, p. 234, tr. Ver Eecke, p. 93; T55, edn. Hyderabad, p. 21.

Synopsis: The enunciation is that there are places (*mawāḍi'*) for the eye from which a magnitude is seen as its half, quarter, or any magnitude it is divided into [i.e., to demonstrate that there is such a locus]. The illustration is that of a magnitude (AB) [drawn horizontally]. The demonstration begins with the geometrical construction of a circle (AEB) around AB, such that AB is not its diameter of the circle and the eye is first at the center (G) and then on the circumference (E), and the ray[s] GA, GB, EA, EB. Then follows, because AEB is half of AGB, AB is seen twice as big from G as from E [def. 4-1]. The conclusion is that because the former angle may become twice or four times as much [or any other multiple] of the latter, the magnitude seen through that angle is seen as half, quarter [and any other fraction accordingly].

Textual Correspondence:[261] This proposition has major structural differences with respect to the Greek. The previous Greek proposition [O48, R48] with the same enunciation as this one contains the problem of how to find such a locus, and there is no correspondence in M to its text and two figures; whereas the demonstration that there is such a locus [O49; R has no corresponding proposition] matches that of Proposition 56 with the same single figure, so the Arabic proposition can be viewed as the combination of the enunciation of O48 [or R48] with the demonstration and figure of O49. [A combined version of O48–49 appears in the form of two different proofs of the same propostition in the Latin *De visu*, V52; M56 would be better grouped with M48–49 as the cases for the circular locus of the eye]. T's abridged demonstration follows M closely, but the figure is slightly different [in the Hyderabad edition, there are two figures, both different from O]. T's demonstration includes the case of the quarter angle in the figure.

Propositions 57–64: Appearances Involving Illusions

The following set of eight propositions deal with appearances involving illusions. The first six (M57–62), which deal with appearances involving motion, can be considered as being specifically concerned with visual illusions insofar as the expression *yuẓannu fī al-ru'ya annahu* (thought through vision that) is used in them [sometimes appearing in the short form *yuẓannu annahu*, as in M25 M31, M33 on the case of spheres, cylinders, and cones, respectively]. Some [e.g., M31] also contain the contrasting expression *yurā bi al-ḥaqīqa* (seen in reality). In contrast to the first six propositions (M57–62) of this set, the last two (M63–64) are concerned with illusions not involving motion, using the term *yurā* (it is seen as) for the [illusioned] convex/concave appearance of certain [three-dimensional] objects (M63) and equal appearance of the sides and diameters of quadrilateral figures (M64).

---

[261] M56 compared to O49, T56: M56, vol. I, pp. 198–201; O48–49, ed. Heiberg, pp. 104–106, tr. Ver Eecke, pp. 44–45; R48, ed. Heiberg, p. 234–236, tr. Ver Eecke, p. 93; T56, edn. Hyderabad, pp. 21–22.

Propositions 57–62: Illusions Involving Motion

Proposition 57–58: Cases for the Relative Position of Objects in Motion

Proposition 57: The Case of Objects Passing with Equal Speed in Front of the Eye but in a Different Plane

Synopsis: The enunciation is that if objects of equal speed are approaching the eye on the same line [i.e., their successive points are on the same vertical line], then the one to the rear is thought (*yuẓannu*) to be seen in front, and when the manner of their being viewed changes (*ikhtalafat ru'yatuhā*) [i.e., once they cross over the direct line of sight], the one ahead is thought to be following and the one following to be ahead. The illustration is that of three magnitudes (AB, GD, EZ) placed in motion, a vertical line (BZ) on which their endpoints lie, and a point (H) for the eye's position. The demonstration begins with the geometrical construction of the ray[s] HZ, HD, HB, where HB is higher than HD and HD higher than HZ, and the assertion that AB is thought to be seen ahead of GD and GD ahead of EZ. Then follows the case of the three parallel magnitudes on the right side of TN (TK, LM, NS), the ray[s] HT, HL, HN and the assertion that [on this side] HN is ahead of [should be higher than] HL, and HL ahead of [should be higher than] HT, so SN is [i.e., thought to be] ahead of ML, and ML with respect to TK. The conclusion is that [in the first case] it is the magnitude to the rear (AB) [with the highest ray HB on it] that is thought to be ahead, and [in this case] it is the magnitude thought to be behind (NS) [i.e., ZE in the first case] that is thought to be ahead.

Textual Correspondence:[262] This proposition represents problems in transmission not only from O and R into M [e.g., M57(8), Ver Eecke, *Euclide: l'Optique et la Catoptrique*, p. 46, n. 1, refers to the problems of the Greek manuscripts with reference to a different part] but also from M to T. Where O and R differ [in enunciation, demonstration, and figure (same lettering)], M is closer to R in the enunciation, where the moving magnitudes are stated as being along the same line [as in R] rather than their endpoints aligned vertically on the same line [as in O], in the demonstration, where ray HB is stated as the most high [in R, the corresponding ray (MT) is the most high *and* the one most to the right; in O it is the one apparent before other rays; see Ver Eecke, *Euclide: l'Optique et la Catoptrique*, p. 46; Burton, "The *Optics* of Euclid," p. 371, has "the one to have diverged (to the right) more than the other rays"], and in the figure, where neither the lines joining B, D, and Z or N, L, and T nor the middle line from point H are drawn [as in R and in contrast to O]. But M also has an element from O: The demonstration does not include any references to rays to

262 M57 as compared to O50, R49, and T57: M57, v. 1 pp. 202–205; O50, ed. Heiberg, pp. 106–110, tr. Ver Eecke, p. 45; R49, ed. Heiberg, p. 236, tr. Ver Eecke, p. 94; T57, edn. Hyderabad, p. 22.

the right and left [as in O, and in contrast to R], though the three magnitudes are stated later in the proposition [M57(13)] as being parallel (*mutawāzī*) and to the right side (*mutayāmin*). The Arabic differs from the Greek in the enunciation, where the expression *(ikhtalafat ru'yatuhā* (the manner of their being viewed changes) is employed with no exact correspondence in O and R [or T] and in the demonstration, where the relative appearance of a moving object is stated in terms of the height of the ray toward it [in R, it is *both* the height and direction, and in O, the prior appearance of the corresponding ray (MT)]. The figure of M, which is closer to R [no joining lines], differs from it in not having the end lines of the moving objects vertically aligned in any of the variants [m2, m4 have a closer alignment than m1, m3, m5 (m4 is incorrectly marked Proposition 56)]. T's abridged version differs from M in the enunciation, where the expression *idhā ikhtalafat ru'yatuhā* (when the manner of their being viewed changes) is replaced by *idhā tawajjahat min aḥad al-jānibayn* (when they turn from one of the two sides of the eye to its front) and where the expression *jāwazat muqābilat al-baṣar* (crosses over in front of the eye) is explicit [as in O], and in the figure, where the joining lines BZ, TN are drawn [as in O, but with no perpendicular line from H].

Proposition 58: The Case of Objects Moving at Different Speeds with Respect to the Eye and in the Same Plane

Synopsis: The enunciation is that for magnitudes moving with unequal speed but one with the same speed as the eye, some are thought (*yuẓannu*) to be fixed and others to be moving ahead or in the opposite direction. The illustration is that of three moving magnitudes [represented as points A, B, G], the middle one (B) equal in speed with the eye (D). The demonstration begins with the geometrical construction of the ray[s] DA, DB, DG. Then follows, since the eye (D) is not stationary, B is thought to be stationary because B and D are on the same line [of the same speed in O], and [again] when the eye is not stationary, point G [at a distance from B and ahead of it] is thought to be moving ahead, while point A [at a distance from B] is thought to be moving in the opposite direction because of the slowness of movement.

Textual Correspondence:[263] In this proposition, where O and R differ in the phraseology of the enunciation and demonstration, M is closer to O in the enunciation, where it starts with the magnitudes rather than the eye [in R] and in the demonstration, where it lacks the part [only in R] where the eye is stated as moving at the same time with all three magnitudes. The Arabic differs from both O and R in minor points such as the lettering [the eye as point D] and in the order and phraseology of the concluding premises, where the case of the

263 M58 compared to O51, R50, T58: M58, v. 1 pp. 206–209; O51, ed. Heiberg, p. 110, tr. Ver Eecke, pp. 46–47; R50, ed. Heiberg, p. 238, tr. Ver Eecke, pp. 94–95; T58, edn. Hyderabad, p. 22.

magnitude moving ahead comes before the one in the opposite direction, and the reason is given as the slowness of movement without further specification. T has a slightly different demonstration: It refers to the fixed appearance of B in terms of ray DB not being fixed [R's reference to the visual ray here is slightly different]. T's final statement adds that "the amount visible" (*al-qadr al-mar'ī*) of their movement is the difference (*al-faḍl*) between the eye's speed from those magnitudes of different speed.

Proposition 59: The Case of a Body Appearing to an Approaching Eye to Grow in Size

Synopsis: The enunciation is that when the eye approaches a [fixed] body (*jism*) [i.e., an object], the body is thought (*yuẓannu*) to be growing or enlarging. The illustration is that of a viewed object (AB) and the eye (point G). The demonstration begins with the geometrical construction of the ray[s] GA, GB so that AB is contained by them, then proceeds to a closer position for the eye (point D) with ray[s] DA, DB. The conclusion is that because angle ADB is greater than angle AGB, AB is thought to be greater from point D than from point G [def. 4-1] [so seen as enlarged as the body comes closer to the eye].

Textual Correspondence:[264] In this proposition [similar in O and R], M matches the Greek closely, but differs from it in the figure, where the lettering of the Arabic is not the same as the Greek, and line GD is drawn [in m2–m5; m1 lacks it as in O and R]. M also has a language in this proposition that is distinct from the rest of the text: In addition to the object of vision being termed "the viewed object" (*manẓūr ilayh*), rather than a visible object (*mubṣar*), it is represented as a body (*jism*) rather than a magnitude (*qadr*) which enlarges (*yazīd*) or grows [*yanmū* in m1, *yanshū* in m2–m5].

Proposition 60: The Case of Objects at Unequal Distances Moving with Equal Speed

Synopsis: The enunciation is that the farthest of magnitudes moving at equal speed is thought (*yuẓannu*) to be the slowest. The illustration is that of two magnitudes [points A and B] moving rectilinearly toward G, A being farther away from the eye (point Z) than B. The demonstration begins with the geometrical construction of the ray[s] ZA, ZB, ZD, ZE and the assertion that because angle BZE is greater than angle AZD, BE is seen as greater than AD [def. 4-1], and since AD and BE move at equal speed, [points] H [and] D are seen by ray ZD [and D is thought to be slower than H]. Then follows the extension of ray ZE [to become ZEG] and the statement that since it [i.e., A (or

---

264 M59 compared to O53, R52, and T59: M59, v. 1, pp. 210–211; O53, ed. Heiberg, p. 112, tr. Ver Eecke, p. 47; R52, ed. Heiberg, p. 240, tr. Ver Eecke, p. 95; T59, edn. Hyderabad, pp. 22–23.

B)] is moving [reading *yataḥarraka* with m1, m2] with equal speed [to B (or A)] or since A and B are moving [reading *taḥarrakat* with m3, m5], at ray ZEG [i.e., when reaching point G] point A falls behind [i.e., viewed by ray ZEG, A at point G is thought to be slower than B at point E]. It ends [without any causal premises] with a repetition of the enunciation that in the case of magnitudes moving with equal speed, the farthest is thought (*yuẓannu*) to be the slowest.

Textual Correspondence:[265] This proposition represents major transmission problems, including structural differences, between Greek and Arabic. In O, the proposition has three parts; R's single treatment corresponds to O54-1. In O54-2 [the part corresponding to M60, with no corresponding section in R53 beyond the enunciation], the Greek is corrupt in a crucial part of the demonstration [Ver Eecke, *Euclide: l'Optique et la Catoptrique*, p. 48, n. 1, suggests that the whole proposition may be an interpolation]. The same place in M contains an additional statement in all copies of M [m4 lacks the text and figure of the whole proposition], which is only comprehensible if a subject such as B [in O, but not in any of the variants of M] is considered as part of the sentence [see M60(13–14), Edition and Translation, v. 1, p. 213]. The treatment in T is also problematic [and not identical in the three published editions]. A proposition with the same number [60 in *abjad* form] has received [together with the next proposition, also partly missing in m4] an extensive treatment [still not fully comprehensible!] in the *Iṣlāḥ al-Manāẓir* (*Correction of Optics*) attributed to al-Kindī, commonly extant as a companion text to many manuscript copies of T [see Edition and Translation, v. 1, p. 226–229]. In it, three cases [approaching, intersecting, and crossing the perpendicular line of the eye respectively] are treated with a corresponding diagram, with specific references to Proposition 57 (O50, R49) [with three similar cases].

Proposition 61: The Case of the Movement of Far and Near Objects with Respect to a Moving Eye

Synopsis: The enunciation is that when the eye is moving, far objects are thought (*yuẓannu*) to be moving with different speeds from those that are close. The illustration is that of two objects [represented as two points, A and G] that are [moving] along straight [lines](AB, GD)[AB being farther from the eye in the figure]. The demonstration begins with the geometrical construction of the ray[s] EG, ED, EA and the extension of ED until it falls on the [farther] object. Then follows, because angle GEB is greater than angle AEB, magnitude AB is seen as smaller than magnitude GD [def. 4-2]. The demonstration ends [without any causal premises] with the conclusion that point A is [i.e., is thought to be] different [in speed from G].

---

[265] M60 compared to O54, R53, and T57: M60, v. 1, pp. 212–213; O54, ed. Heiberg, p. 112, tr. Ver Eecke, pp. 47–48; R53, ed. Heiberg, p. 240 , tr. Ver Eecke, p. 96; T60, edn. Hyderabad, p. 23.

Textual Correspondence:[266] This proposition presents major transmission problems of a different kind, since O and R differ considerably throughout [enunciation, demonstration, figure, and lettering]. M reflects elements from both: Its enunciation is closer to R, while its demonstration, figure, and lettering are all closer to O. M differs from O and R in the enunciation [closer to R], where the more remote object is thought to be different (*mukhtalifa*) [in R, it is thought to be fixed] and the eye is described as moving (*mutaḥarrak*) [in R, its movement is overtaking that of the objects; see Ver Eecke, *Euclide: l'Optique et la Catoptrique*, p. 96, n. 1]; in the figure [closer to O], where the extensions of the two moving points [D and B] are not on the same vertical line with the eye [in O, the extensions are drawn as such]; and in the demonstration, the single statement that point A is different [in O, a long section follows before the concluding premise that magnitude AB is thought to be fixed]. T's abridged demonstration follows M but adds the following statement to the conclusion: "In this manner is the text, so let there be a consideration in it" (*hākadhā fī al-matn, wa-l-yanẓur fīhi*). The "correction" of a proposition with the same number [61 in *abjad* form] in the accompanying *Iṣlāḥ al-Manāẓir,* has again a text and figure [see Edition and Translation, v. 1, pp. 228–229] different from both M and T, this time with no further reference to a specific proposition [as in Proposition 60]. There, in the case of a [moving] eye approaching the two objects [now points A and B], the closer one [A] is stated as having not just a different speed from the farther one [B] but a *varying* speed with respect to it [i.e., faster, slower and equal in speed, presumably depending on the eye's position].

Proposition 62: The Case of Growing Objects Appearing to Approach the Eye

Synopsis: The enunciation is that magnitudes that grow in size are thought (*yuẓannu*) to be approaching the eye. The illustration is that of an object (AB) and a [fixed] point (G) for the eye. The demonstration begins with the geometrical construction of the ray[s] GA, GB, and when AB grows in size to become BD, also ray GD. Then follows, because angle BGD is greater than angle BGA, magnitude BD is greater than magnitude AB. The conclusion is that because objects that are seen from greater angles are thought to be closer to the eye [reverse of prop. 5], BD is thought to be closer to the eye.

---

[266] M61 compared to O55, R54, and T61: M61, v. 1, pp. 214–215; O55, ed. Heiberg, pp. 114–116, tr. Ver Eecke, pp. 48–49; R54, ed. Heiberg, pp. 240–242, tr. Ver Eecke, p. 96; T61, edn. Hyderabad, p. 23.

Textual Correspondence:[267] In this proposition, where O and R differ slightly in the demonstration and lettering of the figure, M has elements of both: in the demonstration, it is closer to R [does not include all the steps of the final conclusion in O], and in the lettering, it is closer to O [Δ, A, B, T; R has Δ, B, T, K]. The Arabic differs from the Greek in the figure, where the triangle made up of the extension of the object and the point of the eye is drawn as isoceles in all varinats of M [in O and R, it is a right triangle], in the demonstration, where the angles are referred to with three letters [in O and R, by the visual rays containing them], and in the conclusion, where it lacks the last line that repeats the validity of the conclusion for the specific case of the illustration. T's abridged demonstration is close to M.

Propositions 63–64: Illusions Not Involving Motion

Proposition 63: The Case of the Convex and Concave Appearance of Objects Whose Middles Are Not Aligned with Their Sides

Synopsis: The enunciation is that when objects that are unequally distant [from the eye] are placed such that the sides are not in line with the middle, then the figure [formed by them] is seen (*yurā*) sometimes as convex and sometimes as concave. The illustration is that of objects (A, B, G), nonaligned points (D, E, Z), and the eye (H). The demonstration begins with the geometrical construction of the ray[s] HB, HA, HG and lines BA, BG, and the statement that when the viewpoint is changed from [the initial] position of the eye (H) to points B, A [and] G, BAG is seen as concave. Then follows the case of the eye at point T, the geometrical construction of TD, TE, TZ, and the statement that when the viewpoint is changed from point T, once to D and once to Z, then DEZ is seen as convex, because points in front of it (*bayna yadayhī*) are not seen.

Textual Correspondence:[268] This proposition is noticeably different from the two Greek versions in both text and figure. And where O and R differ in enunciation and illustration [close in demonstration, figure, and lettering], M is closer to R in the enunciation, where the objects are stated to be "unequally distant" [O has "at the same interval/distance" (*diastêmati*); R has "not at the same interval, and not parallel;" see Ver Eecke, *Euclide: l'Optique et la Catoptrique*, p. 49 and n. 1, and p. 97 and n. 1] and in the illustration, where the objects are represented by three points [as in R; in O, by two lines].

---

[267] M62 compared to O56, R55, and T62: M62, v. 1, pp. 216–217; O56, ed. Heiberg, p. 116, tr. Ver Eecke, p. 49; R55, ed. Heiberg, p. 242, tr. Ver Eecke, pp. 96–97; T62, edn. Hyderabad, p. 23.

[268] M63 compared to O57, R56, and T63: M63, v. 1, pp. 218–221; O57, ed. Heiberg, pp. 116–118, tr. Ver Eecke, pp. 49–50; R56, ed. Heiberg, pp. 242–244, tr. Ver Eecke, p. 97; T63, edn. Hyderabad, pp. 23–24.

But the Arabic differs from the Greek: The enunciation is not identical to R. It lacks the descriptions "nonparallel", and "in the same direction" for the objects. It also has a longer and clearly different demonstration, which includes an additional section where the viewpoint is changed to specific points and a causal premise as the concluding statement. Neither of the two figures of M, which are close in all variants, correspond to O and R in either shape or letters, though in the published versions of T the shape of one of the figures is similar [in t3, both are similar]. T's slightly different demonstration contains expressions such as "viewing together" (*ma'an*) and "combined" (*majmū'*) appearance.

Proposition 64: The Case of the Equal Appearance of the Sides and Diameters of a Quadrilateral Figure

Synopsis: The enunciation is that from a line erected at the intersection of diameters of a quadrilateral figure (*murabba'*), the sides and diameters of the figure are seen (*yurā*) as equal. The illustration is that of a quadrilateral figure (ABGD). The demonstration begins with the geometrical construction of a line (EZ) erected upward at right angles to the assumed plane (*al-saṭḥ al-mawḍū'*) from the intersection point (E) of two of the diameters of the quadrilateral figure to the position of the eye (point Z), from which fall the visual ray[s] ZA, ZB, ZG, ZD. There follows a set of premises on the equality of triangles and the statement that ZG and ZD are equal and so are AB and GD [def. 4-4]. The demonstration continues with the case of two other half-diameters and sides and ends with the conclusion that because the angles under which these pairs are seen are equal, they are seen as equal [def. 4-4].

Textual Correspondence:[269] In this proposition, where O and R slightly differ, M has elements from both: The phraseology of the enunciation is closer to R, and the lettering of the figure is closer to O [m1 is the only variant with a rectangular rather than a square figure]. M differs from both Greek versions in lacking the final reference to the case where the upward line is not perpendicular to the figure's plane [in both O and R, with no demontration or figure]; T's abridged demonstration follows M closely.

---

[269] M64 compared to O58, R57, and T64: M64, v. 1, pp. 222–225; O58, ed. Heiberg, p. 118, tr. Ver Eecke, pp. 50–51; R57, ed. Heiberg, p. 244–246, tr. Ver Eecke, pp. 97–98; T64, edn. Hyderabad, p. 24.

Part III

Outline of Key Points

Outline of Key Points

7 The Versions

The Greek Versions

- As *not* representing single points of reference: Preface, pp. vii–viii.
- As *Euclidis Optica* and *Opticorum recensio Theonis* of Heiberg and Version A and Version B of Jones and Knorr [as represented respectively by Vienna Codex Vindobon. xxxi, 13, 12th century, and the Vatican codex, gr. 204, 10th century]: Preface, p. viii; The Arabic Versions, v. 1, pp. xx–xxi; Textual Transmission, p. xli; The Arabic "Version," v. 1, p. xix, n. 7.
- As designated in the form O and R in this study, from the standpoint of the Arabic tradition: The Arabic "Version," v. 2, p. xix.

The Arabic Versions

- As consisting of early (3rd/9th century) and late (6th/13th century) Arabic versions: a "translation" (*tarjama*) [in two close variants] by Hiliyā ibn Sarjūn (ca. 212/827–828)(M), a recension (*Taḥrīr*) by Ṭūsī (ca. 651/1253) (T) with an attached "Correction" (*Iṣlāḥ*)(Ṣ) attributed to al-Kindī (d. ca. 257/870), and a *Tajrīd* (J) by Ibn Abī Jarāda (ca. 676/1277): The Arabic Versions, v. 1, pp. xix–xxiii.
- As containing 4 definitions and 64 propositions in *all* Arabic versions, in contrast to the 7 definitions and 57 and 58 propositions of the main Greek versions O and R respectively: The Arabic "Version," v. 2, p. xix.
- As representing problems of documentation: Preface, p. viii; The Arabic Version, v. 1, pp. xxiii–xxiv; The Arabic "Version," v. 2, pp. xxi–xxii.
- As evidence against certain misconceptions: The Arabic "Version," v. 2, pp. xx–xxvi.

The Early "Translation"

- As a *translation*: The Arabic Versions, v. 1, p. xix; The Arabic "Version," v. 2, p. xix, xxiii–xxvii; Index of Arabic Terms, v. 2, pp. 125–152.
- As a *translation* by Hiliyā ibn Sarjūn: The Arabic Versions, v. 1, p. xix; The Arabic "Version," v. 2, pp. xxii–xxv, n. 19–23.
- As a translation by a *Sarjūn* of early 3rd/9th-century Baghdad, as distinguished from members of a much earlier family in Damascus: The Arabic "Version," v. 2, p. xxiv, n. 22.
- As an *early* translation: The Arabic Versions, v. 1, p. xix, and p. xxvii; Editorial Procedures and Apparatus, v. 1, p. xxix; The Arabic "Version," v. 2, p. xix, n. 23, and p. xxv; Summary and Commentary: Title, p. 1 and n. 39, p. 2, n. 43; Proposition 9, p. 42, n. 167.
- As a *direct* translation: The Arabic "Version," v. 2, p. xxv and p. xxviii.

- As a *nonstandard* translation: The Arabic "Version," v. 2, pp. xxvi–xxviii; Summary and Commentary: Title, p. 4; Proposition 2, v. 2, p. 15.
- As a translation with different vocabulary from the overlapping part of the Arabic version of the pseudo-Euclidean *De speculis*: The Arabic Versions, v. 1, p. xxii, n. 10; Supplementary Passages, v. 1, p. 243; The Arabic "Version," v. 2, p. xxvi, n. 28; Summary and Commentary: Definition 1, v. 2, pp. 10–11, n. 69, 72; Definition 2, v. 2, pp. 15–17, and n. 89.
- As a translation with a different text from the overlapping treatments of two propositions in *Iṣlāḥ al-Manāẓir*: The Arabic Versions, v. 1, p. xxi, n. 9.
- As a translation with a widely circulated vocabulary: Summary and Commentary: Definition 1, v. 2, p. 7, p. 9, n. 64, p. 10, n. 69, p. 11 and n. 72, pp. 12–14; Definition 2, v. 2, p. 15–18; Proposition 20, v. 2, p. 56.

Manuscript Tradition

- As represented by largely uniform manuscript copies [dated from 600/1203 to at least 750/1349]: Description and Analysis of Manuscripts, v. 1, pp. xxvi–xxix; Edition and Translation [critical apparatus], v. 1, pp. 2–225.
- As represented by close manuscript copies with a striking variance in the opening definition, as in the case of the main two Greek versions: The Arabic Versions, v. 1, p. xx–xxii; Edition and Translation, v. 1, p. 3; Textual Divergence, v. 2, p. xxxv; Variation Sample 1, v. 2, p. xxxviii.
- As represented by two close variants, here presented in the form of odd- and even-numbered sets [m1, m3, m5 vs. m2, m4]: Description and Analysis of Manuscripts, v. 1, p. xxix; Variation Samples 2–4, v. 2, pp. xl–xlv.
- As representing the original more closely through the odd-numbered set, with an apparently older, more authentic, and more widely circulated text: The Arabic Versions, v. 1, p. xxii; Editorial Procedures and Apparatus, v. 1, p. xxix; Summary and Commentary: Definition 1, v. 2, pp. 7–8.
- As presented in the form of a critical edition and English translation (M), primarily based on a member of the odd-numbered manuscripts (m1) considered closest to the original [Editorial Procedures and Apparatus, v. 1, p. xxx] and a critical apparatus for the other manuscript variants [m2–m5]: Edition and Translation, v. 1, pp. 2–225.
- As representing signs of restorations rather than strict revisions: The Arabic Versions, v. 1, pp. xxi–xxii, n. 10; Description and Analysis of Manuscripts, v. 1, p. xxviii; The Arabic "Version," v. 2, xxvi; Summary and Commentary: Definition 1, v. 2, p. 8, n. 63.
- As representing restoration efforts in the same opening section considered defective in Greek: The Arabic Versions, v. 1, p. xxi, n. 10; Summary and Commentary: Definitions; Definition 1, v. 2, p. 6, n. 60; p. 8, n. 63.
- As representing an opening problematic enough for a *De speculis* passing under the name of Euclid (*Kitāb al-Mir'āh li-Uqlīdis*) to have possibly acted as supplement: Summary and Commentary: Definition 1, v. 2, p. 7, n. 61.

Relation to the Greek Tradition

- As represented by two variant sets that do *not* have an exact or consistent correspondence to Greek versions O or R to belong *fully* to either tradition: The Arabic Versions, v. 1, p. xxii, The Arabic "Version," v. 2, p. xxix.
- As represented by two variant sets [m1, m3, m5 and m2, m4] whose odd- and even-numbered manuscripts *both* correspond more closely to O: Summary and Commentary: Definintion 1, v. 1, pp. 8–11.
- As represented by two variant sets [m1, m3, m5 and m2, m4] whose odd- and even-numbered manuscripts *both* correspond more closely to R: Summary and Commentary: Definintion 2, v. 1, p. 18.
- As represented by two variant sets whose even-numbered manuscripts (m2, m4) correspond more closely to R; Summary and Commentary: Definition 1, v. 1, pp. 8–9.
- As represented by two variant sets whose odd-numbered manuscripts (m1, m3, m5) correspond more closely to R; Summary and Commentary: Proposition 35, v. 1, p. 74.
- As represented by two variant sets [m1, m3, m5 and m2, m4] with members from *both* sets [m1, m2, m3, m5] associated with the name of Euclid: The Arabic Versions, v. 1, p. xxii; The Arabic "Version," v. 2, p. xxix.
- As the product of a transmission involving more than one version and passed on as a single voice: Preface, p. vii; The Arabic "Version," v. 2, p. xxix.
- As representing a single text (M) with close correspondence to O in content: Summary and Commentary: Propositions 2, 4–6, 8–10, 12–22, 25–28, 31–34, 36–38, 40–43, 51–53, 55–56, 58, and 61, v. 2, pp. 30–95.
- As representing a single text (M) with close correspondence to O in structure: Summary and Commentary: Propositions 23, 29–30, 43, 52, 53, 56, 60, v. 2, pp. 62–96.
- As representing a single text (M) with close correspondence to R in content: Summary and Commentary: Definitions 1 and 2, v. 2, pp. 7–18, and Propositions 35, 39, 44–50, 54, 57, 62–64, v. 2, pp. 74–99.
- As representing a single text (M) with close correspondence to R in structure: Summary and Commentary: Propositions 6–7, 36–37, 45–47, and 64, v. 2, pp. 38–39, pp. 75–77, 84–86, and p. 99.
- As representing a few cases of closer correspondence to C than to O or R: Summary and Commentary: Proposition 37, p. 77, Proposition 38, p. 79.
- As lacking elements present in both O and R: Summary and Commentary: Definitions, v. 2, p. 6; Proposition 20, p. 55; Proposition 23, p. 62 [corresponding to O48, R44–45, R51].
- As containing elements absent in O: Summary and Commentary: Propositions 46 and 47, v. 2, pp. 85–86.
- As containing elements absent in R: Summary and Commentary: Propositions 30, 52, and 53, v. 2, p. 70 and pp. 89–90.

Relation to the Latin Tradition

- As corresponding to three Arabo-Latin Versions: Textual Transmission, v. 1, p. lxii, n. 54, and p. lxiv; The Arabic "Version," v. 2, p. xx, n. 10.
- As corresponding to the Greco-Latin version V: Textual Transmission, p. lxiv; Correspondence Table 2, v. 2, p. xxxiv; The Arabic "Version," v. 2, p. xxix, n. 32.

The Later "Recension"

- As a better-known and more premeditated form of revision of the same definitions and propositions: The Arabic "Version," v. 2, p. xx.
- As distinguished from the earlier version by longer definitions and shorter or alternative demonstrations: The Arabic Versions, v. 1, p. xxii; Summary and Commentary: Propositions 3, 10, v. 2, p. 35, n. 149, p. 45 and n. 176.
- As presenting problems of documentation including confusion with the earlier version: The Arabic "Version," v. 2, pp. xxi–xxiv.
- As represented by uniform manuscript copies more than ten times as many as the earlier version and almost three times of the *Correction (Iṣlāḥ)* attached to it: The Arabic Versions, v. 1, p. xxiii; List I, v. 1, pp. xxxiv–xliv.
- As containing an opening definition in much closer correspondence to the odd-numbered manuscripts: Variation Sample 1, v. 1, pp. xxxviii–xxxix.
- As representing a text distinct from the attached *Correction* in the treatment of the same two propositions, making such a sequel by Ṭūsī unnecessary: Summary and Commentary: The Arabic Versions, v. 1, p. xxi, n. 9.

8 The "Versions"

The Arabic "Version"

- As representing a "transformation" of the Euclidean texts: Structural Correspondence, v. 1, p. xxvii; Correspondence Tables 1 and 2, v. 1, pp. xxxiii–xxxiv; Summary and Commentary: Definitions, v. 2, pp. 6–21.
- As representing a transformation of the Euclidean terminology: Variation Samples 1–4, v. 1, pp. xxxvii–xlv; Summary and Commentary: Definitions 1–4, v. 2, pp. 7–21; Propositions 9, 20, v. 2, pp. 40–44 and pp. 55–59.
- As representing a transformation of Euclidean visual theory: Summary and Commentary: Definitions 1–4; Propositions 1–3, 9–10, 23, v. 2, pp. 7–65.

The Dependent "Versions"

- As affected by the Greco-Arabic transmission of the Euclidean assumptions: Textual Divergence, v. 2, p. xxxv; Variation Samples 1–4: v. 2, p. xxxvii–xlv; Summary and Commentary: Definitions, v. 2, p. 6–21.

- As affected by the internal transmission of manuscript transcriptions or marginal additions: Variation Sample 1, v. 2, p. xxxviii; Summary and Commentary: Definition 1, p. 10, p. 13, and n. 77–78; Definition 3, p. 18.
- As affected by the internal transmission of the transformed component:s: Summary and Commentary: Definition 1, pp. 7–14; Definition 2, pp. 17–18; Definition 3, p. 18–19.
- As affected by the Arabo-Latin transmission of the transformed components: Summary and Commentary: Definition 1, v. 2, pp. 7–8, and pp. 12–14 and n. 81; Definition 2, v. 2, p. 17 and n. 90; Definition 3, pp. 17–18 and n. 96.

9 The Conclusions

Conclusions Supported by Evidence

That the nature of the Greco-Arabic transmission largely directed the course of developments.

- Relevance of what *was* transmitted

  - The case of the transmitted Euclidean definitions: Summary and Commentary: Definitions 1–4, v. 2, pp. 7–21.
  - The case of propositions involving special problems in appearances: Summary and Commentary: Proposition 1–3, v. 2, pp. 21–36.
  - The case of propositions involving appearances vs. distances: Summary and Commentary: Proposition 9–10, pp. 40–48.
  - The case of propositions involving determination of magnitudes: Summary and Commentary, Propositions 19–22, pp. 54–61.
  - The case of the proposition involving appearance of curvatures: Summary and Commentary: Proposition 23, v. 2, p. 62, n. 220, p. 63. n. 219–220.

- Relevance of what *was not* transmitted

  - The case of two Euclidean definitions associated with linear perspective: Summary and Commentary: Proposition 9, v. 2, p. 44.
  - The case of the intended relationship between appearance and distance in the Euclidean proposition associated with linear perspective: Summary and Commentary, Proposition 9, v. 2, pp. 42–43 and n. 171–172.
  - The case of other parts involving appearance and distance: Summary and Commentary: Proposition 23, v. 2, p. 62, p. 63, n. 217–218.
  - The case of related treatments in Ptolemy's Optics: Summary and Commentary: Proposition 3, v. 2, p. 36 and n. 151.
  - The case of treatments of vision through mediums: The Arabic Tradition, v. 1, p. lvii.

- Relevance of *how* whatever was transmitted, *was* transmitted

  - The case of the Euclidean visual ray: Summary and Commentary, Definition 1, v. 2, pp. 8–10, pp. 12–14.
  - The case of the Euclidean visual cone: Summary and Commentary, Definitions 2, v. 2, pp. 15–18.
  - The case of the conditions for vision: Summary and Commentary, Definition 3, v. 2, pp. 18–19.
  - The case of the size and clarity of appearance: Summary and Commentary: Definition 4, v. 2, pp. 20–21.
  - The case of simultaneous vision: Summary and Commentary: Proposition 1, v. 2, pp. 23–29.
  - The case of clear vision: Summary and Commentary: Proposition 2, v. 2, pp. 30–34.
  - The case of limits of visibility: Summary and Commentary: Proposition 3, v. 2, pp. 34–36.
  - The case of the concept of proportion: Summary and Commentary: Proposition 9, v. 2, pp. 40–44.
  - The case of the appearance of shapes: Summary and Commentary: Proposition 10, v. 2, pp. 44–48; Proposition 23, pp. 62–65.
  - The case of the principle of reflection at a plane mirror: Summary and Commentary: Proposition, 20, v. 2, pp. 55–59 and n. 199.
  - The case of the corrupt Greek propositions on the appearance of motion: Summary and Commentary: Propositions 60–61, v. 2, pp. 95–97.

- Relevance of linguistic elements in textual transmission

  - The case of the flexible sense of the form *kathra*: Summary and Commentary, Definition 1, v. 2, p. 7 and p. 14.
  - The case of the flexible sense of the form *kathīr*: Summary and Commentary, Definition 1, v. 2, p. 7; Definition 4, v. 2, pp. 20–21.
  - The case of the flexible sense of the form *akthar*: Summary and Commentary, Definition 1, v. 2, p. 7; Proposition 2, v. 2, pp. 31–34.
  - The case of the flexible sense of the term *sumūt*: Summary and Commentary, Definition 1, v. 2, pp. 12–14.
  - The case of the flexible sense of the term *shu'ā'*: Summary and Commentary, Definition 1, v. 2, pp. 8–10.
  - The case of the flexible sense of the term *makhrūṭ*: Summary and Commentary, Definition 2, v. 2, pp. 15–17.
  - The case of the nonstandard expression *in'iṭāf*: Summary and Commentary, Proposition 20, v. 2, pp. 55–59.
  - The case of the nonstandard expression *'alā qadr/aqdār*: Summary and Commentary, Definition 9, v. 2, p. 40–44.

- Relevance of orthographic elements in textual transmission

  - The case of the close scripts of *kathīr* (كثير) and *kabīr* (كبير): Summary and Commentary: Definition 4, v. 2, p. 20, n. 100; Proposition 2, v. 2, p. 33, n. 138, n. 140.
  - The case of the close scripts of *ru'ya* (رؤية) and *zāwiya* (زاوية): Summary and Commentary: Proposition 2, v. 2, p. 33, n. 138; Proposition 10, p. 41, n. 163.
  - The case of the close scripts of *in'iṭāf* (انعطاف) and *in'ikās* (انعكاس): Summary and Commentary: Proposition 20, p. 56, n. 197.
  - The case of the close scripts of *yalī*/*'alā al-'ayn* (يلي/على العين) and *fī al-'ayn* (في العين): Summary and Commentary: Definition 2, v. 2, p. 17, n. 92.

- Dominance of one form over another in internal transmission

  - The case of *kathra* (in terms of multitude) over *kathra* (in terms of magnitude): Summary and Commentary: Definition 1, v. 2, p. 14 and n. 82.
  - The case of *samt* (as path/direction) over *samt* (as side) or *bu'd* (distance/interval): Summary and Commentary: Definition 1, v. 2, pp. 12–14, n. 77–78.
  - The case of *al-shu'ā'* (as ray) over *shu'ā' al-baṣar*, *manāẓir* (as *visual* ray[s]), or *quwwa nūriyya* (luminous power): Summary and Commentary, Definition 1, v. 2, p. 9, n. 64; p. 10, n. 69; Proposition 21, v. 2, p. 60.
  - The case of *yakhruj* (as something coming out) over *yanbathth*/*yanba'ith* (as something spreading/being sent out): Summary and Commentary, Definition 1, v. 2, p. 10–11, and n. 72–75.
  - The case of *makhrūṭ* (with undefined base) over *ṣanawbar*: Summary and Commentary, Definition 2, v. 2, p. 16.
  - The case of *ra's* (apex) over *mustaḥadd*/*zujj*: Summary and Commentary, Definition 2, v. 2, p. 17 and n. 91.
  - The case of *yalī al-'ayn* (*next to* the eye) over *fī al-'ayn* (*in* the eye): Summary and Commentary, Definition 2, v. 2, pp. 17–18.

- Transfer of the dominant forms through further transmission

  - The case of the Arabic *kathra* to Latin *multitudo*: Summary and Commentary: Definition 1, v. 2, p. 14.
  - The case of the Arabic *al-shu'ā'* to Latin *radii*: Summary and Commentary: Title, v. 2, pp. 3–4.
  - The case of the Arabic *makhrūṭ* to Latin *pyramis*: Summary and Commentary: Definition 2, v. 2, pp. 16–17 and n. 90.
  - The case of the Arabic *yaqa'u* to Latin *inciderit*: Summary and Commentary: Definition 3, v. 2, p. 19 and n. 96.

Assumptions Contradicted by Evidence

- Relevance of linguistic elements in Arabic *exclusively*

  – The case of the flexible sense of the Greek διάστημα/*diastêma*: Summary and Commentary: Definition 1, v. 2, p. 12.
  – The case of the flexible sense of the Greek πλειόνων/*pleionôn*: Summary and Commentary: Definition 4, v. 2, p. 20.

- Relevance of orthographic elements in Arabic *exclusively*

  – The case of the close scripts of the Greek ὄψεις κατ' εὐθείας γραμμὰς (visual rays in straight lines) and ἐξαγομένας εὐθείας γραμμὰς (straight lines leading out): Summary and Commentary, Definitions, v. 2, p. 6, n. 60.
  – The case of the close scripts of the Greek τι ποιούσας ἀπ' ἀλλήλων (making from each other), and μεγεθῶν μεγάλων (of great magnitudes): Summary and Commentary, Definitions, v. 2, p. 6, n. 60.

- Relevance of linguistic elements in transmission *exclusively*

  – The case of the limited circulation of Ptolemy's *Optics* in Arabic: Preface, p. vii; The Arabic Tradition, v. 1, p. lx, and n. 46, p. lxiii, n. 57.
  – The case of the limited circulation of Ibn al-Haytham's *Optics* in Arabic: The Arabic Tradition, v. 1, p. lx, and n. 46.
  – The case of the wide circulation of Euclid's *Optics* in Arabic: Textual Transmission, v. 1, p. lxv; The Arabic "Version," v. 2, p. xxviii.
  – The case of the wide circulation of Euclid's *Optics* in Latin: Textual Transmission, v. 1, p. lxii–lxiv, and n. 55; The Arabic "Version," v. 2, p. xxviii, and n. 30.
  – The case of the wide circulation of al-Kindī's and Ibn al-Haytham's *Optics* in Latin: Textual Transmission, v. 1, pp. lxiv–lxv, and n. 55.

Generalizations Invalidated by Evidence

- That the nature of transmission determined the course of *all* developments, or certain developments, *exclusively*: Summary and Commentary: Proposition 2–3, v. 2, pp. 30–36; Proposition 9–10, v. 2, pp. 40–48.
- That *language* determined the course of *all* developments, or certain developments, *exclusively*: Summary and Commentary: Propositions 1, 2, 10, 23, 60–61, v. 2, pp. 23–29, pp. 44–48, pp. 62–65, and pp. 95–97.
- That the *Arabic* language determined the course of *all* developments, or certain developments, *exclusively*: The Arabic "Version," v. 2, p. xxviii; Summary and Commentary: Definitions, pp. 6–7; Proposition 2, pp. 30–31.

Bibliography and Indices

Bibliography

Afshār, *Pārs* = Afshār, Iraj, "Kitābkhānih-i Millī-yi Pārs," *Nuskhih'hā-yi khaṭṭī*, v. 5, no. 1039/5, M. T. Dānishpazhūh and Iraj Afshār (eds.), Tehran: Tehran University Press, 1346 (=1967).

Afshār and Dānishpazhūh, *Malik* = Afshār, Iraj, and Dānishpazhūh, M. T. (eds.), *Fihrist-i Kitāb'hā-yi khaṭṭī-yi Kitābkhānih-i Malik* (*vā-bastih bi Āstān-i Quds-i Raḍavī*), 9 vols., Mashhad: *Āstān-i Quds*, 1352–1371 (=1973–1992), v. 2, *Kitāb'hā-yi fārsī* (with the collaboration of M. B. Ḥujjatī and Aḥmad Munzavī), 1354 (=1975); vols. 5–9, *Majmū'ih'hā va jung'hā*, v. 5, 1363 (=1982); v. 6, 1366 (=1985); v. 8, 1369 (=1990); v. 9, 1371 (=1992).

Afshār and Dānishpazhūh, *Iṣfahān* = Afshār, Iraj, and Dānishpazhūh, M. T., "Fihrist-i Nuskhih'hā-yi khaṭṭī-yi Dānishgāh-i Iṣfahān," *Nuskhih'hā-yi khaṭṭī*, v. 11 and 12, Dānishpazhūh and Ḥakīmī (eds.), Tehran: Tehran University Press, 1362 (=1983).

Ahlwardt, *Königlichen Bibliothek* = *Verzeichniss der Arabischen Handschriften der Königlichen Bibliothek zu Berlin von W. Ahlwardt*, Fünfter Band (IX–XIV Buch), Berlin: A. Asher and Co., 1893.

'Āmirī, *al-Qawl fī al-Ibṣār wa al-mubṣar*: see Khalīfāt.

Aristotle, *Problems*, v. 1: Books I–XXI, edited by T. E. Page, E. Capps, and W. H. D. Rouse, with an English translation by W. S. Hett, Loeb Classical Library, Cambridge, Massachusetts: Harvard University Press, and London: William Heinemann Ltd., 1936.

Aristotle, *The Physics*, edited by E. H. Warmington with an English translation by Philip H. Wicksteed and Francis M. Cornford, Cambridge: Harvard University Press, London: William Heinemann Ltd., 1970.

Assemani, *Bibliotheca Medicea* = *Bibliothecae Mediceae Laurentianae et Palatinae Codicum Mss. Orientalium Catalogus,* sub auspiciis regiae celsitudinis serenissimi Francisci III . . . Stephanus Evodius Assemanus, Florence, 1742.

Atiyeh, G. N., *Al-Kindi: The Philosopher of the Arabs,* Rawalpindi: Islamic Research Institute, Publication no. 6, 1966.

Bacharach, Jere L., *A Near East Studies Handbook,* Seattle and London: University of Washington Press, 1974.

Badawī, 'Abdurraḥmān, *Commentaires sur Aristote perdus en Grec et Autres Épîtres*, Publiés et annotés par Dr. 'Abdurraḥmān Badawī, *Recherche Publiées sous la direction de l'Institut de Lettres Orientales de Beyrouth,* Nouvelle Série, Tome I, Beyrouth: Dar El-Machreq Éditeurs, 1971.

Bahādur, *Riāsat* = Ṣāhib Bahādur, M., *Fihrist-i Kutub-i 'arabī-yi Mawjūdih-i Kutubkhānih-i Riāsat-i Rāmpūr,* Rampur: The Nigar Book Agency, 1902.

Bakhtīshū', 'Ubaydullāh, *al-Rawḍa al-ṭibbiyya* = *Ar-Raoudat at-tibbiyya*, ed. Paul Sbath, Cairo: H. Friedrich and Co., 1927.

– The prefixes *al-* and *de* have been disregarded for alphabetizing the entries.

de Beaurecueil, *Manuscrits d'Afghanistan* = *Manuscrits d'Afghanistan* par S. de Laugier de Beaurecueil, O. P., Cairo: Institut Français d'Archéologie Orientale du Caire, *Recherches d'Archéologie, de Philologie et d'Histoires*, 26, 1964 [previously published as extrait de MIDEO: *Mélanges d'Institut Dominicain d'Etudes Orientales*, 3, 1956].

Bihrūzī and Faqīrī, *Fārs* = Bihrūzī, A. N., and Faqīrī, Ṣādiq, *Fihrist-i Kutub-i khaṭṭī-yi Kitābkhānih-i Millī-yi Fārs*, v. 2, Shiraz: Publication of the Society for Public Libraries of Shiraz, 1351 (=1972).

Bīrūnī, *Al-Tafhīm li awā'il al-ṣinā'at al-tanjīm*: see Wright, *The Book of Instruction in the Elements of the Art of Astrology by Abu'l-Rayḥān Muḥammad ibn Aḥmad al-Bīrūnī*.

Björnbo and Vogl, *Alkindi, Tideus und Pseudo-Euklid* = *Alkindi, Tideus und Pseudo-Euklid: Drei optische Werke*, herausgegeben und erklärt von A. Björnbo und Sebastian Vogl, *Abhandlungen zur Geschichte der mathematischen Wissenschaften*, Leipzig/Berlin, 1912, 26, 3: 1–176.

Blochet, *Bibliothèque Nationale* = *Bibliothèque Nationale, Catalogue des Manuscrits Arabes des Nouvelles Acquisitions* (1884–1924) par E. Blochet, Paris: Éditions Ernest Leroux, 1925.

Boyer, Carl B., "Aristotelian References to the Law of Reflection," *Isis*, 1946, 36, 104: 92–95.

Brockelmann, *GAL* = Brockelmann, Carl, *Geschichte der arabischen Litteratur*, 2nd edition, Leiden: E. J. Brill, 2 vols. and 3 suppl., 1937–1949 (*GAL I* = vol. I, 1943; *GAL II* = vol. II, 1949; *GALS I* = Suppl. I, 1937; *GALS II* = Suppl. II, 1938; *GALS III* = Suppl. III, 1943).

Brockelmann, Carl, "al-Ya'kūbī," *EI*[1], 1987, 8: 1152–1153.

Brownson, C. D., "Euclid's *Optics* and Its Compatibility with Linear Perspective," *Archive for History of Exact Sciences*, 1981, 24, 3: 165–194.

Burton, H. E., "The *Optics* of Euclid," *Journal of the Optical Society of America*, 1945, 35: 357–372.

Carmody, *Arabic Astronomical Sciences* = Carmody, Francis J., *Arabic Astronomical and Astrological Sciences in Latin Translation. A Critical Bibliography*, Berkeley and Los Angeles: University of California Press, 1956.

Cohen and Drabkin, *Source Book* = Cohen, Morris R., and Drabkin, I. E, *A Source Book in Greek Science*, Cambridge: Harvard University Press, 1958.

Daiber, Hans, *Aetius Arabus: Die Vorsokratiker in arabischer Überlieferung*, Wiesbaden, 1980.

Dānishpazhūh, *Mishkāt* = *Catalogue méthodique, descriptif, et raisonné des manuscrits scientifiques (médicine, mathématique, sciences occultes et histoire de philosophie) de la Bibliothèque de l'Université de Téhéran (Don de M. le Professeur Meshkāt)* par M. T. Daneche-Pajouh, (*Fihrist-i Kitābkhanih-i ihdā'ī-yi Āghā-yi Seyyed Muḥammad-i Mishkāt bi Kitābkhānih-i Dānishgāh-i Tehran*), Tehran: Tehran University Press, v. 3, part. 2, no. 181, 1332=1953; v. 3, part 5, no. 533, 1338=1959.

Dānishpazhūh, *Markazī* = *Catalogue méthodique, descriptif, et raisonné des manuscrits de la Bibliothèque Centrale de l'Université de Téhéran* par M. T. Daneche-Pajouh (*Fihrist-i Kitābkhānih-i Markazī-yi Dānishgāh-i Tehran*), Tehran: Tehran University Press, 18 vols., v. 8, no. 665, 1339=1960; v. 9, no. 669, 1340=1961; v. 12, no. 693, 1340=1961; v. 15, no. 1096, 1345=1966; v. 16, 1357 =1978.

Dānishpazhūh, *Mikrufilm'hā* = Dānishpazhūh, M. T., *Fihrist-i Mikrufilm'hā-yi Kitābkhānih-i Markazī-yi Dānishqāh-i Tehran*, Tehran University, 3 vols, v. 1, 1348 (=1969); v. 2, 1353 (=1974); v. 3, 1363 (=1984).

Dānishpazhūh and Munzavī, *Sipahsālār* = *Catalogue of Manuscripts in Sepahsālār-Library* by M. T. Daneshpazhuh, and A. N. Monzavi *(Fihrist-i Kitābkhānih-i Sipahsālār)*, 3 vols. (v. 3–5), Tehran: Iranian Council for Philosophy and Humanistic Studies, v. 3, ed. Munzāvī, 1340=1962, v. 5, ed. Dānishpazhūh, 1356 =1977.

Dānishpazhūh and Afshār, *Nuskhih'hā-yi khaṭṭī* = Dānishpazhūh, M. T., and Afshār, Iraj (eds.), *Nuskhih'hā-yi khaṭṭī*, Tehran: Tehran University Press, v. 4, 1345 (=1966); v. 5, 1346 (=1967).

Dānishpazhūh and 'Ilmī-yi Anvārī, *Majlis* = Dānishpazhūh, M. T., and 'Ilmī-yi Anvārī, B., *Fihrist-i Kutub-i khaṭṭī-yi Kitābkhānih-i Majlis-i Shawrā-yi Islāmī:* no. 2 *(Sinā-yi sābiq), Fārsī, 'Arabī, Turkī*, Tehran: Majlis Library Press, v. 2., 1359 (=1980).

Derenbourg, *Bibliothèque Nationale* = Derenbourg, H., *Les Manuscrits arabes de la Collection Schefer à la Bibliothèque Nationale*, Paris: Imprimerie Nationale, 1901.

De Young, Gregg R., *The Arithmetic Books of Euclid's Elements in the Arabic Tradition: An Edition, Translation and Commentary*, Ph.D. Thesis, Harvard University, 1981.

De Young, Gregg R., "The Arabic Textual Traditions of Euclid's *Elements*," *Historia Mathematica*, 1984, 11: 147–160.

Dorn, *Chanykov* = Dorn, B., "Die vordem Chanykov'sche, jetzt der Kaiserl. Öffentlichen Bibliothek zugehörige Sammlung von morgenländischen Handschriften," *Mélanges Asiatiques, tirés du Bulletin de l'Académie Impériale des Sciences de St. Pétersbourg*, 1865, 5: 221–313.

DSB = *Dictionary of Scientific Biography*, ed. C. C. Gillispie, 16 vols (including Supplement and Index), New York: Charles Scribner's Sons, 1970–1980.

EI[1] = *First Encyclopaedia of Islam*, 1st edition, Leiden: E. J. Brill, 1913–1936.

EI[2] = *Encyclopaedia of Islam*, 2nd edition, Leiden: E. J. Brill, 1960–

Engroff, John W., Jr., *The Arabic Tradition of Euclid's Elements, Book V*, Ph.D. Thesis, Harvard University, 1980.

Euclid, *Optics*: see Heiberg, *Euclidis Optica*; Burton, "The Optics of Euclid;" Ver Eecke, *Euclide: l'Optique et la Catoptrique*.

Euclid, *De visu*: see Theisen, "*Liber de visu*."

Euclid, *Elements of Geometry*: see Heath, *The Thirteen Books of Euclid's Elements*.

Fārābī, Abū Naṣr, *Iḥṣā' al-'ulūm*, ed. 'Uthmān Amīn [= Osman Amine], Cairo: Librairie Anglo-Égyptienne, 1968 [first edition, Cairo, 1931].

Fārisī, Kamāl al-Dīn, *Tanqīḥ al-Manāẓir li-dhawī al-abṣār wa al-baṣā'ir*, 2 vols., Hyderabad, 1347–1348 (=1928–1930).

Galen, *De usu Partium*, Arabic translation: *Kitāb al-Manāfi' al-a'ḍā'*, *Bibliothèque Nationale*: MS ar. 2583; English translation, see Margaret May.

Gartz, J. C., *De Interpretibus et explanatoribus Euclidis arabicis*, Halle, 1823.

Goldstein, Bernard (ed.) "The Arabic Version of Ptolemy's *Planetary Hypotheses*," *Transactions of the American Philosophical Society*, 1976, 57, 4: 3–35.

Grant, Edward (ed.), *A Source Book in Medieval Science*, Cambridge, Massachusetts: Harvard University Press, 1974.

Gulchīn Ma'ānī, *Āstān-i Quds* = *A Catalogue of Manuscripts in the Astan Quds Razavi Library* (*Fihrist-i Kutub-i khaṭṭī-yi Kiṭabkhānih-i Āstān-i Quds-i Raḍavī*) by Ahmad Golchin Maani, v. 8, Mashhad: A publication of the Cultural and Library Affairs, no. 6, 1350=1971.

H. P. Kraus Catalogue 137 = *Important Works in the Field of Science*, New York: H. P. Kraus, 1974.

Ḥā'irī, *Majlis* = A *Catalogue of the Manuscripts in the Parliament Library: (Persian-Arabic-Turkish)* by Abdol Hossein Haeri (*Fihrist-i Kitābkhānih-i Majlis-i Shawrā-yi Millī*), v. 19, Tehran: *Majlis Library*, 1350=1972; v. 22–23 = *Catalogue of Iranian Parliament Library Manuscripts* (*Fihrist-i Kitābkhānih-i Majlis-i Shawrā-yi Islāmī*) *Collection of Mohammad Sadegh Tabatabai* by AbdolHossein Haeri, Tehran: Iranian Majles Library and Documentation Center (IMALDOC), v. 2, 1374=1996; v. 23, 1376=1997.

Ḥājjī Khalīfa, *Kashf al-ẓunūn* = *Kashf al-ẓunūn 'an asāmī al-kutub wa al-funūn*, *Lexicon Bibliographicum et Encyclopaedicum a Mustafa Ben Abdallah. . Ḫaji Khalfa*, ed. Gustav Flügel, v. 5, London: The Oriental Translation Fund, 1850.

Hava, J. G., *Al-Farā'id:Arabic-English Dictionary*, Beirut: Catholic Press, 1964.

Heath, Thomas L., *The Thirteen Books of Euclid's Elements, Translated from the Text of Heiberg, with Introduction and Commentary*, Second Edition, Revised with Additions, 3 vols., Cambridge: Cambridge University Press, 1926 [reprinted New York: Dover Publications, 1956].

Heiberg, J. L., *Litterargeschichtliche Studien über Euklid*, Leipzig: Teubner, 1882.

Heiberg, *Euclidis Optica* = Heiberg, J. L., *Euclidis Optica, Opticorum recensio Theonis, Catoptrica, cum scholiis antiquis*, edidit Heiberg, *Euclidis Opera Omnia*, ediderunt J. L. Heiberg et H. Menge, vol. VII, Leipzig: Teubner, 1895: 1–121.

Heiberg, *Opticorum recensio Theonis* = Heiberg, J. L., *Euclidis Optica, Opticorum recensio Theonis, Catoptrica, cum scholiis antiquis*, edidit Heiberg, *Euclidis Opera Omnia*, ediderunt J. L. Heiberg et H. Menge, vol. VII, Leipzig: Teubner, 1895: 143–247.

Hultsch, *Pappus Alexandrinus Collectio* = *Pappi Alexandrini Collectionis: quae supersunt*. E libris manu scriptis edidit Latina interpretatione et commentariis instruxit Fridericus Hultsch, 3 vols., Berlin: 1876–1878, volumen II, reprinted Amsterdam: Hakkert, 1965.

Ḥunayn Ibn Isḥāq, *Kitāb al-'Ashar maqālāt fī al-'ayn*, see Meyerhof, *The Book of the Ten Treatises of the Eye Ascribed to Hunain ibn Is-hâq (809–877 A.D.)*.

Ḥusaynī and Mar'ashī, *Mar'ashī* = Ḥusaynī, S. A., and Mar'ashī, S. M., *Fihrist-i Nuskhih'hā-yi khaṭṭī-yi Kitābkhānih-i Haḍrat-i Āyatullāh al-'Uẓmā Najāfī-yi Mar'ashī . . .*, v. 11, 1364 (=1985); v. 12, 1365 (=1986).

Ibn Abī Uṣaybi'a, *'Uyūn al-anbā' fī Ṭabaqāt al-aṭibbā'*, ed. August Müller, 2 vols., Cairo, 1299 (=1882); Königsberg, 1884.

Ibn al-Haytham, *Maqāla* (or *Qawl*) *fī al-Ḍaw'*, *Majmū' al-rasā'il*, Hyderabad, 1357 (=1938–39); see also Rashed, "Le 'Discours de la Lumière' d'Ibn al-Haytham."

Ibn al-Haytham, *al-Shukūk 'alā Baṭlamyūs*: see Sabra and Shehaby, *Alhazen's al-Shukūk 'alā Baṭlamyūs*.

Ibn al-Haytham, *Kitāb al-Manāẓir*: see Sabra (ed.), *Kitāb al-Manāẓir: Books I–II–III*, and Sabra, *The Optics of Ibn al-Haytham*, 2 vols..

Ibn al-Haytham, *Risāla fī Ru'yat al-kawākib*: see Sabra and Heinen, "On Seeing the Stars."

Ibn Khaldūn, *Muqaddimah*, ed. 'A. 'A. Wāfī, Cairo: Lajnat al-Bayān al-'Arabī, v. 3, 1957 (translated from the Arabic by Franz Rosenthal, *The Muqaddimah*, 3 vols., Princeton, N. J., 1967).

Ibn al-Nadīm, *Kitāb al-Fihrist*, ed. Flügel = Ibn al-Nadīm, Muḥammad ibn Isḥāq, *Kitāb al-Fihrist*, edited by Gustav Flügel, 2 vols., Leipzig: Verlag von F. C. W. Vogel, 1871–1872; reprinted, Beirut: Khayyāṭ, 1964.

Ibn al-Nadīm, *Kitāb al-Fihrist*, tr. Dodge = *The Fihrist of al-Nadīm: A Tenth-Century Survey of Muslim Culture*, Bayard Dodge (editor and translator), 2 vols. (Records of Civilization: Sources and Studies no. 83), New York and London: Columbia University, 1970.

Ibn al-Qifṭī, *Ta'rīkh al-ḥukamā'* = *Ibn al-Qifṭī's Ta'rīḫ al-ḥukamā'*, ed. J. Lippert, Leipzig: Dieterich'sche Verlagsbuchhandlung, 1903.

Ibn Sīnā, *Al-Shifā*, *Al-Ṭabī'iyyāt*, 6: *Kitāb al-Nafs*, ed. G. C. Anawati et S. Zayed, Cairo: al-Hay'at al-Miṣriyya, 1394=1974.

'Irfāniān, *Āstān-i Quds* = 'Irfāniān, Gh., *A Descriptive Catalogue of the Persian, Arabic & Turkish MSS. Preserved in the Astan Quds Library (Fihrist-i Kutub-i khaṭṭī-yi Kiṭabkhānih-i Āstān-i Quds-i Raḍavī)*, vol X, including Algebra, Arithmetic, Astronomical-tables, Astronomy, Geometry, etc., Mashhad: A. Q. Library Publications, no. 8, 1362=1984.

Iskandar, *University of California* = Iskandar, A. Z., *A Descriptive List of Manuscripts on Medicine and Science at the University of California*, Leiden: E. J. Brill, 1984.

I'tiṣāmī, *Majlis* = I'tiṣāmī, Y. (I'tiṣām al-Mulk), *Catalogue des manuscrits Persans et Arabes de la Bibliothèque du Madjless par Y. Etessami (Fihrist-i Kitābkhānih-i Majlis-i Shawrā-yi Millī)*, Publication de la Bibliothèque, vol. II, Téhéran: Imprimérie du Madjless, 1311=1933.

Ivanow, *Curzon Collection* = Ivanow, Wladimir, *Concise Descriptive Catalogue of Persian Manuscripts in the Curzon Collection*, Calcutta: Asiatic Society of Bengal, 1926.

Jāḥiẓ, *Kitāb al-Tarbī' wa al-tadwīr*: see Pellat, *Le Kitāb al-tarbī' wa-t-tadwīr de Ǵāḥiẓ*.

Jones, Alexander, "On Some Borrowed and Misunderstood Problems in Greek Catoptrics," *Centaurus*, 1987, 30, 1: 1–17.

Jones, "Peripatetic and Euclidean Theories" = Jones, Alexander, "Peripatetic and Euclidean Theories of the Visual Ray," *Physis*, 1994, 31, 1: 47–76.

de Jong and de Goeje, *Bibliotheca Academiae Lugduno Batavae* = *Catalogus Codicum Orientalium Bibliothecae Academiae Lugduno Batavae, Auctoribus* P. De Jong et M. J. De Goeje, volumen Tertium, Lugduni Batavorum: E. J. Brill, 1865.

Karatay, *Topkapı Sarayı* 12 = Karatay, Fehmi Edhem, *Topkapı Sarayı Müzesi Kütüphanesi Farsça Yazmalar Kataloğu*, Istanbul Topkapı Sarayı Müzesi Yayinlari, Istanbul: Topkapı Sarayı Müzesi, no. 12, 1961.

Karatay, *Topkapı Sarayı* 15 = Karatay, Fehmi Edhem, *Topkapı Sarayı Müzesi Kütüphanesi Arapça Yazmalar Kataloğu*, Istanbul Topkapı Sarayı Müzesi Yayinlari, Istanbul: Topkapı Sarayı Müzesi, no. 15, 1966.

Khalīfāt, *'Āmirī, Rasā'il* = *Rasā'il-i Abu al-Ḥasan-i 'Āmirī bā muqaddamih va taṣḥīḥ-i Saḥbān Khalīfāt, tarjumih-i muqqadamih, Mihdī Tadayyun*, Tehran: Markaz-i Nashr-i Dānishgāhī (University Publications), 1375=1996 [earlier edition, *Rasā'il Abī al-Ḥasan al-'Āmirī wa-shadharātuhu al-falsafīyah: dirāsah wa-nuṣūs*, Amman: al-Jāmi'ah al-Urdunīyah, 1988].

Kheirandish, Elaheh, *The Medieval Arabic Tradition of Euclid's Optika*, Ph.D. Thesis, Harvard University, 1991.

Kheirandish, Elaheh, "The Arabic 'Version' of Euclidean Optics: Transformations as Linguistic Problems in Transmission," *Tradition, Transmission, Transformation*: 227–247, see Ragep and Ragep (eds.).

Kheirandish, Elaheh, "Optics in the Islamic World," *Encyclopaedia of the History of Science, Technology, and Medicine in Non-Western Cultures*, Dordrecht: Kluwer Academic Publishers, 1997: 795–799.

Kheirandish, Elaheh, "The 'Manāẓir' Tradition through Persian Sources," *La sciences dans la monde iranien,* Tehran: Institut Français de Recherche en Iran (IFRI), forthcoming.

Kheirandish, Elaheh, "Optics: Highlights from Islamic Lands," *Science in Islam,* Paris: UNESCO, forthcoming.

Kindī, *De aspectibus*: see Björnbo and Vogl, *Alkindi, Tideus und Pseudo-Euklid: Drei optische Werke.*

King, "Notes on Sources" = King, David A., "Notes on the Sources for the History of Islamic Mathematics," *Journal of the American Oriental Society,* 1979, 99.3: 450–459.

King, *Cairo Catalogue* = King, David A., *A Catalogue of the Scientific Manuscripts in the Egyptian National Library,* Cairo: General Egyptian Book Organization in collaboration with The American Research Center in Egypt, and the Smithsonan Institution, Part II, 1986 [Part I, 1981].

King, *Cairo Survey* = King, David A., *A Survey of the Scientific Manuscripts in the Egyptian National Library,* published for The American Research Center in Egypt, Catalogs, v. 5, Winona Lake, Indiana: Eisenbrauns, 1986.

King, David A., "al-Marrākushī," EI^2, 1989, 6, fasc. 107–108: 598.

Klamroth, "Schriftstellern bei al-Ja'qūbī" = Klamroth, Martin, "Ueber die Auszüge aus griechischen Schriftstellern bei al-Ja'qūbī," *Zeitschrift der Deutschen Morgenländischen Gesellschaft* (ZDMG), 1888, 42: 1–44.

Knorr, Wilbur, "Archimedes and Pseudo-Euclidean *Catoptrics:* Early Stages in the Ancient Geometric Theory of Mirrors," *Archives Internationales d'Histoire des Sciences,* 1985, 35, 114–115: 28–105.

Knorr, Wilbur, "On the Principle of Linear Perspective in Euclid's *Optics,*" *Centaurus,* 1991, 34: 193–210.

Knorr, Wilbur, "When Circles Don't Look Like Circles: An Optical Theorem in Euclid and Pappus," *Archive for History of Exact Sciences,* 1992, 44, 4: 287–329.

Knorr, "Pseudo-Euclidean Reflections in Ancient Optics" = Wilbur R., "Pseudo-Euclidean Reflections in Ancient Optics: A Re-Examination of Textual Issues Pertaining to the Euclidean *Optica* and *Catoptrica,*" *Physis,* 1994, 31, 1: 1–45.

Krause, "Stambuler Handschriften" = Krause, Max, "Stambuler Handschriften islamischer Mathematiker," *Quellen und Studien zur Geschichte der Mathematik, Astronomie und Physik,* Berlin, 1936, Band 3, Heft 4: 437–532.

Kunitzsch, *Der Almagest* = Kunitzsch, Paul, *Der Almagest: Die Syntaxis Mathematica des Claudius Ptolemäus in arabisch-lateinischer Überlieferung,* Wiesbaden: Otto Harrassowitz, 1974.

Kunitzsch, "Ibn aṣ-Ṣalāḥ" = Kunitzsch, Paul, "Ibn aṣ-Ṣalāḥ Zur Kritik der Koordinatenüberlieferung im Sternkatalog des Almagest", *Abhandlungen der Akademie der Wissenschaften in Göttingen,* Philologisch-Historische Klasse Dritte Folge, Nr. 94, Göttingen: Vandenhoeck & Ruprecht, 1975.

Lejeune, *Euclide et Ptolémée* = Lejeune, Albert, *Euclide et Ptolémée, deux stades de l'optique géométrique grecque*, Louvain: Université de Louvain, *Recueil de travaux d'histoire et de philologie*, 3e série, 31e fasc., 1948.

Lejeune, *L'Optique de Claude Ptolémée* = Lejeune, Albert, *L'Optique de Claude Ptolémée dans la version latine d'après l'arabe de l'émir Eugène de Sicile: Édition critique et exégétique augmentée d'une traduction française et de complément*, Leiden: E. J. Brill, *Collection de Travaux de l'Académie Internationale d'Histoire des Sciences*, no. 31, 1989 [previously published as: *L'Optique de Claude Ptolémée dans la version latine d'après l'arabe de l'émir Eugène de Sicile*, Louvain: Université de Louvain, *Recueil de travaux d'histoire et de philologie*, 4. sér., fasc. 8, 1956].

Lejeune, Albert, *Recherches sur la catoptrique greque, d'après les sources antiques et médiévales*, Brussels: Académie Royale des Sciences, des Lettres et des Beaux-arts, 1957.

Liddell and Scott, *A Greek-English Lexicon*, compiled by Henry George Liddell and Robert Scott, with a Supplement, Oxford: Clarendon Press, 1968.

Lindberg, David C., "Alkindi's Critique of Euclid's Theory of Vision," *Isis*, 1971, 62: 469–489 [reprinted in Lindberg, *Studies in the History of Medieval Optics*, Variorum Reprints, London, 1983, II: 469–489].

Lindberg, David C., *A Catalogue of Medieval and Renaissance Optical Manuscripts*, Toronto: The Pontifical Institute of Mediaeval Studies, Subsidia Mediaevalia, IV, 1975.

Lindberg, *Theories of Vision* = Lindberg, David C., *Theories of Vision from Al-Kindi to Kepler*, Chicago and London: The University of Chicago Press, 1976.

Lindberg, David, C., "The Intromission–Extramission Controversy in Islamic Visual Theory: Alkindi versus Avicenna," *Interrelations in the History and Philosophy of Science*, edited by Peter K. Machame and Robert Turnbull, Columbus: Ohio State University Press,1978 [reprinted in Lindberg, *Studies in the History of Medieval Optics*, London: Variorum Reprints, 1983, IV: 137–159].

Lindberg, David C., *Roger Bacon and the Origins of Perspectiva in the Middle Ages: A critical edition and English translation of Bacon's Perspectiva with Introduction and Notes*, Oxford: Clarendon Press, 1996.

Mach, *Garrett Collection* = *Catalogue of Arabic Manuscripts (Yahuda Section) in the Garrett Collection*, Princeton University Library by Rudolph Mach, Index by Robert D. McChesney, Princeton: Princeton University Press, 1977.

May, Margaret T., Galen *On the Usefulness of the Parts of the Body: Translated from the Greek with an Introduction and Commentary*, 2 vols., Ithaca: Cornell University Press, 1968–1969.

Meyerhof, Max, "Die Optik der Araber," *Zeitschrift für ophthalmologische Optik*, 1920, 8:16–29, 42–54, 86–90.

Meyerhof, Max, *The Book of the Ten Treatises on the Eye Ascribed to Hunain ibn Is-hâq (809–877 A.D.)*, Cairo: Government Press, 1928.

Mingana, *John Rylands* = Mingana, A., *Catalogue of the Arabic Manuscripts in the John Rylands Library*, Manchester: The Manchester University Press, Deansgate, 1934.

Mudarris-i Raḍavī, *Aḥvāl va āthār* = Mudarris-i Raḍavī, M. T., *Aḥvāl va athār-i Ustād-i Bashar . . . mulaqqab bi Khwāja Naṣīr al-Dīn*, Tehran: Tehran University Press, no. 282, 1334 (=1955) [reprinted Tehran: Bunyād-i Farhang-i Īrān, 1354 (=1975), 1370 (=1991)].

Mudarrisī, *Sarguzasht* = Mudarrisī-yi Zanjānī, M., *Sarguzasht va 'aqāyid-i falsafī-yi Khāwaja Naṣīr al-Dīn-i Ṭūsī*, Tehran: Tehran University Press, no. 309, 1335 (=1956) [reprinted, Tehran: Amīr Kabīr, 1363 (=1984)].

Munzavī, *Persian Manuscripts* = *A Catalogue of Persian Manuscripts* (*Fihrist-i Nuskhih'hā-yi khaṭṭī-yi fārsī*) by Ahmad Monzavi, Tehran: Regional Cultural Institute, publication, v. 1, no. 14, 1348 (=1969).

Murdoch, John E., "Euclid," *DSB*, 4: 438–443.

Mushār, *Mu'allifīn* = Mushār, Khān Bābā, *Mu'allifīn-i kutub-i chāppī-yi fārsī va 'arabī*, Tehran, v. 4, 1342 (=1963); v. 5, 1343 (=1964).

Mushār, *Chāppī-yi 'arabī* = Mushār, Khān Bābā, *Fihrist-i Kitāb'hā-yi chāppī-yi 'arabī*, Tehran, 1344 (=1965).

Mushār, *Chāppī-yi fārsī* = Mushār, Khān Bābā, *Fihrist-i Kitāb'hā-yi chāppī-yi fārsī*, Tehran, v. 2, 1351 (=1972).

Nicoll and Pusey, *Bibliotheca Bodleiana* = *Bibliothecae Bodleianae codicum manuscriptorum orientalium catalogi, partis secundae volumen secundum arabicos complectens confecit Alexander Nicoll, J. C. D., edidit et catalogum Urianum aliquatenus emendavit* E. B. Pusey, S. T .B., Oxford, 1835.

O'Leary, De Lacy, *How Greek Science Passed to the Arabs*, London: Routledge and Kegan Paul Ltd., 1949.

Pappus of Alexandria, *Collection*: see Hultsch, *Pappus Alexandrinus Collectio*, and Ver Eecke, *Pappus: La Collection*.

Pellat, Charles, *Le Kitāb al-tarbī' wa-t-tadwīr de Ǵāḥiẓ*, Damascus: Institut Français de Damas, 1955.

Pellat, Charles, *The Life and Works of Jāḥiẓ: Translations of selected texts*, Berkeley and Los Angeles: University of California Press, 1969.

Peters, F. E., *Aristoteles Arabus: The Oriental Translations and Commentaries on the Aristotelian Corpus*, Leiden: E. J. Brill, 1968.

Petraitis, Casimir, *The Arabic Version of Aristotle's Meteorology: A Critical Edition with an Introduction and Greek-Arabic Glossaries*, Beyrouth: Dar El-Machreq Éditeurs, 1967.

Plooij, E. B., *Euclid's Concept of Ratio and His Definition of Proportional Magnitudes as Criticized by Arabian Commentators*, Rotterdam: Uitgeverij W. J. van Hengel, 1950.

Ptolemy, *Optica*: see Lejeune, *L'Optique de Claude Ptolémée*, and Smith, "Ptolemy's Theory of Visual Perception: An English Translation of the *Optics* with Introduction and Commentary."

Ptolemy, *Planetary Hypotheses*: see Goldstein, "The Arabic Version of Ptolemy's *Planetary Hypotheses*."

Ragep, F. J., *Naṣīr al-Dīn al-Ṭūsī's Memoir on Astronomy: al-Tadhkira fī 'ilm al-hay'a, Edition, Translation and Commentary*, 2 vols., New York: Springer-Verlag, *Sources in the History of Mathematics and Physical Sciences*, no. 12 (ed. G. J. Toomer), 1993.

Ragep and Ragep (eds.), *Tradition, Transmission, Transformation* = *Tradition, Transmission, Transformation: Proceedings of Two Conferences on Pre-modern Science Held at the University of Oklahoma*, eds. F. Jamil Ragep and Sally P. Ragep with Steven Livesey, Leiden: E. J. Brill, *Collection de travaux de l'Académie Internationale d'Histoire des Sciences*, Tome 37 (ed. John D. North), 1996.

Rashed, Roshdi, "Le 'Discours de la Lumière' d'Ibn al-Haytham (Alhazen)," *Revue d'Histoire des Sciences*, 1968, 21:197–224 [reprinted in Rashed, *Optique et mathématiques*, 1992, V: 197–224].

Rashed, Roshdi, "A Pioneer in Anaclastics: Ibn Sahl on Burning Mirrors and Lenses," *Isis*, 1990, 81, 308: 464–491 [reprinted in Rashed, *Optique et mathématiques*, 1992, VI: 464–491].

Rashed, *Optique et mathématiques* = Rashed, Roshdi, *Optique et mathématiques: Recherches sur l'histoire de la pensée scientifique en arabe*, London: Variorum Reprints (Collected Studies Series no. 388), 1992.

Rashed, Roshdi, *Géometrie et dioptrique au Xe siècle: Ibn Sahl, al-Qūhī et Ibn al-Haytham*, *Collection Sciences et Philosophie Arabes: Textes et Études*, Paris, les Belles Lettres, 1993.

Ritter and Plessner, "Stambuler Bibliotheken" = Ritter, H. and Plessner, M., "Schriften Ja'qūb ibn Isḥāq al-Kindī's in Stambuler Bibliotheken," *Archiv Orientalni*, 1932, 4: 363–372.

Rosenthal, Franz, "On the Knowledge of Plato's Philosophy in the Islamic world," *Islamic Culture*, 1940, 14: 412–416.

Rosenthal, Franz, "al-Kindî and Ptolemy," *Studi orientalistici in onore di Giorgio Levi della Vida*, Rome: Istituto per l'Oriente, 1956, 2: 436–456.

Rosenthal, Franz, *Das Fortleben der Antike im Islam*, Zurich, 1965, translated as *The Classical Heritage in Islam*, by E. and J. Marmorstein, London: Routledge and Kegan, Paul, 1975.

Sabra, A. I., "The Astronomical Origin of Ibn al-Haytham's Concept of Experiment," *Actes XIIe Congrès International d'Histoire des Science*, Paris: Albert Blanchard, 1971: 133–136 [reprinted in Sabra, *Optics, Astronomy and Logic*, 1994, VI: 133–136].

Sabra, A. I., "Ibn al-Haytham," *DSB*, 1972, 6: 189–210 [reprinted in Sabra, *Optics, Astronomy and Logic*, 1994, II: 189–210].

Sabra, A. I., "The Physical and the Mathematical in Ibn al-Haytham's Theory of Light and Vision," reproduced with corrections from *The Commemoration Volume of Bīrūnī International Congress in Tehran*, Tehran: High Council of Culture and Arts, publication no. 38, 1976: 439–478 [reprinted in Sabra, *Optics, Astronomy and Logic*, 1994, VII: 1–20].

Sabra, A. I., "A Note on Codex Biblioteca Medicea Laurenziana Or. 152," *Journal for the History of Arabic Science*, 1977, 1, 2: 276–283.

Sabra, A. I., "Ibn al-Haytham and the Visual-Ray Hypothesis," *Ismā'īlī Contributions to Islamic Culture*, ed. S. H. Nasr, Tehran: Imperial Iranian Academy, 1977: 189–205.

Sabra, A. I., "Sensation and Inference in Alhazen's Theory of Visual Perception," in Machamer and Turnball (eds.) *Studies in Perception: Interrelations in the History and Philosophy of Science*, Columbus: Ohio State University Press, 1978: 160–185 [reprinted in Sabra, *Optics, Astronomy and Logic*, 1994, X: 160–185].

Sabra (ed.), *Kitāb al-Manāẓir* = Sabra, A. I., *Al-Ḥasan Ibn al-Haytham, Kitāb al-Manāẓir: Books I–II–III <On Direct Vision>: Edited with Introduction, Arabic-Latin Glossaries and Concordance Tables by Abdelhamid I. Sabra*, Kuwait: The National Council for Culture, Arts and Letters, 1983.

Sabra, A. I., "Manāẓir, or 'Ilm al-Manāẓir," *EI*[2], 1987, 6, fasc. 103–104: 376–377.

Sabra, A. I., "Optics, Islamic," *Dictionary of the Middle Ages* edited by Joseph R. Strayer, New York: Charles Scribner's Sons, 1987, 9: 240–247.

Sabra, A. I. "Psychology vs. Mathematics: Ptolemy and Alhazen on the Moon Illusion," *Mathematics and its Applications to Science and Natural Philosophy in the Middle Ages: Essays in the Honor of Marshall Clagett*, ed. Edward Grant and John E. Murdoch, Cambridge: Cambridge University Press, 1987: 217–247 [reprinted in Sabra, *Optics, Astronomy and Logic*, 1994, IX: 217–247].

Sabra, *The Optics of Ibn al-Haytham* = Sabra, A. I., *The Optics of Ibn al-Haytham: Books I–III On Direct Vision, Translated with Introduction and Commentary by A. I. Sabra*, 2 vols. (Studies of the Warburg Institute edited by J. B. Trapp, vols. 40 i–ii), London: The Warburg Institute, University of London, 1989.

Sabra, A. I., "Form in Ibn al-Haytham's Theory of Vision," *Zeitschrift für Geschichte der arabisch-islamischen Wissenschaften*, 1989, 5:115–140 [reprinted in Sabra, *Optics, Astronomy and Logic*, 1994, XI: 115–140].

Sabra, *Optics, Astronomy and Logic* = Sabra, A. I., *Optics, Astronomy and Logic: Studies in Arabic Science and Philosophy*, London: Variorum Reprints (Collected Studies Series no. 444), 1994.

Sabra, A. I., and Shehaby, Nabil (eds.), *Alhazen's al-Shukūk 'alā Baṭlamyūs (Dubitationes in Ptolemaeum)*, Edition and Introduction by A. I. Sabra and Nabil Shehaby, Cairo: National Library, 1971.

Sabra and Heinen, "On Seeing the Stars" = Sabra, A. I., and Heinen, Anton, "On Seeing the Stars: Edition and Translation of Ibn al-Haytham's *Risāla fī Ru'yat al-kawākib*," *Zeitschrift für Geschichte der arabisch-islamischen Wissenschaften*, 1991/92, 7: 31–72.

Şeşen, *Maktabāt Turkiyya* = Seşen, R., *Nawādir al-makhṭūṭāt al-'arābiyya fī maktabāt Turkiyya*, Beirut 3 vols., 1975–1982, v. 1, 1975.

Şeşen, *Köprülü* = Seşen, R., *Catalogue of Manuscripts in the Köprülü Library*, Istanbul: OIC Research Center for Islamic History, Art and Culture, v. 1, 1986.

Sezgin, *GAS* = Sezgin, Fuat, *Geschichte der arabischen Schrifttums*, Leiden: E. J. Brill, 1970–1979 (*GAS V*: Mathematik, 1974; *GAS VI*: Astronomie, 1978; *GAS VII*: Astrologie, Meteorologie, und Verwandtes, 1979).

de Slane, *Bibliothèque Nationale* = Slane, William, Le Baron de, *Bibliothèque Nationale: Département des manuscrits, catalogue des manuscrits arabes*, Paris: Imprimerie Nationale, 1883–1895.

Smith, Mark A., "Ptolemy's Search for a Law of Refraction," *Archive for History of Exact Sciences*, 1982, 26, 3: 221–240.

Smith, "*Ptolemy's Theory of Visual Perception*" = Smith, A. Mark, "Ptolemy's Theory of Visual Perception: An English Translation of the *Optics* with Introduction and Commentary," *Transactions of the American Philosophical Society Held at Philadelphia for Promoting Useful Knowledge*, 1996, 86, 2: 1–270.

Steinschneider, "Die mittleren Bücher" = Steinschneider, Moritz, "Die 'mittleren' Bücher der Araber und ihre Bearbeiter," *Zeitschrift für Mathematik und Physik*: Leipzig: B. G. Teubner, 1865, 10: 456–498.

Steinschneider, Moritz, "Euklid bei den Arabern," *Zeitschrift für Mathematik und Physik*: Historische-Literarische Abtheilung, 1886, 31, 3: 81–109.

Steinschneider, "Die arabischen Uebersetzungen" = Steinschneider, Moritz, "Die arabischen Uebersetzungen aus dem Griechischen," *Zeitschrift der Deutschen Morgenländischen Gesellschaft* (*ZDMG*), Leipzig, 1896, 50, 2: 161–417.

Storey, *Persian Literature* = Storey, Charles, *Persian Literature: A Biobibliographical Survey*, v. 2, part 2, London: Luzac and Co., 1958; v. 2 part 3, Leiden: The Royal Asiatic Society of Great Britain and Ireland, 1977.

Suter, "*Die Mathematiker*" = Suter, Heinrich, "Die Mathematiker und Astronomer der Araber und ihre Werke," *Abhandlungen zur Geschichte der Mathematischen Wissenschaften*, X, Leipzig, 1900.

Tahānawī, *Kashshāf iṣṭilāḥāt al-funūn*: Ta'līf al-Shaikh Muḥammad ibn 'Alī ibn 'Alī al-Tahānawī, ed. 1318 (=1900).

Takahashi, Ken'ichi, *The Medieval Latin Traditions of Euclid's Catoptrica*, A Critical Edition of *De speculis* with an Introduction, English Translation and Commentary, Higashi-ku: Kyushu University Press, 1992.

Theisen, *The Mediaeval Tradition* = Theisen, Wilfred R., *The Mediaeval Tradition of Euclid's Optics*, Ph.D. Thesis, University of Wisconsin, 1972, facsimile, Ann Arbor: University Microfilms International, 1984.

Theisen, "*Liber de visu*" = Theisen, Wilfred R., "*Liber de visu*: The Greco-Latin Translation of Euclid's *Optics*" [edited with introduction and notes], *Mediaeval Studies*, Toronto: Pontifical Institute of Mediaeval Studies, 1979, 41: 44–105.

Ṭihrānī, *Al-Dharī'a* = Ṭihrānī, Sheykh Āghā Buzurg, *Al-Dharī'a ilā taṣānīf al-Shī'a*: Najaf: Maṭba'at al-ghurā, v. 3, 1357 (=1938), v. 5, 1363 (=1944); v. 9, part 3, *Bibliography of Shi'ite's Literary Works*, Tehran: Danshgahe Tehran, 1383=1964, v. 17, Tehran: Islamiyya, 1387 = 1967.

Tobin, Richard, "Ancient Perspective and Euclid's Optics," *Journal of the Warburg and Courtauld Institutes*, 1990, 53: 14–42.

Toomer, G. J., *Diocles on Burning Mirrors: The Arabic Translation of the Lost Greek Original*, Edited with English Translation and Commentary by G. J. Toomer, New York: Springer-Verlag, *Sources in the History of Mathematics and Physical Sciences*, no. 1 (eds. M. J. Klein ad G. J. Toomer), 1976.

Toomer, G. J., "Ptolemy," *DSB*, 1976, 11: 186–206.

Toomer, G. J., *Apollonius, Conics Books V to VII. The Arabic Translation of the Lost Greek Original in the Version of Banū Mūsā*, Edited with Translation and Commentary, 2 vols., NewYork: Springer-Verlag, 1990.

Ṭūsī, Naṣīr al-Dīn, *Taḥrīr al-Manāẓir, Majmū'at al-Ukarāt wa al-Mutawassiṭāt*, Tehran: Lithograph Publication, 1304 (= 1886–1887).

Ṭūsī, Naṣīr al-Dīn, *Taḥrīr al-Manāẓir, Majmū' al-rasā'il*, Hyderabad: Uthmāniyya, part 1, 1358 (=1939), 5: 2–24.

Ṭūsī, Naṣīr al-Dīn, *Taḥrīr al-Manāẓir*, ed. A. S. Dimirdash, *Revue de l'Institut des Manuscrits Arabes (RIMA)*, 1963, 9: 251–290.

Ṭūsī, Naṣīr al-Dīn, *Talkhīs al-Muḥaṣṣal*, ed. A. Nūrānī, Mc Gill University, Montreal Canada: Institute of Islamic Studies, Tehran Branch, in collaboration with Tehran University, Tehran, 1980.

Tytler, John, "Analysis and Specimens of a Persian Work on Mathematics and Astronomy," *The Journal of the Royal Asiatic Society of Great Britain and Ireland*, v. 4 , London: J. W. Parker, West Strand, 1837: 254–272.

Ullmann, Manfred, *Wörterbuch der Klassischen Arabischen Sprache*, Band I, Begründet von Jörg Kraemer und Helmut Gätje, in Verbindung mit Anton Spitaler bearbeit von Manfred Ullmann, Weisbaden: Otto Harrassowitz, 1970.

Ullmann, Manfred, "Die Natur- und Geheimwissenschaften im Islam," *Handbuch der Orientalistik*, Ergänzungsband VI, Zweiter Abschnitt, Leiden: E. J. Brill, 1972.

Uri, *Bibliotheca Bodleiana* = Uri, Joannes, *Bibliothecae Bodleianae codicum manuscriptorum orientalium, videlicet Hebraicorum, Chaldaicorum, Syriacorum, Aethiopicorum, Arabicorum, Persicorum, Turcicorum, Copticorumque Catalogus, jussu curatorum preli academici a Joanne Uri confectus*, Oxford: Pars primus, 1787.

Vajda, *Bibliothèque Nationale* = Vajda, G., *Index général des manuscrits arabes musulmans de la Bibliothèque Nationale de Paris*, Paris: Editions du Centre National de la Recherche Scientifique (CNRS), 1953.

Vander, *Biblioteca Publica Universitatis Lugduno-Batavae* = *Catalogus Librorum tam Impressorum quam manuscriptorum Bibliotecae Publicae Universitatis Lugduno-Batavae*, sumptibus Petri Vander, Lugduni apud Batavos, 1716.

Veltman, Kim H., *Optics and Perspective: A Study in the Problems of Size and Distance*, Ph.D. Thesis, London: The Warburg Institute, 1975.

Ver Eecke, *Pappus: La Collection* = *Pappus d'Alexandrie: La Collection Mathématique*, Œuvre traduite pour la première fois du grec en français avec une Introduction et des Notes par Paul Ver Eecke, Tome Second, Ouvrage publié sous les auspices de la Fondation Universitaire de Belgique, Paris, Bruges: Desclée de Brouwer et Cie., 1933.

Ver Eecke, *Euclide: l'Optique et la Catoptrique* = *Euclide: l'Optique et la Catoptrique*, Œuvres traduites pour la première fois du grec en français, avec une Introduction et des Notes par Paul Ver Eecke, Nouveau Tirage, Paris: Librairie Scientifique et Technique Albert Blanchard, 1959.

Voorhoeve, *Handlist* = Voorhoeve, P., *Codices Manuscripti VII: Handlist of Arabic Manuscripts in the Library of the University of Leiden and other Collections in the Netherlands*, Lugduni Batavorum, 1957.

Wenrich, *De Auctorum Graecorum* = Wenrich, J. G., *De Auctorum Graecorum versionibus et commentariis syriacis arabicis armeniacis persicisque*, Leipzig, 1842.

Wiedemann, Eilhard, "Über die Reflexion und Umbiegung des Lichtes von Naṣīr al Dīn al Ṭūsī," *Jahrbuch für Photographie und Reproduktionstechnik*, Halle, 1907, 21: 38–44.

Winter, "British Isles" = Winter, H. J. J., "The Arabic Optical MSS in the British Isles: Mathematical and Physical Optics," *Centaurus*, 1956, 5, 1: 73–88.

Winter, H. J. J. and 'Arafat, W., "A Statement on Optical Reflection and 'Refraction' Attributed to Naṣīr ud-Dīn aṭ-Ṭūsī," *Isis*, 1951, 42: 138–142.

Wolfson, H. A. "The Internal Senses in Latin, Arabic and Hebrew Philosophical Texts," *Harvard Theological Review*, 1935, 28: 69–133.

Wright, Ramsay R., *The Book of Instruction in the Elements of the Art of Astrology by Abu'l-Rayḥān Muḥammad ibn Aḥmad al-Bīrūnī* [facsimile ed. with facing English translation], London: Luzac and Co., 1934.

Ya'qūbī, *Ta'rīkh* = Ibn-Wādhih qui dicitur Al-Ja'qubī, *Historiae*, pars prior historiam ante-Islamicam continens, edidit indicesque adjecit M. Th. Houtsma, 2 vols., Leiden: E. J. Brill, 1883.

Index of Arabic Terms

– The index includes all textual occurrences and forms without the article (*al-*).
Key: M9(27) = text M, Proposition 9, line 27; S(4) = text S, line 4.

ص

ك

ل

ن

General Index

Sources in the History of Mathematics and Physical Sciences

Continued from page ii

MIX
Papier aus verantwortungsvollen Quellen
Paper from responsible sources
FSC® C105338

If you have any concerns about our products,
you can contact us on
ProductSafety@springernature.com

In case Publisher is established outside the EU,
the EU authorized representative is:
Springer Nature Customer Service Center GmbH
Europaplatz 3, 69115 Heidelberg, Germany

Printed by Libri Plureos GmbH
in Hamburg, Germany